Advanced Molecular Quantum Mechanics

STUDIES IN CHEMICAL PHYSICS

General Editor

A. D. Buckingham, Professor of Chemistry, University of Cambridge

Series Foreword

The field of science known as 'Chemical Physics' has greatly expanded in recent years. It is an essential part of both physics and chemistry and now impinges on biology, crystallography, the science of materials and even on astronomy. The aim of this series is to present short, authoritative and readable books on different topics in chemical physics at a level that is appreciated by the non-specialist and yet is of prime interest to the expert—in fact, the type of book that we all welcome and enjoy.

I was grateful to be given the opportunity to help plan this series, and warmly thank the authors and publishers whose efforts have brought it into being.

A. D. Buckingham,
University Chemical Laboratory,
Cambridge, U.K.

Other titles in the series

Electronic Transitions and the High Pressure Chemistry and Physics of Solids, H. G. Drickamer and C. W. Frank
Chemical Applications of Molecular Beam Scattering, M. A. D. Fluendy and K. P. Lawley

Advanced Molecular Quantum Mechanics

An Introduction to Relativistic Quantum Mechanics and the Quantum Theory of Radiation

R. E. MOSS
Lecturer in Chemistry
University of Southampton

CHAPMAN AND HALL · LONDON

First published 1973
by Chapman and Hall Ltd
11 New Fetter Lane, London EC4P 4EE

Set by E. W. C. Wilkins Ltd, London N12 0EH
and printed in Great Britain by
Redwood Press Limited, Trowbridge, Wiltshire

SBN *412 10490 3*

Distributed in the U.S.A.
by Halsted Press, a Division
of John Wiley & Sons, Inc.
New York

TO SHIRLEY
AND JONATHAN

Preface

This book is primarily intended for graduate chemists and chemical physicists. Indeed, it is based on a graduate course that I give in the Chemistry Department of Southampton University.

Nowadays undergraduate chemistry courses usually include an introduction to quantum mechanics with particular reference to molecular properties and there are a number of excellent textbooks aimed specifically at undergraduate chemists. In valence theory and molecular spectroscopy physical concepts are often encountered that are normally taken on trust. For example, electron spin and the anomalous magnetic moment of the electron are usually accepted as postulates, although they are well understood by physicists. In addition, the advent of new techniques has led to experimental situations that can only be accounted for adequately by relatively sophisticated physical theory. Relativistic corrections to molecular orbital energies are needed to explain X-ray photoelectron spectra, while the use of lasers can give rise to multiphoton transitions, which are not easy to understand using the classical theory of radiation. Of course, the relevant equations may be extracted from the literature, but, if the underlying physics is not understood, this is a practice that is at best dissatisfying and at worst dangerous. One instance where great care must be taken is in the use of spectroscopically determined parameters to test the accuracy of electronic wave functions. The practice of employing classical analogies to add small terms to the molecular Hamiltonian, as and when they are needed to explain new phenomena, is no longer satisfactory. It is necessary to start with as complete a molecular Hamiltonian as possible to make sure that important contributions to spectroscopic parameters are not omitted.

However, the physics literature that deals with such topics as relativistic quantum mechanics and the quantum theory of electromagnetic radiation usually assumes a broad physical background that can only be acquired by reading a number of physics texts. In addition, the pertinent knowledge is often embedded in a considerable amount of information that is not directly relevant to molecular physics. The aim of this book is, therefore, to present sufficient theoretical physics that the theory needed to understand some of the recent developments in molecular physics may be better appreciated.

Only the quantum mechanics and mathematics that, in my experience, is usually covered in undergraduate chemistry courses is assumed. Even so, the relevant parts are summarized in the early chapters, although vector algebra is considered in some detail, since there are a number of useful sophistications that do not appear to be covered in the usual texts. Classical mechanics and special relativity are then introduced with particular reference to a system of particles and this theory is used to develop electromagnetic theory in such a way that no knowledge of these subjects is necessary beyond that taught in schools. With this background the Dirac equation for the electron is considered in detail; this theory leads naturally to the concept of electron spin and the correct value for the spin magnetic dipole moment of the electron. The extension to many electrons and to molecules is then discussed, the end product being a Hamiltonian containing all the terms of (chemical) interest. Finally, the classical theory of electromagnetic fields is quantized and it is shown how quantum field theory may be used to describe the interaction of radiation with matter. This is but a brief review of the topics covered, but reference to the list of contents may be made for a more detailed summary.

I do not pretend that the theory described in this book is entirely adequate to account for all molecular properties. There are, for example, radiative corrections that can only be discussed satisfactorily in terms of quantum electrodynamics, a subject that is beyond the scope of this book. Nevertheless, an attempt is made to give a qualitative description of the processes responsible for such phenomena as the Lamb shift and the radiative correction to the g factor of the electron.

In a way the subject matter reflects my own interests, since it represents what I have found necessary to learn, since graduating as a chemist, in order to pursue my current research interest in high resolution molecular spectroscopy. There is another personal stamp on this book, since I have a dislike for appendices and these have been kept to a minimum. The appendix containing a summary of the useful vector relations derived in Chapter 2 requires no comment. Of the other two, one is a review of the SI system of units, which is used throughout the book. For those who, despite international agreement, still prefer to use the mixed (or Gaussian) system, the transition to the SI system is not trivial. However, no problems should be encountered here, since electromagnetism, where the main difficulties occur, is developed from first principles and since tables for the conversion of units and equations are given. In the remaining appendix a general bibliography is presented; this comprises a list of the books that are referred to most frequently in the bibliographies at the ends of the individual chapters. I have made no attempt to provide a complete bibliography and indeed this would be an almost impossible task. The sources that are cited are those that I have found useful; where appropriate I have given a brief description of the material covered by the individual references. Anyone wishing to consult other sources can always find more references in the articles and books that are mentioned in the bibliographies.

It will be apparent that no problems as such are given, but the reader will often find results quoted, the verification of which provides ample material for exercises. This is not intended as a textbook, its purpose is to provide sufficient background that the more recent research literature may be approached with confidence.

Finally, I must thank a number of people for their part in the realization of this book. Professor A. Carrington first turned my attention to relativistic effects and Professor A. D. Buckingham suggested the inclusion of the quantum theory of radiation. I am also grateful to these and other colleagues, in particular Dr J. M. Brown, Dr B. J. Howard, Dr I. J. Ketley and Mr A. J. Perry, for reading all or parts of the manuscript and helping to remove inaccuracies. Needless to say the remaining errors are my responsibility alone and I will appreciate them being pointed out to me. Thanks are also due to Mrs Jean Gordon and Mrs Christine Croucher, who between them expertly typed the bulk of the manuscript.

Richard E. Moss

Southampton
July, 1972

Contents

CHAPTER ONE

Non-relativistic Quantum Mechanics

This introductory chapter has two purposes. The first is to provide a brief résumé of the quantum theory that is assumed to be familiar to the reader and, in particular, those results that are used later. A deep discussion of the foundations of quantum mechanics will not be given and in this chapter the results will be quoted without proof; the justification for this policy is that there are a number of excellent texts available. Nevertheless, some of the mathematics employed in this presentation is discussed in more detail later. For example, an acquaintance with vector and matrix algebra is assumed here, although Chapter 2 is devoted to this subject.

The other objective of this chapter is to point out the unsatisfactory features of non-relativistic quantum mechanics. This theory is based on a number of postulates, the justification of which is that the results derived from them are in agreement with experiment. However, in a relativistic theory some of these postulates are unnecessary and here we are thinking in particular of electron spin and the phenomena associated with it.

1.1 Formal quantum mechanics

For convenience just one particle is considered here, but the extension to many-particle systems is readily made. Moreover, it should be noted that the present section is not intended to be a logical development of quantum theory.

The state of a one-particle system is described by a wave function $\Psi(\mathbf{r}, t)$, which is a function of particle coordinates and time. The probability of the particle being found in a small volume $\mathrm{d}\tau$ at position $\mathbf{r}$ and time t is given by $\Psi(\mathbf{r}, t)^*\Psi(\mathbf{r}, t)\mathrm{d}\tau$, where the asterisk indicates complex conjugation. Since the particle must be somewhere in space:

$$\int \Psi(\mathbf{r}, t)^* \Psi(\mathbf{r}, t) \mathrm{d}\tau = \int |\Psi(\mathbf{r}, t)|^2 \mathrm{d}\tau = 1 \tag{1.1}$$

and a wave function satisfying this condition is said to be normalized. Since the wave function is to be regarded as a probability amplitude, it must be finite, single-valued and continuous; in addition, it must be possible to perform the integration in Equation 1.1. Subscripts may be used to distinguish wave functions representing different states of a system. Two different wave functions, Ψ_m and Ψ_n, are then said to be orthogonal if $\int \Psi_m^* \Psi_n \mathrm{d}\tau$ vanishes.

It is often convenient to use an alternative notation to describe the state of a system. This is the $\langle \text{bra}|c|\text{ket}\rangle$ notation due to Dirac, in which the ket $|n\rangle$ is used instead of Ψ_n and the bra $\langle n|$ in place of Ψ_n^*. In this scheme Equation 1.1 is written as $\langle n|n\rangle = 1$ and the orthogonality condition is $\langle m|n\rangle = 0$ $(m \neq n)$; in each case integration over all space is implied by the notation. (It is tempting to say that Ψ_n and $|n\rangle$ are equal, but this is not strictly accurate. The Ψ_n are functions of coordinates, while the $|n\rangle$ are vectors in so-called Hilbert space and are independent of the coordinate system used. In addition, in the Dirac notation a quantity such as $\langle n|n\rangle$ is a scalar product of vectors rather than an integration over space.)

Observables or measurable properties of a system are represented by operators. If a state, which is characterized by Ψ_n, has a definite value α_n for the observable represented by the operator A, then Ψ_n is an eigenfunction of A with eigenvalue α_n and the equation:

$$A\Psi_n = \alpha_n \Psi_n \tag{1.2}$$

is satisfied. By multiplying this equation on the left by Ψ_n^*, integrating over all space and using Equation 1.1 α_n may be obtained:

$$\alpha_n = \int \Psi_n^* A \Psi_n \mathrm{d}\tau. \tag{1.3}$$

This is the expectation value of A in the state n and in the Dirac notation is written as $\langle n|A|n\rangle$. In general, expressions of the type $\int \Psi_m^* A \Psi_n \mathrm{d}\tau$ or $\langle m|A|n\rangle$ are referred to as matrix elements of the operator A between states m and n.

In quantum mechanics the operators are all linear so that the result of operating on a linear combination of wave functions is:

$$A(c_m\Psi_m + c_n\Psi_n) = c_m A\Psi_m + c_n A\Psi_n. \tag{1.4}$$

For an operator to represent a physical observable all its eigenvalues must be real and this imposes the restriction that the operator be Hermitian. The condition for the Hermiticity of A is that the equation:

$$\int \Psi_m^* A \Psi_n \mathrm{d}\tau = \int (A^* \Psi_m^*) \Psi_n \mathrm{d}\tau = \int (\Psi_n^* A \Psi_m)^* \mathrm{d}\tau \tag{1.5}$$

should always hold or, in the alternative notation, $\langle m|A|n\rangle$ should always equal $\langle n|A|m\rangle^*$. In addition, the eigenfunctions corresponding to different (real) eigenvalues of an Hermitian operator are orthogonal to one another.

A general normalized function Φ can be written as a combination of a complete set of orthogonal functions. This set might be all the eigenfunctions of the Hermitian operator A:

$$\Phi = \sum_n c_n \Psi_n. \tag{1.6}$$

However, Φ is not necessarily itself an eigenfunction of A, but a determination of the physical property associated with A will always give one of the α_n as the result; the probability that a particular result α_m is obtained being $c_m^* c_m$. The expectation value of A in the state represented by Φ may still be determined, but it will be a weighted average of all the eigenvalues α_n of A.

A rule is still needed to construct the operator for a particular observable. It is usual to take the classical expression for the observable and to replace the position $\mathbf{r}$ and linear momentum $\mathbf{p}$ by the operators $\mathbf{r}$ and $\mathbf{p} = -i\hbar(\partial/\partial\mathbf{r})$, respectively, where the symbol i represents $\sqrt{-1}$ and $\hbar$ is related to Planck's constant h by $\hbar = h/2\pi$; the vector operator $(\partial/\partial\mathbf{r})$ has components $(\partial/\partial x)$, $(\partial/\partial y)$ and $(\partial/\partial z)$. If both $\mathbf{r}$ and $\mathbf{p}$ appear in the classical expression then care must be taken that the resulting operator is Hermitian. In addition, the total energy E of a system may also be replaced by the operator $i\hbar(\partial/\partial t)$.

1.2 The Schrödinger equation

Classically the energy of a system is the sum of its kinetic and potential energies. For a particle of mass m the former is $p^2/2m$ and the latter is a function of position and time:

$$E = p^2/2m + V(\mathbf{r}, t). \tag{1.7}$$

On quantization the right-hand side of this equation becomes the Hamiltonian operator:

$$\mathcal{H} = -(\hbar^2/2m)(\partial^2/\partial x^2 + \partial^2/\partial y^2 + \partial^2/\partial z^2) + V(\mathbf{r}, t), \tag{1.8}$$

where x, y, z are the components of the position vector $\mathbf{r}$. When this operates on a wave function it must give the same result as the energy operator $i\hbar(\partial/\partial t)$, so that

$$\mathcal{H}\Psi = i\hbar(\partial\Psi/\partial t); \tag{1.9}$$

this is the time-dependent Schrödinger equation.

If the potential energy of the particle is time independent, the Hamiltonian is also and the space and time variables in Equation 1.9 may be separated. The solution may then be written:

$$\Psi(\mathbf{r}, t) = \exp(-iEt/\hbar)\,\psi(\mathbf{r}), \tag{1.10}$$

where E and ψ are given by the time-independent Schrödinger equation:

$$\mathcal{H}\psi = E\psi; \tag{1.11}$$

E is now the energy of the system and is independent of time. In addition, Equation 1.10 shows that $|\Psi(\mathbf{r}, t)|^2 = |\psi(\mathbf{r})|^2$, so that these wave functions describe a stationary state of the system.

Only in very simple cases is it possible to solve the time-independent Schrödinger equation 1.11 exactly. In general, approximate methods must be employed and two approaches are commonly used.

In the first use is made of the variation principle, which states that, if E_0 is the lowest eigenvalue of the particular Hamiltonian, then for any normalized trial solution ψ' of Equation 1.11:

$$\int \psi'^{*}\mathcal{H}\psi'\mathrm{d}\tau \geqslant E_0. \tag{1.12}$$

Using this fact a particular form for ψ' may be optimized by varying adjustable parameters to minimize $\int \psi'^{*}\mathcal{H}\psi'\mathrm{d}\tau$. In particular, ψ' may be taken as a linear combination of known normalized functions ϕ_n:

$$\psi' = \sum_n c_n\phi_n, \tag{1.13}$$

and the best approximation of this form to the true lowest eigenvalue of $\mathcal{H}$ is obtained by the variation of the coefficients c_n. This leads to a set of coupled equations:

$$\sum_n (\mathcal{H}_{mn} - ES_{mn})c_n = 0, \tag{1.14}$$

where $\mathcal{H}_{mn}$ is the matrix element of $\mathcal{H}$ between ϕ_m and ϕ_n and S_{mn} is the overlap integral $\int \phi_m^{*}\phi_n\mathrm{d}\tau$, which is unity for $m = n$ but is only zero for $m \neq n$ if the ϕ_n are orthogonal to one another.

The other method involves perturbation theory, in which the Hamiltonian is split into two parts:

$$\mathcal{H} = \mathcal{H}_0 + \mathcal{H}'. \tag{1.15}$$

It is assumed that the eigenfunctions of $\mathcal{H}_0$ are known:

$$\mathcal{H}_0\psi_n^0 = E_n^0\psi_n^0 \tag{1.16}$$

and that the eigenfunctions ψ_n of the total Hamiltonian may be written as a linear combination of them. We then have:

$$\psi_n = \psi_n^0 + \sum_{m \neq n} \psi_m^0 \mathcal{H}'_{mn}/(E_n^0 - E_m^0) + \ldots \tag{1.17}$$

and

$$E_n = E_n^0 + \mathcal{H}'_{nn} + \sum_{m \neq n} \mathcal{H}'_{nm}\mathcal{H}'_{mn}/(E_n^0 - E_m^0) + \ldots; \tag{1.18}$$

provided $\mathcal{H}'$ is a relatively small perturbation these series converge.

1.3 Heisenberg's uncertainty principle and related topics

For the wave function describing a state to be a simultaneous eigenfunction of two operators, A and B, then these operators must commute. That is, AB must equal BA and this condition is usually expressed as:

$$[A, B] = AB - BA = 0. \tag{1.19}$$

If A and B do commute, the observables they represent can in principle be measured simultaneously and precisely. However, if they do not, there is an uncertainty in the determined values of the observables and this is expressed by Heisenberg's uncertainty principle:

$$(\Delta A)(\Delta B) \geqslant |\langle [A, B] \rangle / 2i|; \tag{1.20}$$

this relates the uncertainties, ΔA and ΔB, to the expectation value of the commutator of the operators A and B. In particular, the x components of the position and momentum operator for a particle do not commute and $(\Delta x)(\Delta p_x) \geqslant \hbar/2$. Similarly, the energy operator $i\hbar(\partial/\partial t)$ does not commute with t, so that $(\Delta E)(\Delta t) \geqslant \hbar/2$ and the precision with which the energy of a state may be determined depends upon its lifetime.

In this connection we may note that we are using the Schrödinger picture in which operators such as **p** and **r** are time independent and any time dependence of the system is contained in the wave functions, which may evolve with time. In an alternative formulation of quantum mechanics it is the operators that are time dependent and the wavefunctions that are independent of time; this is the Heisenberg picture. It is occasionally useful to know about the way in which an operator, or one of its expectation values, varies with time. This section is concluded with a derivation of Heisenberg's equation of motion, which can provide this information.

The Heisenberg and Schrödinger pictures of quantum mechanics may be related to one another by insisting that matrix elements of an operator be the same in the two pictures:

$$\int\Psi^*(t)A(0)\Psi'(t)\mathrm{d}\tau = \int\Psi^*(0)A(t)\Psi'(0)\mathrm{d}\tau. \tag{1.21}$$

Now, an alternative way of writing Equation 1.10 is:

$$\Psi(t) = \exp(-i\mathcal{H}t/\hbar)\Psi(0) \tag{1.22}$$

and substitution of this and a similar expression for $\Psi'(t)$ into Equation 1.21 shows that:

$$A(t) = \exp(i\mathcal{H}t/\hbar)A(0)\exp(-i\mathcal{H}t/\hbar). \tag{1.23}$$

Since $A(0)$ is time independent, the derivative of this equation with respect to time gives:

$$\mathrm{d}A/\mathrm{d}t = (i/\hbar)(\mathcal{H}A - A\mathcal{H}) = (i/\hbar)[\mathcal{H},A], \tag{1.24}$$

which is Heisenberg's equation of motion.

As an example of the use of this equation, the velocity operator for a particle may be determined by setting A equal to $\mathbf{r}$. The appropriate Hamiltonian is given in Equation 1.8, so that $\mathrm{d}\mathbf{r}/\mathrm{d}t = \mathbf{p}/\mathrm{m}$ which is what one would expect by analogy with classical mechanics.

Finally, it may be noted that for any operator A that commutes with the Hamiltonian its time derivative $\mathrm{d}A/\mathrm{d}t$, and the expectation value of this, must vanish, so that A is a constant of motion. Consequently, the energy and the observable corresponding to A may be determined precisely and this is in agreement with the earlier discussion.

1.4 Angular momentum

Classically the angular momentum of a particle is given by $\mathbf{r}_\wedge \mathbf{p}$, the vector product of its position and its linear momentum. The corresponding operator in quantum mechanics is given by the same expression and may be written as:

$$\hbar\mathbf{l} = \mathbf{r}_\wedge\mathbf{p} = -i\hbar\mathbf{r}_\wedge(\partial/\partial\mathbf{r}), \tag{1.25}$$

so that $\mathbf{l}$ is a dimensionless operator. The x component is given by:

$$\hbar l_x = -i\hbar(y\,\partial/\partial z - z\,\partial/\partial y) \tag{1.26}$$

and the other components may be obtained by a cyclic permutation of x, y and z.

The components of the operator $\mathbf{l}$ do not commute:

$$[l_x, l_y] = il_z; \quad [l_y, l_z] = il_x; \quad [l_z, l_x] = il_y, \tag{1.27}$$

so that a wave function cannot be a simultaneous eigenfunction of l_x, l_y and l_z. However, the operator $l^2 = (l_x^2 + l_y^2 + l_z^2)$ does commute with these components and it is usual to employ wave functions that are eigenfunctions of l^2 and l_z. Their explicit form may be determined by taking l^2 and l_z to be operators involving partial derivatives with respect to spatial coordinates as in Equation 1.26 and solving the resulting partial differential equations. The resulting eigenfunctions are angular functions, or spherical harmonics, and are characterized by two quantum numbers l and m_l. The equations they satisfy are:

$$l^2\psi_{lm_l} = l(l+1)\psi_{lm_l}; \quad l_z\psi_{lm_l} = m_l\psi_{lm_l}, \tag{1.28}$$

where $l = 0, 1, 2, \ldots$ and $m_l = l, l-1, \ldots, -l$.

Thus an atomic electron is said to possess an orbital angular momentum which is an integer in units of $\hbar$. This is in agreement with observations made in atomic spectroscopy. In addition, experiment shows that associated with the orbital motion of an electron is a magnetic moment $-e\hbar\mathbf{l}/2m$, where $-e$ and m are the charge and mass of the electron, respectively. (In this book SI units are employed and these are summarized in Appendix A; in addition, those units associated with electromagnetic phenomena are discussed in more detail in Chapter 5.) This orbital magnetic moment of an electron arises quite naturally in non-relativistic quantum theory when the interaction with electromagnetic fields is introduced and this is demonstrated in Chapter 7.

Of particular interest is the so-called g factor which is the magnitude of the ratio of an angular momentum and its associated magnetic moment, the ratio being measured in units of $e/2m$. Thus the orbital g factor is just 1 and this is not surprising, since classical theory gives the same result. For an electron moving with speed v in a circle of radius r the angular momentum is mvr. The associated magnetic moment is $-evr/2$, which is just the product of the current $(-ev/2\pi r)$ flowing round the circle and its area (πr^2). The ratio of interest is $-e/2m$, so that classically the orbital g factor is unity.

Equations 1.28 are obtained by taking the eigenfunctions of l^2 and l_z to be functions of the electron's coordinates. However, a general angular momentum $\hbar\mathbf{j}$ may be defined, the components of which satisfy similar commutation relations (Equations 1.27) to $\hbar\mathbf{l}$:

$$[j_x, j_y] = ij_z; \quad [j_y, j_z] = ij_x; \quad [j_z, j_x] = ij_y. \tag{1.29}$$

Without making any assumptions about the form of the simultaneous eigenfunctions ψ_{jm_j} of the operators j^2 and j_z, these commutation relations may be used to show that:

$$j^2\psi_{jm_j} = j(j+1)\psi_{jm_j}; \quad j_z\psi_{jm_j} = m_j\psi_{jm_j}. \tag{1.30}$$

These results are analogous to those for l^2 and l_z except that in this case the values that the quantum numbers may adopt are $j = 0, 1/2, 1, 3/2, \ldots$ and

$m_j = j, j-1, \ldots, -j$. Thus half-in ral angular momentum is possible. For integer values of j the eigenfunctions ay be written as spherical harmonics, but it is not possible to give the eigenfunction an analytic form when j is a half integer.

1.5 Electron spin

The existence of orbital angular momentum and its associated magnetic moment does not explain all the features of the fine structure of atomic spectra and, in particular, Zeeman splittings. In non-relativistic quantum mechanics it is necessary to postulate that in addition the electron possesses an intrinsic angular momentum or 'spin' of $\frac{1}{2}$ in units of $\hbar$ together with a spin magnetic moment of magnitude $e\hbar/2m$.

The spin angular momentum corresponds to the case $j = \frac{1}{2}$ mentioned at the end of the previous section and by combining spin and orbital angular momentum the other half-integer values for j may be obtained. The spin angular momentum operator is usually deignasted $\hbar\mathbf{s}$ and the quantum numbers corresponding to j and m_j are $s(=\frac{1}{2})$ and $m_s = \pm\frac{1}{2}$. The eigenfunctions of the operators s^2 and s_z may then be written as $|s, m_s\rangle$ or just $|m_s\rangle$, since s is always $\frac{1}{2}$; in addition, the notation $|\alpha\rangle$ and $|\beta\rangle$ is often used in place of $|+\frac{1}{2}\rangle$ and $|-\frac{1}{2}\rangle$, respectively.

In what follows it is often more convenient to use the operator $\boldsymbol{\sigma} = 2\mathbf{s}$ instead of $\mathbf{s}$ itself. The commutation relations for the components of $\boldsymbol{\sigma}$ are then:

$$[\sigma_x, \sigma_y] = 2i\sigma_z; \quad [\sigma_y, \sigma_z] = 2i\sigma_x; \quad [\sigma_z, \sigma_x] = 2i\sigma_y, \qquad (1.31)$$

and the way in which these operators act on the spin eigenfunctions $|\alpha\rangle$ and $|\beta\rangle$ is given by:

$$\begin{aligned} \sigma_x|\alpha\rangle &= |\beta\rangle; & \sigma_x|\beta\rangle &= |\alpha\rangle; \\ \sigma_y|\alpha\rangle &= i|\beta\rangle; & \sigma_y|\beta\rangle &= -i|\alpha\rangle; \\ \sigma_z|\alpha\rangle &= |\alpha\rangle; & \sigma_z|\beta\rangle &= -|\beta\rangle. \end{aligned} \qquad (1.32)$$

From these relations the effect of operators constructed from the components of $\boldsymbol{\sigma}$ may readily be determined. In particular, it may be shown that $|\alpha\rangle$ and $|\beta\rangle$ are both eigenfunctions of the operators σ_x^2, σ_y^2 and σ_z^2 with the same eigenvalue 1. Thus we may write:

$$\sigma_x^2 = \sigma_y^2 = \sigma_z^2 = 1. \qquad (1.33)$$

As we know the commutator of two operators A and B is $[A,B] = (AB - BA)$. The anticommutator of these operators may also be defined; this is just $(AB + BA)$ and is written in the abbreviated form $[A, B]_+$. The anti-commutation properties of the components of $\boldsymbol{\sigma}$ may now be found and as an illustration we evaluate $[\sigma_x, \sigma_y]_+$ by using Equations 1.31:

$$\begin{aligned} [\sigma_x, \sigma_y]_+ &= \sigma_x \sigma_y + \sigma_y \sigma_x, \\ &= -\tfrac{1}{2}i([\sigma_y, \sigma_z]\sigma_y + \sigma_y[\sigma_y, \sigma_z]), \\ &= -\tfrac{1}{2}i[\sigma_y^2, \sigma_z]. \end{aligned} \tag{1.34}$$

Now, since σ_y^2 can be replaced by unity (Equation 1.33), it must commute with σ_z and $[\sigma_x, \sigma_y]_+$ vanishes. Similarly, the other pairs of components of $\boldsymbol{\sigma}$ may be shown to anticommute so that:

$$[\sigma_x, \sigma_y]_+ = [\sigma_y, \sigma_z]_+ = [\sigma_z, \sigma_x]_+ = 0. \tag{1.35}$$

It should be noted that in general the components of the angular momentum operator $\mathbf{j}$ do not satisfy the relations of Equations 1.33 and 1.35 and that these properties are peculiar to the case when $j = \frac{1}{2}$.

It is possible to construct a matrix representation of the components of $\boldsymbol{\sigma}$, that is, a set of three matrices which have the same commutation (Equations 1.31) and anticommutation (Equations 1.35) properties and thus mimic the behaviour of σ_x, σ_y and σ_z. This may be done using two-by-two matrices one such representation being the so-called Pauli spin matrices:

$$\sigma_x = \begin{pmatrix} 0 & 1 \\ 1 & 0 \end{pmatrix}; \quad \sigma_y = \begin{pmatrix} 0 & -i \\ i & 0 \end{pmatrix}; \quad \sigma_z = \begin{pmatrix} 1 & 0 \\ 0 & -1 \end{pmatrix}. \tag{1.36}$$

These matrices also have the property that the square of each is equal to the two-by-two unit matrix in agreement with Equation 1.33.

With this formalism the spin functions $|\alpha\rangle$ and $|\beta\rangle$ are written as two-component column vectors:

$$|\alpha\rangle = \begin{pmatrix} 1 \\ 0 \end{pmatrix}; \quad |\beta\rangle = \begin{pmatrix} 0 \\ 1 \end{pmatrix}, \tag{1.37}$$

the complex conjugates of these being the corresponding row vectors. Using these with Equations 1.36, the relationships of Equations 1.32 may be recovered and indeed the elements of the matrices in Equations 1.36 are just the matrix elements of the corresponding operators between the spin functions.

Electron spin can now be incorporated into the quantum mechanical description of the electron by using classical analogies to add appropriate spin-dependent terms to the Hamiltonian and using wave functions which are products of spatial and spin functions. The resulting Schrödinger equation can then be expressed as one equation using the formalism employed in Equations 1.32 or as a pair of coupled equations using the matrix representation of the spin operators (Equations 1.36) and spin functions (Equations 1.37)

It must be emphasized that although this theory is largely successful it is based on the postulate of a 'spinning' electron with angular momentum $\hbar \mathbf{s}$ and associated magnetic moment $-e\hbar \mathbf{s}/m$. The electron spin g factor, which is the magnitude of the ratio of these two quantities measured in units of $e/2m$, is thus 2. This result is contrasted by the orbital g factor of 1 and consequently the spin g factor and the spin magnetic moment are often said to be anomalous. The reason for this is that, unlike an orbiting electron, it is not possible to construct a satisfactory classical model for a spinning electron.

Any model in which the electron has its mass and charge distributed over a finite volume explains the existence of both the angular momentum and the magnetic moment when the electron spins about an axis. However, the g factor cannot be explained so easily. The simplest model consists of a sphere with uniform mass and charge densities, but any model in which the mass and charge are distributed in the same way leads to a spin g factor of 1 just as it does for the orbital motion of an electron. Of course, if the electron consists of conducting material the charge might be expected to reside on the surface of the sphere and this modification of the model leads to a spin g factor of 5/3. Although this is not the required result, it does suggest that models may be devised with $g = 2$ and indeed this is so; for illustration we quote just one such model. In this the electron is a solid cylinder of uniform density and all the charge resides on the curved surface. The angular momentum is $mr^2\omega/2$, where m is the mass, r the radius and ω the angular velocity of the cylinder, and the magnitude of the associated magnetic moment is $er^2\omega/2$, so that $g = 2$ as required.

Nevertheless, this model, and others like it that give a classical g factor of 2, are rather far-fetched and cannot be taken as a justification for the introduction of the concept of electron spin. In non-relativistic quantum mechanics the existence of electron spin angular momentum and its associated magnetic moment must be regarded as postulates.

All these remarks apply specifically to the electron, but nuclei may be described in a similar fashion. They too possess spin angular momenta and magnetic moments, although the spin is not restricted to $\frac{1}{2}$ as it is for the electron, and these properties must also be incorporated in the theory in a phenomenological manner.

Finally, we might mention a further postulate that is necessary when dealing with systems of many identical particles. This is the Pauli exclusion principle, which states that wave functions must be antisymmetric with respect to the interchange of like particles with half-integer spin and symmetric with respect to interchange of like particles with zero or integer spin. The former category of particles, which includes the electron, are called fermions and obey Fermi-Dirac statistics, while the latter are bosons and are subject to Bose-Einstein statistics.

1.6 The need for a relativistic theory

The velocity of the electron in the first Bohr orbit of a one-electron atom is $Z\alpha c$, where Z is the atomic number, $\alpha(\simeq 1/137)$ is the fine structure constant and c is the velocity of light. In the hydrogen atom the relativistic corrections expected are small although they are observable. However, in heavier atoms these corrections can become quite large for inner electrons despite the shielding of the nuclear charge by other electrons. For this reason alone a relativistic theory is needed particularly as techniques such as X-ray photoelectron spectroscopy are now being used to study the inner electrons in atoms and molecules.

At this point it might be mentioned that we will only be interested in special relativity. The effects of general relativity, which is concerned with gravitation, are negligible in atoms and molecules. This may be seen by considering the hydrogen atom. In the first Bohr orbit the total energy of the electron is approximately 2×10^{-18} J, whereas the potential energy due to the gravitational attraction of the electron by the proton is of the order of 2×10^{-57} J.

As it happens the incorporation of special relativity in quantum mechanics gives rise to far greater benefits than might be expected. In particular, it leads quite naturally to the concept of electron spin, so that this need no longer be regarded as a postulate. Indeed, both the spin angular momentum and the spin magnetic moment are predicted by the theory and the spin g factor is found to be exactly 2. (In fact the spin g factor of the electron differs slightly from 2 and this discrepancy can only be explained by quantum electrodynamics, which is beyond the scope of this book, although an indication of its origin is given in Chapter 9).

Morevoer, the spin-dependent terms in the Hamiltonian arise directly from a relativistic theory and it is not necessary to use dubious classical analogies to construct them as in the non-relativistic approach. Here we might cite the spin-orbit interaction, for which classical analogy actually leads to a term in the Hamiltonian that is twice as large as is needed to explain experimental observations. It is possible to explain the necessary correction factor classically, but only if special relativity is included. A section in Chapter 5 is devoted to this topic, but no problems arise if relativistic quantum mechanics is used from the outset.

Not only are terms in the Hamiltonian with classical analogues explained readily in the relativistic theory, but also there are some terms that have no classical analogues and only appear if special relativity is allowed for.

The theory to be developed in this book applies specifically to particles of spin $\frac{1}{2}$, and in particular to the electron. However, it turns out that nuclei may be included adequately in atomic and molecular Hamiltonians by treating them as electrons but with anomalous masses, charges, spins and magnetic moments.

Finally we note that the Pauli exclusion principle is derivable from relativistic theory, but only by using quantum field theory and as far as electrons are concerned this is too advanced a topic to be covered in this text, although a qualitative discussion is given in Chapter 12.

Having made all these assertions about the results that may be obtained by using relativistic quantum theory, we have now to demonstrate them. However, before this may be done a certain amount of background theory must be developed and the next few chapters are devoted to this aim.

Bibliography

Quantum mechanics

Atkins, P. W. (1970), *Molecular Quantum Mechanics*, Clarendon Press, Oxford.
Dirac, P. A. M. (1958), *Quantum Mechanics*, Clarendon Press, Oxford.
Eyring, H., Walter, J., and Kimball, G. E. (1944), *Quantum Chemistry*, Wiley, New York.
Landau, L. D., and Lifshitz, E. M. (1959), *Quantum Mechanics*, Pergamon Press, Oxford.
Murrell, J. N., Kettle, S. F. A., and Tedder, J. M. (1965), *Valence Theory*, Wiley, New York.
Pauling, L., and Wilson, E. B., Jr., (1935), *Introduction to Quantum Mechanics,* McGraw-Hill, New York.

Electron spin

Uhlenbeck, G. E., and Goudsmit, S. A. (1926), 'Spinning electrons and the structure of spectra', *Nature*, **117**, 264: the paper in which electron spin was postulated.
Kusch, P. (1966), 'The electron dipole moment – a case history', *Physics Today*, **19** (2), 23: a review of the experiments that showed that the *g* factor of the electron differed slightly from 2.
Crane, H. R. (1968), 'The *g* factor of the electron', *Scientific American*, **218** (1), 72: a description of the experiments now used to measure the electron's *g* factor together with classical models of the electron.
McWeeny, R. (1970), *Spins in Chemistry*, Academic Press, New York.
Rose*: Chapter 1.

*References consisting of the author's name only are given in full in the General Bibliography in Appendix C.

CHAPTER TWO

Vector and Matrix Algebra

It has already been remarked that the reader is assumed to be acquainted with vectors and matrices, indeed it is unlikely that he will have survived the first chapter without a rudimentary knowledge of these subjects. However, the techniques of vector and matrix algebra are used widely in this book and a summary of these subjects is desirable. This summary could have been relegated to an appendix, but there are a number of sophistications that may well be unfamiliar, since they do not appear to be covered in elementary mathematical textbooks. The advantages of these sophisticated techniques are that they provide us with an abbreviated notation and can considerably reduce the manipulations involved particularly in vector theory.

We start by recalling the elementary theory of three-dimensional vectors. With this background, the repeated subscript summation convention can be introduced together with the Kronecker delta, δ_{ij}, and the ϵ_{ijk} notation; it is the introduction of the latter that leads to the greatest economies. These techniques are used to review the multiplication and differentiation of vectors. The properties of matrices and determinants are then summarized using the new notation where appropriate. Finally the possibility of having vectors with four or more components is mentioned.

2.1 Vectors and vector multiplication

A vector is a quantity that has two properties, magnitude and direction. In three-dimensional space a vector **A** may be specified completely by its three components, A_x, A_y and A_z, in three orthogonal directions x, y and z. If unit vectors in these directions are denoted by $\mathbf{e}_x$, $\mathbf{e}_y$ and $\mathbf{e}_z$, then **A** may be expressed as

$$\mathbf{A} = A_x\mathbf{e}_x + A_y\mathbf{e}_y + A_z\mathbf{e}_z. \tag{2.1}$$

Situations arise where it is expedient to change the axis system by rotating the axes to a new orientation; the vector is unchanged by this rotation, but of course its components in the new axis system will in general differ from those in the original axis system.

Two types of vector multiplication will be considered, the scalar product and the vector product. As their names imply the first results in a scalar quantity, while the second gives a new vector.

The scalar (or dot) product of two vectors **A** and **B** is written as **A.B** and is defined by:

$$\mathbf{A}.\mathbf{B} = A_xB_x + A_yB_y + A_zB_z. \tag{2.2}$$

If A denotes the length of vector **A**, then the scalar product of **A** with itself gives A^2:

$$\mathbf{A}.\mathbf{A} = A_x^2 + A_y^2 + A_z^2 = A^2. \tag{2.3}$$

An equivalent way of expressing a scalar product is:

$$\mathbf{A}.\mathbf{B} = AB\cos\theta, \tag{2.4}$$

where θ is the angle between the two vectors. This angle is clearly unchanged by a rotation of the axis system, so that the scalar product of two vectors has the important property of being invariant to rotation. From Equation 2.4 we note that the scalar product of two perpendicular vectors is zero; two such vectors are said to be orthogonal. The scalar products of the unit orthogonal vectors among themselves are thus:

$$\mathbf{e}_x.\mathbf{e}_x = \mathbf{e}_y.\mathbf{e}_y = \mathbf{e}_z.\mathbf{e}_z = 1; \tag{2.5a}$$

$$\mathbf{e}_x.\mathbf{e}_y = \mathbf{e}_y.\mathbf{e}_z = \mathbf{e}_z.\mathbf{e}_x = 0. \tag{2.5b}$$

The vector (or cross) product of **A** and **B** is written as $\mathbf{A}\wedge\mathbf{B}$. It gives a new vector **C**, which is orthogonal to both **A** and **B** ($\mathbf{C}.\mathbf{A} = \mathbf{C}.\mathbf{B} = 0$) and whose components are given by:

$$\begin{aligned} C_x &= (\mathbf{A}\wedge\mathbf{B})_x = A_yB_z - A_zB_y \\ C_y &= (\mathbf{A}\wedge\mathbf{B})_y = A_zB_x - A_xB_z \\ C_z &= (\mathbf{A}\wedge\mathbf{B})_z = A_xB_y - A_yB_x. \end{aligned} \tag{2.6}$$

This definition may be expressed more compactly using a determinant:

$$\mathbf{C} = \mathbf{A}_\wedge \mathbf{B} = \begin{vmatrix} \mathbf{e}_x & \mathbf{e}_y & \mathbf{e}_z \\ A_x & A_y & A_z \\ B_x & B_y & B_z \end{vmatrix}; \tag{2.7}$$

on expansion this gives:

$$\mathbf{C} = \mathbf{e}_x(A_yB_z - A_zB_y) - \mathbf{e}_y(A_xB_z - A_zB_x) + \mathbf{e}_z(A_xB_y - A_yB_x), \tag{2.8}$$

which is equivalent to Equation 2.6. We note that, unlike the scalar product, the order in which the vectors appear in a vector product is important, since:

$$\mathbf{A}_\wedge \mathbf{B} = -\mathbf{B}_\wedge \mathbf{A}. \tag{2.9}$$

Any one of the four preceding equations shows immediately that the vector product of a vector with itself vanishes, while in general the magnitude of the vector $\mathbf{A}_\wedge \mathbf{B}$ is given by $AB \sin\theta$, where again θ is the angle between **A** and **B**. The vector products of the unit orthogonal vectors among themselves are thus:

$$\begin{aligned} &\mathbf{e}_x \wedge \mathbf{e}_x = \mathbf{e}_y \wedge \mathbf{e}_y = \mathbf{e}_z \wedge \mathbf{e}_z = 0; \\ &\mathbf{e}_x \wedge \mathbf{e}_y = \mathbf{e}_z; \quad \mathbf{e}_y \wedge \mathbf{e}_z = \mathbf{e}_x; \\ &\mathbf{e}_z \wedge \mathbf{e}_x = \mathbf{e}_y. \end{aligned} \tag{2.10}$$

2.2 The repeated subscript convention for summation

Equation 2.2, which defines the scalar product of two vectors, may be written in the alternative form:

$$\mathbf{A.B} = \sum_i A_iB_i, \tag{2.11}$$

where the summation is over $i = x, y, z$. This equation can be abbreviated still further:

$$\mathbf{A.B} = A_iB_i, \tag{2.12}$$

where the sign indicating summation over i has been omitted. The question arises as to when summation over a subscript is intended and when it is not. This problem may be solved by introducing the convention that, if a subscript is repeated, that is, if it occurs twice in an expression, then summation over all the possible values of the subscript is implied; if a subscript appears only once in the expression, then we do not sum over its possible values. This is the repeated subscript convention for summation. It will be seen in the examples in the remainder of this chapter that it is not possible for a subscript to appear more than twice; if it did the meaning of the expression would be ambiguous and in fact a mistake would have been made. Other checks can be applied to equations, since a subscript that appears once in one expression must appear once and only once in all the other expressions in the equation. In addition, a subscript that appears twice in one expression must appear twice or not at all in the other terms of the equation. All this should become clearer as examples are met in the rest of this chapter.

This section will be concluded by using the convention to abbreviate some of the equations that have been encountered already; in doing this a number of dangers associated with the convention will be met. For example Equation 2.1 may be written as:

$$\mathbf{A} = A_i\mathbf{e}_i \tag{2.13}$$

and Equation 2.3 appears as:

$$\mathbf{A}.\mathbf{A} = A_iA_i. \tag{2.14}$$

The first example presents no problems, but in the second the right-hand side could have been written as A_i^2, which obscures the fact that the subscript i is repeated; it is wise to leave it in the expanded form used in Equation 2.14. Finally, it is tempting to rewrite Equation 2.5a as $\mathbf{e}_i.\mathbf{e}_i = 1$, but this is not correct, since the subscript i is repeated and summation over the possible values of i must be performed:

$$\mathbf{e}_i.\mathbf{e}_i = \mathbf{e}_x.\mathbf{e}_x + \mathbf{e}_y.\mathbf{e}_y + \mathbf{e}_z.\mathbf{e}_z = 3. \tag{2.15}$$

2.3 The Kronecker delta δ_{ij}

We cannot write Equation 2.5a using the general subscript i for x, y and z, because repeating the subscript i automatically implies summation. However, Equation 2.5a could be written as:

$$\mathbf{e}_i.\mathbf{e}_j = 1 \quad \text{for } i = j; \tag{2.16a}$$

Equation 2.5b would then appear as:

$$\mathbf{e}_i.\mathbf{e}_j = 0 \quad \text{for } i \neq j. \tag{2.16b}$$

These two equations could be written as just one equation:

$$\mathbf{e}_i.\mathbf{e}_j = \delta_{ij}, \tag{2.17}$$

if the new symbol, the Kronecker delta, is defined in the following way:

$$\begin{aligned} \delta_{ij} &= 1 \quad \text{for } i = j; \\ \delta_{ij} &= 0 \quad \text{for } i \neq j. \end{aligned} \tag{2.18}$$

The Kronecker delta can appear in expressions in a number of different ways. For example:

$$\delta_{ij}A_j = A_i, \tag{2.19}$$

since summation over j is implied. Alternatively, we could encounter the product of two Kronecker deltas having a subscript in common:

$$\delta_{ij}\delta_{kj} = \delta_{ik}; \tag{2.20}$$

this sort of relationship may be referred to as a sum rule. However, once again a note of caution should be sounded, since δ_{ii} is not equal to unity, but instead:

$$\delta_{ii} = \delta_{xx} + \delta_{yy} + \delta_{zz} = 3, \tag{2.21}$$

since we are now committed to the repeated subscript convention.

Another example of the use of the Kronecker delta is in writing the anti-commutation relations for the components of the spin vector $\boldsymbol{\sigma}$, which were encountered in the first chapter; Equations 1.33 and 1.35 can now be written with economy as:

$$[\sigma_i, \sigma_j]_+ = 2\delta_{ij}. \tag{2.22}$$

However, the great utility of the Kronecker delta in vector algebra will only be appreciated when one more symbol has been introduced and this is done in the next section.

2.4 The ϵ_{ijk} notation

The definition of ϵ_{ijk}, which is known as the permutation symbol or the unit antisymmetric tensor in three dimensions, is introduced immediately; it will then be seen how it can be used to simplify equations that have been met already. If ijk is an even permutation of xyz, then ϵ_{ijk} takes on the value one, while an odd permutation gives ϵ_{ijk} the value minus one:

$$\begin{aligned} \epsilon_{xyz} &= \epsilon_{yzx} = \epsilon_{zxy} = 1; \\ \epsilon_{xzy} &= \epsilon_{yxz} = \epsilon_{zyx} = -1. \end{aligned} \tag{2.23a}$$

There is still the possibility that a pair of the subscripts may be identical, in which case ϵ_{ijk} has the value zero:

$$\epsilon_{ijk} = 0 \quad \text{if any pair of } i, j, k \text{ are identical.} \tag{2.23b}$$

It can now be seen that Equation 2.6 for the vector product of two vectors may be written as:

$$C_i = (\mathbf{A}_\wedge \mathbf{B})_i = \epsilon_{ijk}A_jB_k. \tag{2.24}$$

This is because the subscripts j and k are both repeated on the right-hand side of the equation so that, for example:

$$\begin{aligned} C_x &= \epsilon_{xxx}A_xB_x + \epsilon_{xxy}A_xB_y + \epsilon_{xxz}A_xB_z \\ &\quad + \epsilon_{xyx}A_yB_x + \epsilon_{xyy}A_yB_y + \epsilon_{xyz}A_yB_z \\ &\quad + \epsilon_{xzx}A_zB_x + \epsilon_{xzy}A_zB_y + \epsilon_{xzz}A_zB_z \\ &= \epsilon_{xyz}A_yB_z + \epsilon_{xzy}A_zB_y \\ &= A_yB_z - A_zB_y \end{aligned} \tag{2.25}$$

by Equations 2.23. Alternatively, Equation 2.7 could be rewritten:

$$\mathbf{C} = (\mathbf{A}_\wedge \mathbf{B}) = \mathbf{e}_i\epsilon_{ijk}A_jB_k, \tag{2.26}$$

where summation over i must be performed as well.

The anticommutation of two vectors under the operation of taking a vector product (Equation 2.9) may easily be demonstrated using this new notation:

$$\mathbf{A}_\wedge\mathbf{B} = \mathbf{e}_i\epsilon_{ijk}A_jB_k = -\mathbf{e}_i\epsilon_{ikj}B_kA_j = -\mathbf{B}_\wedge\mathbf{A}, \tag{2.27}$$

since interchange of any pair of subscripts in ϵ_{ijk} must change its sign. In addition, Equations 2.10, which state the result of taking the vector product of a pair of unit orthogonal vectors, can now be written as one simple equation:

$$\mathbf{e}_i\,{}_\wedge\,\mathbf{e}_j = \epsilon_{ijk}\mathbf{e}_k. \tag{2.28}$$

So far we have only met products involving two vectors. The triple scalar product, which is written as $\mathbf{A}.(\mathbf{B}_\wedge\mathbf{C})$, can be expanded using the new notation:

$$\mathbf{A}.(\mathbf{B}_\wedge\mathbf{C}) = \epsilon_{ijk}A_iB_jC_k. \tag{2.29}$$

From its definition cyclic permutation of the subscripts i, j, k does not alter the value of ϵ_{ijk}, so that cyclic permutation of the vectors **A**, **B** and **C** does not alter their triple scalar product:

$$\mathbf{A}.(\mathbf{B}_\wedge\mathbf{C}) = \mathbf{B}.(\mathbf{C}_\wedge\mathbf{A}) = \mathbf{C}.(\mathbf{A}_\wedge\mathbf{B}). \tag{2.30}$$

The other important type of product involving three vectors is the triple vector product $\mathbf{A}_\wedge(\mathbf{B}_\wedge\mathbf{C})$. This too can be expressed in the new notation:

$$\begin{aligned} \mathbf{A}_\wedge(\mathbf{B}_\wedge\mathbf{C}) &= \mathbf{e}_i\epsilon_{ijk}A_j(\mathbf{B}_\wedge\mathbf{C})_k \\ &= \mathbf{e}_i\epsilon_{ijk}A_j\epsilon_{klm}B_lC_m. \end{aligned} \tag{2.31}$$

The combination of two permutation symbols in one expression has not been encountered before. In addition, the subscript k is repeated in the product

$\epsilon_{ijk}\epsilon_{klm}$ ($= \epsilon_{kij}\epsilon_{klm}$), so that the sum rules involving the permutation symbols must be considered before Equation 2.31 may be simplified.

2.5 The ϵ_{ijk} sum rules

The permutation symbol ϵ_{ijk} has been used to express the vector product of two vectors in terms of their components (Equation 2.26). This was also accomplished in Equation 2.7 using a determinant. It would not be surprising if ϵ_{ijk} could be expressed as a determinant and indeed it is possible, one way being:

$$\epsilon_{ijk} = \begin{vmatrix} \delta_{ix} & \delta_{iy} & \delta_{iz} \\ \delta_{jx} & \delta_{jy} & \delta_{jz} \\ \delta_{kx} & \delta_{ky} & \delta_{kz} \end{vmatrix} . \tag{2.32}$$

This may easily be checked. From the rules relating to the interchange of rows in determinants, it can be seen that a cyclic permutation of the subscripts i,j,k does not alter the value of the determinant, but a non-cyclic one changes its sign; thus explicit consideration of, say, ϵ_{xyz} shows that this definition is consistent with Equation 2.23a. Agreement with Equation 2.23b may be demonstrated by noting that if any pair of subscripts are identical, then the determinant will be zero, since two rows are identical. (Any reader unfamiliar with the manipulations of determinants is reminded that these are summarized in the section on matrices later in this chapter).

Using the rule for multiplying two determinants together (similar to that for matrices), the general product of two permutation symbols may be expressed as one determinant:

$$\epsilon_{ijk}\epsilon_{lmn} = \begin{vmatrix} \delta_{ix} & \delta_{iy} & \delta_{iz} \\ \delta_{jx} & \delta_{jy} & \delta_{jz} \\ \delta_{kx} & \delta_{ky} & \delta_{kz} \end{vmatrix} \begin{vmatrix} \delta_{lx} & \delta_{ly} & \delta_{lz} \\ \delta_{mx} & \delta_{my} & \delta_{mz} \\ \delta_{nx} & \delta_{ny} & \delta_{nz} \end{vmatrix}$$

$$= \begin{vmatrix} \delta_{il} & \delta_{im} & \delta_{in} \\ \delta_{jl} & \delta_{jm} & \delta_{jn} \\ \delta_{kl} & \delta_{km} & \delta_{kn} \end{vmatrix} ; \tag{2.33}$$

this result is best obtained by transposing the second determinant before multiplication. Although each subscript appears three times in the above determinant, expansion leads to a sum of six terms, each of which contains a particular subscript once and only once, so there is no inconsistency with the repeated subscript convention.

However, if the product $\epsilon_{ijk}\epsilon_{imn}$ is under consideration, summation over i is implied and we have:

$$\epsilon_{ijk}\epsilon_{imn} = \begin{vmatrix} \delta_{ii} & \delta_{im} & \delta_{in} \\ \delta_{ji} & \delta_{jm} & \delta_{jn} \\ \delta_{ki} & \delta_{km} & \delta_{kn} \end{vmatrix}$$

$$\begin{aligned} &= \delta_{ii}(\delta_{jm}\delta_{kn} - \delta_{jn}\delta_{km}) \\ &\quad - \delta_{im}(\delta_{ji}\delta_{kn} - \delta_{jn}\delta_{ki}) \\ &\quad + \delta_{in}(\delta_{ji}\delta_{km} - \delta_{jm}\delta_{ki}). \end{aligned} \tag{2.34}$$

In evaluating this expression we recall that $\delta_{ii} = 3$ (Equation 2.21) and note that the Kronecker delta sum rule (Equation 2.20) gives, for example, $\delta_{im}\delta_{ji} = \delta_{mj}$. Substitution leads to the first sum rule:

$$\epsilon_{ijk}\epsilon_{imn} = \delta_{jm}\delta_{kn} - \delta_{jn}\delta_{km}. \tag{2.35}$$

Before using this result on the vector triple product, the problem that initiated this discussion of sum rules, two other sum rules will be established. The first arises when there are two pairs of identical subscripts in the product of two permutation symbols, for example $\epsilon_{ijk}\epsilon_{ijn}$. Equation 2.35 gives immediately:

$$\begin{aligned} \epsilon_{ijk}\epsilon_{ijn} &= \delta_{jj}\delta_{kn} - \delta_{jn}\delta_{kj} \\ &= 3\delta_{kn} - \delta_{kn} \\ &= 2\delta_{kn}. \end{aligned} \tag{2.36}$$

Finally, for three pairs of identical subscripts:

$$\epsilon_{ijk}\epsilon_{ijk} = 2\delta_{kk} = 6. \tag{2.37}$$

Returning to the problem of the triple vector product in Equation 2.31, use of Equations 2.35, 2.19 and 2.12 gives:

$$\begin{aligned} \mathbf{A}_\wedge(\mathbf{B}_\wedge\mathbf{C}) &= \mathbf{e}_i\epsilon_{kij}\epsilon_{klm}A_jB_lC_m \\ &= \mathbf{e}_i(\delta_{il}\delta_{jm} - \delta_{im}\delta_{jl})A_jB_lC_m \\ &= \mathbf{e}_iA_j(B_iC_j - B_jC_i) \\ &= (\mathbf{A}.\mathbf{C})\mathbf{B} - (\mathbf{A}.\mathbf{B})\mathbf{C}. \end{aligned} \tag{2.38}$$

2.6 Examples I

In this section we consider a number of examples of the use of the notation that has been introduced. The reader is advised to attempt these problems as exercises before looking at the solutions.

(a) Show that the quadruple vector product $(\mathbf{A}_\wedge \mathbf{B}).(\mathbf{C}_\wedge \mathbf{D})$ may be expanded as:

$$(\mathbf{A}_\wedge \mathbf{B}).(\mathbf{C}_\wedge \mathbf{D}) = (\mathbf{A}.\mathbf{C})(\mathbf{B}.\mathbf{D}) - (\mathbf{A}.\mathbf{D})(\mathbf{B}.\mathbf{C}). \tag{2.39}$$

Solution:

$$\begin{aligned}(\mathbf{A}_\wedge \mathbf{B}).(\mathbf{C}_\wedge \mathbf{D}) &= \epsilon_{ijk}A_jB_k\epsilon_{imn}C_mD_n \\ &= (\delta_{jm}\delta_{kn} - \delta_{jn}\delta_{km})A_jB_kC_mD_n \\ &= A_jB_kC_jD_k - A_jB_kC_kD_j \\ &= (\mathbf{A}.\mathbf{C})(\mathbf{B}.\mathbf{D}) - (\mathbf{A}.\mathbf{D})(\mathbf{B}.\mathbf{C}).\end{aligned} \tag{2.40}$$

(b) Express as one equation the commutation relations in Equations 1.31 for the components of the spin vector $\boldsymbol{\sigma}$:

$$[\sigma_x, \sigma_y] = 2i\sigma_z; \quad [\sigma_y, \sigma_z] = 2i\sigma_x; \quad [\sigma_z, \sigma_x] = 2i\sigma_y.$$

Solution:

$$[\sigma_i, \sigma_j] = 2i\epsilon_{ijk}\sigma_k \tag{2.41}$$

or multiplying by $\frac{1}{2}\epsilon_{ijn}\mathbf{e}_n$, expanding the commutator and using Equations 2.26 and 2.36:

$$\boldsymbol{\sigma}_\wedge \boldsymbol{\sigma} = 2i\boldsymbol{\sigma}. \tag{2.42}$$

(c) Prove the following identity for any two vectors, **A** and **B**, that commute with the spin vector $\boldsymbol{\sigma}$, but not necessarily with each other:

$$(\boldsymbol{\sigma}.\mathbf{A})(\boldsymbol{\sigma}.\mathbf{B}) = (\mathbf{A}.\mathbf{B}) + i\boldsymbol{\sigma}.(\mathbf{A}_\wedge \mathbf{B}). \tag{2.43}$$

Solution:

In many of the previous examples, it has been assumed without comment that all the components of all the vectors commute with one another. In most of the situations that will be encountered in this book, and in the present example in particular, this assumption is not justified and we must keep track of the order of the vector components. The commutation and anticommutation relations for the components of $\boldsymbol{\sigma}$ are:

$$\boldsymbol{\sigma}_\wedge \boldsymbol{\sigma} = 2i\boldsymbol{\sigma} \tag{2.42}$$

and

$$[\sigma_i, \sigma_j]_+ = 2\delta_{ij}. \tag{2.22}$$

To prove Equation 2.43 we note that, since $\boldsymbol{\sigma}$ commutes with **A** and **B**, the term $(\boldsymbol{\sigma}.\mathbf{A})(\boldsymbol{\sigma}.\mathbf{B})$ is identical to the first term on the right-hand side of Equation 2.39, provided **C** and **D** are both set equal to $\boldsymbol{\sigma}$. Equation 2.39 cannot be used directly, since the order of vector components was not preserved in its proof, but we can start with the third line of Equation 2.40:

$$(\mathbf{A}_\wedge \mathbf{B}).(\boldsymbol{\sigma}_\wedge\boldsymbol{\sigma}) = A_jB_k\sigma_j\sigma_k - A_jB_k\sigma_k\sigma_j.$$

Substitution of Equations 2.42 and 2.22 gives:

$$\begin{aligned} 2i\boldsymbol{\sigma}.(\mathbf{A}_\wedge \mathbf{B}) &= \sigma_j\sigma_kA_jB_k - (-\sigma_j\sigma_k + 2\delta_{jk})A_jB_k \\ &= 2\sigma_j\sigma_kA_jB_k - 2A_jB_j. \end{aligned}$$

On rearrangement and division by 2:

$$(\boldsymbol{\sigma}.\mathbf{A})(\boldsymbol{\sigma}.\mathbf{B}) = (\mathbf{A}.\mathbf{B}) + i\boldsymbol{\sigma}.(\mathbf{A}_\wedge \mathbf{B});$$

this is an important relationship and will be used a number of times in Chapter 8 onwards.

2.7 The vector operator ∇

So far two types of quantity have been encountered, scalars and vectors. Both scalars and vectors can be functions of time and space coordinates or other variables. (These two types of functions are usually called scalar fields and vector fields. An example often given for a scalar field is the temperature, since this can vary in space and time, but is represented by a scalar quantity at each point, while an example of a vector field is heat flow, which needs a vector at each point in space and time to specify it.) Situations frequently arise where it is necessary to differentiate a scalar or a vector. Differentiation with respect to time presents no problems and will not be considered here, but differentiation with respect to spatial coordinates leads to the vector operator ∇.

Let us consider a scalar function $\phi(x, y, z)$. For infinitesimal increases in x, y and z the corresponding change in ϕ is given by:

$$\mathrm{d}\phi = (\partial\phi/\partial x)\mathrm{d}x + (\partial\phi/\partial y)\mathrm{d}y + (\partial\phi/\partial z)\mathrm{d}z. \tag{2.44}$$

This expression can be thought of as the scalar product of two vectors with components $((\partial\phi/\partial x), (\partial\phi/\partial y), (\partial\phi/\partial z))$ and $(\mathrm{d}x, \mathrm{d}y, \mathrm{d}z)$. If the point x, y, z is specified by the vector **r** then the second vector may be thought of as d**r**, while the first vector consists of the scalar ϕ being operated on by a vector operator with components $((\partial/\partial x), (\partial/\partial y), (\partial/\partial z))$. This operator is called 'del' and has the symbol ∇:

$$\nabla = \mathbf{e}_x \partial/\partial x + \mathbf{e}_y \partial/\partial y + \mathbf{e}_z \partial/\partial z$$

$$= \mathbf{e}_i \partial/\partial r_i = \mathbf{e}_i \nabla_i, \tag{2.45}$$

where the repeated subscript convention has again been introduced.

When the operator ∇ operates on a scalar function ϕ, there is only one possible result, a vector $\nabla\phi$, which is called the gradient of ϕ. However, two possibilities arise when ∇ is allowed to operate on a vector $\mathbf{A}$, since either the scalar product or the vector product between ∇ and $\mathbf{A}$ may be taken. The first possibility, $\nabla.\mathbf{A}$, is a scalar quantity and is known as the divergence of $\mathbf{A}$, while the second, $\nabla \wedge \mathbf{A}$, is a vector called the curl of $\mathbf{A}$. The reason for the names, gradient, divergence and curl, will become clearer when their physical significance is considered in the following sections.

2.8 The gradient

The gradient of a scalar function $\phi(x, y, z)$ is a vector and from Equation 2.45 may be written:

$$\nabla\phi = \mathbf{e}_i(\partial\phi/\partial r_i) = \mathbf{e}_i \nabla_i \phi. \tag{2.46}$$

For a small change $\mathbf{dr}$ in the position vector $\mathbf{r}$ of the point (x, y, z) the corresponding change in ϕ is, from Equation 2.44:

$$\mathrm{d}\phi = (\nabla\phi).\mathbf{dr}. \tag{2.47}$$

To see what $\nabla\phi$ means physically we note that, if k is a constant, the equation $\phi(x,y,z) = k$ represents a surface. If $\mathbf{dr}$ is taken to be tangential to the surface, that is the change in position $\mathbf{dr}$ takes us to a new point in the same surface, then $\mathrm{d}\phi = 0$ and

$$(\nabla\phi).\mathbf{dr} = 0. \tag{2.48}$$

Since $\mathbf{dr}$ is tangential to the surface and can be in any arbitrary direction in the surface, $\nabla\phi$ must be perpendicular to the surface. Then again, if $\mathbf{dr}$ were not tangential to the surface, but were to take us to some new surface, $\phi = k'$, Equation 2.47 indicates that the increase in ϕ will be greatest when $\mathbf{dr}$ is parallel to $\nabla\phi$, that is perpendicular to the surface. Thus $\nabla\phi$ represents in magnitude and direction, the maximum rate of increase of the scalar function ϕ.

Suppose Equation 2.47 were to be integrated between two points 1 and 2 along some line joining the points. That is for each infinitesimal element $\mathbf{dr}$ on the line, we calculate $(\nabla\phi).\mathbf{dr}$ and then add the contributions from all the elements of the line. The result would be:

$$\phi_2 - \phi_1 = \int_1^2 (\nabla\phi).\mathbf{dr}. \tag{2.49}$$

The left-hand side of this equation depends only on the location of the two points and is independent of the line along which the integral is performed. It follows that the line integral on the right-hand side of the equation must also be independent of the path taken. In the special case where the integration is over a closed path and point 2 is coincident with point 1:

$$\int_1^1 (\nabla\phi).\mathbf{dr} = \oint (\nabla\phi).\mathbf{dr} = 0. \tag{2.50}$$

2.9 The divergence

The divergence of the vector **A** has already been defined as the scalar $\nabla.\mathbf{A}$; in terms of components:

$$\nabla.\mathbf{A} = (\partial A_i/\partial r_i) = \nabla_i A_i. \tag{2.51}$$

Let us consider the way in which electric charge (or heat etc.) flows out of a volume element $d\tau = dxdydz$ centred on the point x, y, z (Fig. 2.1). If the

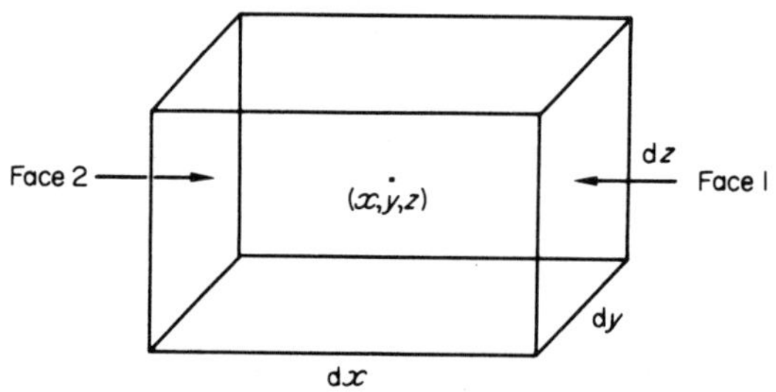

Fig. 2.1. The volume element $dxdydz$ used to demonstrate the significance of the divergence of a vector.

vector $\mathbf{j} = \mathbf{j}(x, y, z)$ represents the current density (flow of charge, or current, per unit area perpendicular to the direction of **j**), then the charge loss per unit time from the volume element through face 1 is:

$$j_x(x + \tfrac{1}{2}dx, y, z)dydz \simeq [j_x + \tfrac{1}{2}(\partial j_x/\partial x)dx]\,dydz, \tag{2.52}$$

while for face 2 the rate of gain of charge is:

$$j_x(x - \tfrac{1}{2}dx, y, z)dydz \simeq [j_x - \tfrac{1}{2}(\partial j_x/\partial x)dx]\,dydz. \tag{2.53}$$

The combined charge loss rate through the two faces is thus $(\partial j_x/\partial x)\mathrm{d}x\mathrm{d}y\mathrm{d}z$. The charge flow across the other faces of the volume element may be found in the same way giving for the total charge loss rate from the volume element:

$$[(\partial j_x/\partial x) + (\partial j_y/\partial y) + (\partial j_z/\partial z)]\mathrm{d}x\mathrm{d}y\mathrm{d}z = (\nabla.\mathbf{j})\mathrm{d}\tau \qquad (2.54)$$

or a loss rate per unit volume of $\nabla.\mathbf{j}$. Thus, since $\mathbf{j}$ is itself a rate, the divergence of a vector represents a loss per unit volume.

If the charge density associated with the volume element under consideration is $\rho(x, y, z)$, this loss rate per unit volume could also be expressed as $-(\partial\rho/\partial t)$ so that the equation:

$$\nabla.\mathbf{j} + (\partial\rho/\partial t) = 0 \qquad (2.55)$$

expresses mathematically the law of conservation of charge, that is charge can be neither created nor destroyed and any charge lost by a volume element must go through the surface of the element.

The rate of loss of charge can also be thought of as an integral over the surface of the volume element. An infinitesimal surface element may be represented by ds, which is a vector normal to the surface element and with magnitude equal to the area of the element. Thus the total loss rate from the volume element is $\int \mathbf{j}.\mathrm{d}\mathbf{s}$, so that from Equation 2.54:

$$\int \mathbf{j}.\mathrm{d}\mathbf{s} = \nabla.\mathbf{j}\,\mathrm{d}\tau. \qquad (2.56)$$

This equation applies to an infinitesimal volume element $\mathrm{d}\tau$, but it may be extended to a finite irregularly shaped volume by dividing this volume into infinitesimal volume elements and integrating. Except at the boundaries of the volume under consideration any face of a volume element is shared with another. The loss from one volume element through a shared face is the gain of the other volume element through the same face and the two cancel on integration, so that Equation 2.56 generalizes to:

$$\int \mathbf{j}.\mathrm{d}\mathbf{s} = \int \nabla.\mathbf{j}\,\mathrm{d}\tau \qquad (2.57)$$

for an extended region of space; the integral on the left is a surface integral, while that on the right is a volume integral, the whole equation being known as Gauss' theorem.

2.10 The curl

The curl of a vector **A**, written as $\nabla_\wedge \mathbf{A}$, is itself a vector and from Equation 2.26 may be expanded in terms of the components:

$$\nabla_\wedge \mathbf{A} = \mathbf{e}_i \epsilon_{ijk} \nabla_j A_k. \tag{2.58}$$

Its physical significance is not as easy to see as in the case of the gradient or the divergence, but consideration of an area element is instructive. For convenience it is assumed to be parallel to the xy plane (Fig. 2.2), but generalization to other orientations is possible. The circulation of the vector **A** round

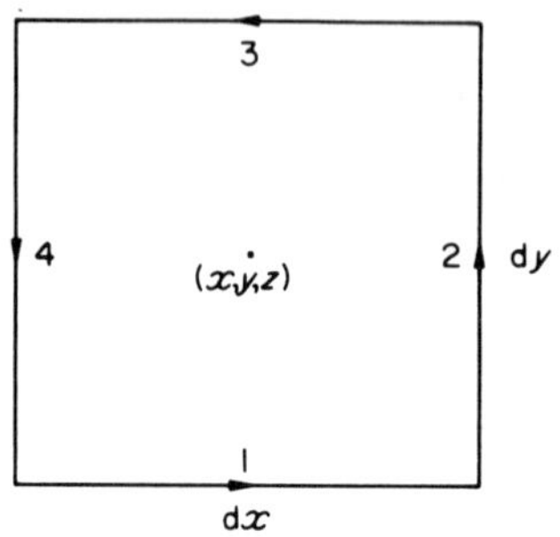

Fig. 2.2. The surface element dxdy used to demonstrate the significance of the curl of a vector.

the boundary of the surface element is the line integral round this closed loop of the component **A** along the line element, that is $\oint \mathbf{A}.\mathbf{dr}$. This integral may be broken down into contributions from the sides of the area element, in the same way that the contributions from each face of a volume element were considered when the divergence of a vector was discussed. If the integration round the boundary of the area element is carried out in the direction of the arrows in the figure, the contribution from side 1 is:

$$A_x(x, y - \tfrac{1}{2}\mathrm{d}y, z)\mathrm{d}x \simeq [A_x - \tfrac{1}{2}(\partial A_x/\partial y)\mathrm{d}y]\mathrm{d}x, \tag{2.59}$$

while that from side 3 is:

$$-A_x(x, y + \tfrac{1}{2}\mathrm{d}y, z)\mathrm{d}x \simeq -[A_x + \tfrac{1}{2}(\partial A_x/\partial y)\mathrm{d}y]\mathrm{d}x. \tag{2.60}$$

The combined contribution from these two sides is $-(\partial A_x/\partial y)\mathrm{d}x\mathrm{d}y$. The sides 2 and 4 may be treated in the same manner to give for the total line integral around the boundary of the area element:

$$\oint \mathbf{A}.\mathbf{dr} = [(\partial A_y/\partial x) - (\partial A_x/\partial y)]\mathrm{d}x\mathrm{d}y. \tag{2.61}$$

The term in square parentheses is just the z component of $\nabla_\wedge \mathbf{A}$, while dxdy is the magnitude of the area element. If as before the area element is represented by a vector ds normal to it, Equation 2.61 may be generalized to:

$$\oint \mathbf{A}.\mathbf{dr} = (\nabla_\wedge \mathbf{A}).\mathbf{ds}. \tag{2.62}$$

Thus the circulation of a vector round the boundary of an infinitesimal surface element is the component of the curl of the vector perpendicular to the element times the area of the element.

Equation 2.62 may be extended to a general closed loop by taking any surface bounded by the loop and splitting the surface up into infinitesimal surface elements. Except at the boundary of the surface each edge of a surface element is shared and the contributions to the line integral from two neighbouring elements cancel, since the direction of circulation is taken to be anticlockwise in all elements. Thus:

$$\oint \mathbf{A}.\mathrm{d}\mathbf{r} = \int (\boldsymbol{\nabla}_\wedge \mathbf{A}).\mathrm{d}\mathbf{s}; \qquad (2.63)$$

the left-hand side is a line integral round a closed loop while the right-hand side is a surface integral over any surface bounded by the loop. Equation 2.63 is known as Stokes' theorem.

2.11 Examples II

In this section are given a number of useful relationships involving the vector differentiation of products; these are also listed in Appendix B for convenience. If **A** and **B** are vectors and ϕ is a scalar, then:

$$\boldsymbol{\nabla}.(\phi\mathbf{A}) = (\boldsymbol{\nabla}\phi).\mathbf{A} + \phi(\boldsymbol{\nabla}.\mathbf{A}); \qquad (2.64)$$

$$\boldsymbol{\nabla}_\wedge(\phi\mathbf{A}) = (\boldsymbol{\nabla}\phi)_\wedge\mathbf{A} + \phi(\boldsymbol{\nabla}_\wedge\mathbf{A}); \qquad (2.65)$$

$$\boldsymbol{\nabla}(\mathbf{A}.\mathbf{B}) = (\mathbf{B}.\boldsymbol{\nabla})\mathbf{A} + (\mathbf{A}.\boldsymbol{\nabla})\mathbf{B} + \mathbf{B}_\wedge(\boldsymbol{\nabla}_\wedge\mathbf{A}) + \mathbf{A}_\wedge(\boldsymbol{\nabla}_\wedge\mathbf{B}); \qquad (2.66)$$

$$\boldsymbol{\nabla}.(\mathbf{A}_\wedge\mathbf{B}) = (\boldsymbol{\nabla}_\wedge\mathbf{A}).\mathbf{B} - \mathbf{A}.(\boldsymbol{\nabla}_\wedge\mathbf{B}); \qquad (2.67)$$

$$\boldsymbol{\nabla}_\wedge(\mathbf{A}_\wedge\mathbf{B}) = \mathbf{A}(\boldsymbol{\nabla}.\mathbf{B}) - (\mathbf{A}.\boldsymbol{\nabla})\mathbf{B} + (\mathbf{B}.\boldsymbol{\nabla})\mathbf{A} - (\boldsymbol{\nabla}.\mathbf{A})\mathbf{B}. \qquad (2.68)$$

In these relationships it has been assumed that ϕ and the components of **B** all commute with **A** so that their order has not been preserved when a simpler expression results from rearrangement. Consequently care must be used in applying Equations 2.66 and 2.68 in some situations.

The proofs of these relationships should be regarded as exercises. However, the proofs of Equations 2.64 and 2.67 will be given here as examples.

(a) Proof of Equation 2.64:

$$\boldsymbol{\nabla}.(\phi\mathbf{A}) = \nabla_i(\phi A_i) = (\nabla_i\phi)A_i + \phi(\nabla_i A_i)$$

$$= (\boldsymbol{\nabla}\phi).\mathbf{A} + \phi(\boldsymbol{\nabla}.\mathbf{A}). \qquad (2.64)$$

(b) Proof of Equation 2.67:

$$\begin{aligned}\nabla.(\mathbf{A}\wedge\mathbf{B}) &= \nabla_i(\epsilon_{ijk}A_jB_k) = \epsilon_{ijk}[(\nabla_iA_j)B_k + A_j(\nabla_iB_k)]\\ &= \epsilon_{kij}(\nabla_iA_j)B_k - \epsilon_{jik}A_j(\nabla_iB_k)\\ &= (\nabla\wedge\mathbf{A}).\mathbf{B} - \mathbf{A}.(\nabla\wedge\mathbf{B}).\end{aligned} \tag{2.67}$$

2.12 Second derivatives in vector calculus

There are six possible ways in which ∇ can operate twice on a scalar ϕ or a vector **A**: $\nabla.(\nabla\phi)$; $(\nabla.\nabla)\mathbf{A}$; $\nabla(\nabla.\mathbf{A})$; $\nabla\wedge(\nabla\phi)$; $\nabla.(\nabla\wedge\mathbf{A})$; $\nabla\wedge(\nabla\wedge\mathbf{A})$. There are no meaningful combinations of the symbols, that are not included in this list.

The first case, $\nabla.(\nabla\phi)$, may also be written without ambiguity as $(\nabla.\nabla)\phi$ or even $\nabla^2\phi$ and the scalar operator ∇^2 expanded in terms of components:

$$\nabla^2 = \nabla_i\nabla_i = (\partial^2/\partial x^2) + (\partial^2/\partial y^2) + (\partial^2/\partial z^2); \tag{2.69}$$

it occurs frequently in physics and is known as the Laplacian. Since it is a scalar, the Laplacian can also operate on a vector and an example of this possibility appears second in the original list; however, this possibility is of little interest as is the third possibility, $\nabla(\nabla.A)$.

The next two cases both vanish as can be seen by rewriting them in alternative but equivalent ways:

$$\nabla\wedge(\nabla\phi) = (\nabla\wedge\nabla)\phi = 0 \tag{2.70}$$

and

$$\nabla.(\nabla\wedge\mathbf{A}) = (\nabla\wedge\nabla).\mathbf{A} = 0; \tag{2.71}$$

in each case the operator $(\nabla\wedge\nabla)$ vanishes. These two identities have an interesting use. Suppose that the curl of some vector **B** vanishes, that is $(\nabla\wedge\mathbf{B}) = 0$. Then Equation 2.70 tells us that, if **B** is expressed as the gradient of an appropriate scalar, the condition that its curl is zero is automatically satisfied. Similarly, from Equation 2.71 it can be seen that, if a divergenceless vector is written as the curl of another vector, its divergence is guaranteed to be zero.

Finally, we note that the remaining possibility, $\nabla\wedge(\nabla\wedge\mathbf{A})$, may be expanded using Equation 2.68 provided it is remembered that ∇ and **A** do not commute and their order must be preserved:

$$\nabla\wedge(\nabla\wedge\mathbf{A}) = \nabla(\nabla.\mathbf{A}) - \nabla^2\mathbf{A}. \tag{2.72}$$

2.13 The Dirac delta function

So far we have been referring to general vectors **A**, **B** etc. In the remainder of this book the position vector **r**:

$$\mathbf{r} = \mathbf{e}_x r_x + \mathbf{e}_y r_y + \mathbf{e}_z r_z = \mathbf{e}_i r_i \tag{2.73}$$

will often be encountered. In general the result of operating on a component of **r** with a component of ∇ is given by:

$$\nabla_i r_j = (\partial r_j / \partial r_i) = \delta_{ij}. \tag{2.74}$$

This equation, together with Equation 2.21, immediately tells us that the divergence of **r** is equal to three:

$$\boldsymbol{\nabla}.\mathbf{r} = \nabla_i r_i = \delta_{ii} = 3. \tag{2.75}$$

On the other hand the curl of **r** vanishes:

$$\boldsymbol{\nabla} \wedge \mathbf{r} = \mathbf{e}_i \epsilon_{ijk} \nabla_j r_k = \mathbf{e}_i \epsilon_{ijk} \delta_{jk} = \mathbf{e}_i \epsilon_{ijj} = 0. \tag{2.76}$$

Another simple result that is occasionally useful is:

$$(\mathbf{A}.\boldsymbol{\nabla})\mathbf{r} = A_i \nabla_i \mathbf{e}_j r_j = \mathbf{e}_j A_i \delta_{ij} = \mathbf{e}_j A_j = \mathbf{A}, \tag{2.77}$$

where **A** is any vector.

A more complicated example, that will be encountered later, is $\boldsymbol{\nabla}(1/r)$, where $r\ (=(r_i r_i)^{\frac{1}{2}})$ is the magnitude of the position vector **r**:

$$\begin{aligned}\boldsymbol{\nabla}(1/r) &= \mathbf{e}_i \nabla_i [(r_j r_j)^{-\frac{1}{2}}] = \mathbf{e}_i [-r^{-3}/2](\nabla_i r_j r_j) \\ &= -\mathbf{e}_i [r^{-3}/2][2 r_j \delta_{ij}] = -\mathbf{e}_i r^{-3} r_i \\ &= -\mathbf{r}/r^3.\end{aligned} \tag{2.78}$$

We could proceed in the same way to evaluate $\nabla^2(1/r)$, which will also be met later. From Equation 2.78 this is:

$$\begin{aligned}\nabla^2(1/r) &= \boldsymbol{\nabla}.(-\mathbf{r}/r^3) = -\nabla_i(r_i/r^3) \\ &= -r^{-3}(\nabla_i r_i) - r_i(\nabla_i r^{-3}) \\ &= -r^{-3}\delta_{ii} - r_i(-3r^{-5} r_i) \\ &= -3r^{-3} + 3r^{-3} = 0.\end{aligned} \tag{2.79}$$

This suggests that $\nabla^2(1/r)$ vanishes everywhere, but is this true when $\mathbf{r} = 0$ and $(1/r)$ is infinite? Equation 2.78 shows that $\boldsymbol{\nabla}(1/r)$, which is expected to vary roughly as r^{-2}, also becomes infinite at $\mathbf{r} = 0$ and thus poses no problems. But

$\nabla^2(1/r)$ is expected to behave roughly as r^{-3}, so that it should also be infinite at $\mathbf{r} = 0$. In fact, Equation 2.79 is incomplete, since it is not true when $\mathbf{r}$ vanishes, although it is correct for non-zero $\mathbf{r}$.

To demonstrate how Equation 2.79 needs to be modified, the correct result will first be quoted, since a new symbol needs to be introduced. Equation 2.79 should read:

$$\nabla^2(1/r) = -\nabla.(\mathbf{r}/r^3) = -4\pi\delta(\mathbf{r}), \tag{2.80}$$

where $\delta(\mathbf{r})$ is the Dirac delta function (not to be confused with the Kronecker delta), which is zero everywhere except when $\mathbf{r}$ is zero. Now $\delta(\mathbf{r})$ is a three-dimensional delta function and can be factorized into a product of one-dimensional delta functions:

$$\delta(\mathbf{r}) = \delta(r_x)\delta(r_y)\delta(r_z) = \delta(x)\delta(y)\delta(z), \tag{2.81}$$

since $\mathbf{r}$ is only zero when x, y and z all vanish. We will concentrate for the moment on the one-dimensional Dirac delta function $\delta(x)$, since we can easily generalize to the three-dimensional case using Equation 2.81.

The function $\delta(x)$ is zero everywhere except when x vanishes, but what happens to $\delta(x)$ when $x = 0$? From what has been said already one might expect $\delta(x = 0)$ to be infinite and indeed it is, but it is defined so that it is infinite in a special way. If $f(x)$ is any function of x then $\delta(x)$ must satisfy:

$$\int_{-\infty}^{\infty} f(x)\delta(x)\mathrm{d}x = f(0). \tag{2.82}$$

In particular if $f(x) = 1$ then:

$$\int_{-\infty}^{\infty} \delta(x)\mathrm{d}x = 1, \tag{2.83}$$

so that $\delta(x)$ is zero everywhere except at the origin where it is infinite in such a way that the area under it is unity.

Thus $\delta(x)$ has very strange properties compared with the mathematical functions usually encountered. However, it can be related to the usual type of functions in the following way. The function:

$$(n/\pi)^{\frac{1}{2}} \exp(-nx^2), \tag{2.84}$$

is plotted in Fig. 2.3 for a number of positive values of n. If this function is integrated with respect to x between $\pm\infty$, then no matter what the value of n is, the result is unity because there is a standard integral:

$$\int_{-\infty}^{\infty} \exp(-a^2x^2)\mathrm{d}x = \sqrt{\pi}/a. \tag{2.85}$$

From the figure it can be seen that as n increases the function (Equation 2.84) gets spikier at $x = 0$, and in the limit when n becomes infinite the function has the same properties as the Dirac delta function $\delta(x)$ in that it is zero everywhere except at $x = 0$, where it is infinite in such a way that the area under it is unity and Equations 2.82 and 2.83 are satisfied. Thus an alternative definition of $\delta(x)$ is:

$$\delta(x) = \underset{n\to\infty}{Lt} (n/\pi)^{\frac{1}{2}} \exp(-nx^2). \tag{2.86}$$

Other limits may be used to represent $\delta(x)$ and one of these will be encountered in Chapter 13.

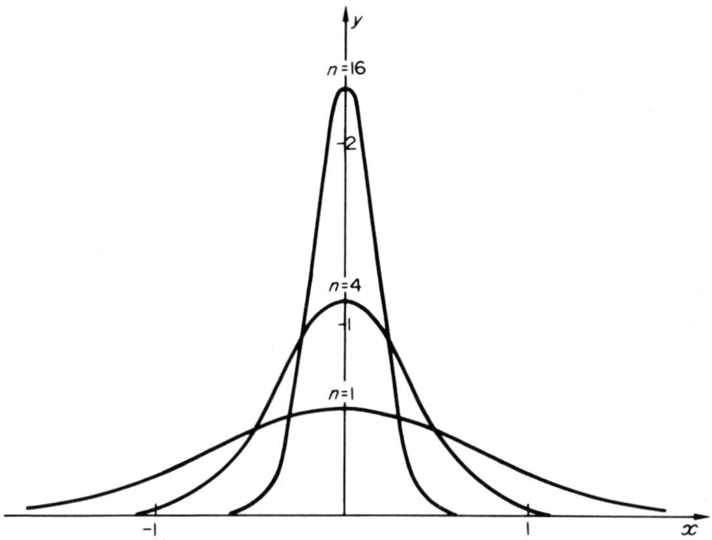

Fig. 2.3. A plot of $y = (n/\pi)^{\frac{1}{2}} \exp(-nx^2)$ for $n = 1$, 4 and 16.

It should be noted that because of its definition the Dirac delta function can only be used in integrals as in Equation 2.82; however, in this book the delta function usually arises in quantum mechanical operators, the matrix elements of which involve integration. Consequently, an operator expression such as $x\delta(x)$ may be replaced by zero:

$$x\delta(x) = 0, \tag{2.87}$$

since Equation 2.82 tells us that an integration involving it will vanish. In addition, Equation 2.82 may be used to show that:

$$\delta(cx) = \delta(x)/c, \tag{2.88}$$

where c is a constant.

It is now possible to demonstrate the validity of Equation 2.80. To do this some of the results of Chapter 5 are anticipated, but the pertinent ones should be familiar. It will be recalled that Gauss' law, which is one of Maxwell's equations, may be written in SI units as:

$$\boldsymbol{\nabla}.\mathbf{E} = \rho/\epsilon_0, \tag{2.89}$$

where $\mathbf{E}$ is the electric field, ρ the charge density and ϵ_0 the permittivity in a vacuum. Now in electrostatics the electric field may be expressed as $-\boldsymbol{\nabla}\phi$, where ϕ is a scalar potential, so that:

$$\nabla^2\phi = -\rho/\epsilon_0. \tag{2.90}$$

In the special case of a point charge q at the origin, the scalar potential at a point $\mathbf{r}$ is given by:

$$\phi = q/4\pi\epsilon_0 r. \tag{2.91}$$

Substitution into Equation 2.90 then gives:

$$\nabla^2(1/r) = -4\pi\rho/q. \tag{2.92}$$

The left-hand side of this equation is the quantity we wish to evaluate; it only remains to interpret ρ, the charge density due to a point charge at the origin. This is zero everywhere except at $\mathbf{r} = 0$ where it is infinite; in addition the total charge in space $\int \rho d\tau$, must be just q so that:

$$\rho = q\delta(\mathbf{r}) \tag{2.93}$$

is an appropriate description. Substitution into Equation 2.92 gives Equation 2.80, which is the desired result.

In a later chapter the expression $\nabla_i(r_j/r^3)$ will be met. If $i \neq j$, reference to Equation 2.79 shows that it may be expressed as:

$$\nabla_i(r_j/r^3) = -3r_ir_j/r^5 \qquad (i \neq j), \tag{2.94}$$

but if $i = j$ Equation 2.80 shows that

$$\nabla_i(r_i/r^3) = 4\pi\delta(\mathbf{r}), \tag{2.95}$$

where summation over i is of course implied. These two equations may be written as one:

$$\nabla_i(r_j/r^3) = (4\pi/3)\delta(\mathbf{r})\delta_{ij} + \delta_{ij}/r^3 - 3r_ir_j/r^5, \tag{2.96}$$

since $\delta_{ii} = 3$ (Equation 2.21) and by symmetry the contributions from each of the three terms in the summation on the left-hand side of Equation 2.95 are equal.

2.14 Matrices and determinants: a summary

Although a knowledge of matrices is assumed, it will be useful to summarize some of the more pertinent properties of matrices using where possible the repeated subscript convention, the Kronecker delta and the permutation symbol.

A matrix is merely a rectangular array of numbers. An $m \times n$ matrix **A** has m rows and n columns, a typical element being A_{ij} $(i = 1, \ldots m; j = 1, \ldots n)$. Provided the dimensions are the same, two matrices may be added or subtracted by adding or subtracting corresponding elements; these operations are associative and commutative. However, the more important operation of matrix multiplication is in general non-commutative, although it is associative. Again the sizes of the matrices must be compatible in that an $m \times n$ matrix **A** can only be multiplied by an $n \times p$ matrix **B** to give an $m \times p$ matrix **C**(= **AB**) as the product. The elements of these matrices are related by:

$$C_{ik} = A_{ij}B_{jk}, \tag{2.97}$$

where summation over $j(= 1, \ldots n)$ is implied by the repeated subscript notation.

The transpose $\widetilde{\mathbf{A}}$ of an $m \times n$ matrix **A** is defined as the $n \times m$ matrix with the elements:

$$\widetilde{A}_{ij} = A_{ji}. \tag{2.98}$$

It may be shown by a simple application of Equation 2.97 that the transpose of the product of two matrices is the product of the transposes of the two matrices but taken in reverse order:

$$(\widetilde{\mathbf{AB}}) = \widetilde{\mathbf{B}}\widetilde{\mathbf{A}}. \tag{2.99}$$

There are a number of specific types of $n \times n$ or square matrices. A diagonal matrix is a square matrix in which only the elements on the major diagonal are non-zero; a special case of a diagonal matrix is the unit matrix **I**, in which all the diagonal elements are equal to unity:

$$I_{ij} = \delta_{ij}. \tag{2.100}$$

The adjoint $\mathbf{A}^+$ of the matrix **A** is obtained by transposing the elements and taking the complex conjugate:

$$\mathbf{A}^+ = \widetilde{\mathbf{A}}^* \quad \text{or} \quad A^+_{ij} = A^*_{ji}. \tag{2.101}$$

A matrix is said to be Hermitian if it is self-adjoint:

$$\mathbf{A}^+ = \mathbf{A}. \tag{2.102}$$

A square matrix **A** has an inverse if there exists a matrix $\mathbf{A}^{-1}$ such that:

$$\mathbf{A}\mathbf{A}^{-1} = \mathbf{A}^{-1}\mathbf{A} = \mathbf{I}, \tag{2.103}$$

and is said to be orthogonal if this inverse is equal to its transpose:

$$\mathbf{A}^{-1} = \widetilde{\mathbf{A}}; \tag{2.104}$$

on the other hand **A** is unitary if its inverse is the same as its adjoint:

$$\mathbf{A}^{-1} = \mathbf{A}^{+}. \tag{2.105}$$

As for the transpose the inverse of the product of two square matrices is the product of the inverses taken in reverse order:

$$(\mathbf{AB})^{-1} = \mathbf{B}^{-1}\mathbf{A}^{-1} \tag{2.106}$$

as may be checked by substitution into Equation 2.103.

$Tr(\mathbf{A})$ is the notation used for the trace of a square matrix, that is the sum of its diagonal elements:

$$Tr(\mathbf{A}) = A_{ii}. \tag{2.107}$$

It is easy to show that the trace of the product of two matrices $Tr(\mathbf{AB})$ is unchanged by reversing the order of multiplication of the matrices:

$$Tr(\mathbf{AB}) = (\mathbf{AB})_{ii} = A_{ij}B_{ji} = B_{ji}A_{ij} = (\mathbf{BA})_{jj} = Tr(\mathbf{BA}). \tag{2.108}$$

Each $n \times n$ matrix **A** has associated with it a number, the determinant $|A|$, which is a sum of products of the elements:

$$|A| = \Sigma\,(-1)^h A_{1i}A_{2j}\ldots A_{nk}, \tag{2.109}$$

where the summation is over all permutations of $i, j, \ldots, k$ and h is the number of interchanges of pairs of these subscripts necessary to restore the natural order $1, 2, \ldots, n$. Recalling the definition of ϵ_{ijk} we note that a 3×3 determinant can be written as:

$$|A| = \epsilon_{ijk}A_{1i}A_{2j}A_{3k}, \tag{2.110}$$

if i, j, k are interpreted as $1, 2, 3$ rather than x, y, z. The connection between the determinental definition (Equation 2.7) of the vector product of two vectors and the definition (Equation 2.26) using ϵ_{ijk} should now be apparent. From the definition (Equation 2.109) of a determinant a number of useful properties may be proved: interchange of two rows (or columns) will change its sign; if two rows (or columns) are identical its value is zero; adding one row (or column) to another row (or column) does not alter its value etc. Finally, it may be noted that determinants may be multiplied in the same way as square matrices and that the determinant of the product of two matrices is the product of the determinants of the individual matrices.

A set of n simultaneous equations:

$$A_{ij}x_j = y_i \qquad (i, j = 1, \ldots n) \tag{2.111}$$

can be written in matrix form as:

$$\mathbf{Ax} = \mathbf{y}, \tag{2.112}$$

where $\mathbf{A}$ is an $n \times n$ matrix and both $\mathbf{x}$ and $\mathbf{y}$ are $n \times 1$ matrices or column vectors, for example:

$$\mathbf{x} = \begin{pmatrix} x_1 \\ x_2 \\ \cdot \\ \cdot \\ \cdot \\ x_n \end{pmatrix}, \tag{2.113}$$

their transposes being $1 \times n$ matrices or row vectors. In fact, a vector in three-dimensional space can be thought of as a 3×1 matrix, and for this reason no confusion should arise from using heavy type for both vectors and matrices. The only way in which difficulties might arise is in knowing what size matrix is intended by a particular symbol. Since the elimination of this problem would necessitate a clumsy notation, it must always be remembered what size matrix is denoted by a particular symbol in a particular context.

The formal solution of Equation 2.112 is:

$$\mathbf{x} = \mathbf{A}^{-1}\mathbf{y}, \tag{2.114}$$

and we recall that the inverse of a matrix is found by finding for each element A_{ij} its cofactor, which is $(-1)^{i+j}$ times the determinant obtained by removing row i and column j, transposing the resulting matrix and dividing each element by the determinant of $\mathbf{A}$. That is

$$(\mathbf{A}^{-1})_{ij} = \hat{A}_{ji}/|A|, \tag{2.115}$$

where $\hat{A}_{ij}$ is the cofactor of element A_{ij}. If the determinant of the matrix $\mathbf{A}$ vanishes, $\mathbf{A}$ is said to be singular and can possess no inverse.

If $\mathbf{y} = \lambda\mathbf{x}$ were substituted into Equation 2.112, it would read:

$$(\mathbf{A} - \lambda\mathbf{I})\mathbf{x} = 0, \tag{2.116}$$

which is the general form of the eigenvalue equation found in quantum mechanics; the non-trivial solutions of this set of n equations consists of n values of the eigenvalue λ each with its corresponding eigenvector $\mathbf{x}$. The possible eigenvalues are found by solving the polynomial equation:

$$|\mathbf{A} - \lambda\mathbf{I}| = 0 \tag{2.117}$$

and the eigenvectors $\mathbf{x}_r$ corresponding to each eigenvalue λ_r are determined by back-substitution into Equation 2.116 followed by solution of the resulting simultaneous equations. The eigenvectors $\mathbf{x}_r$ are column vectors, which can be written side by side as a square matrix $\mathbf{X}$. This matrix $\mathbf{X}$ is such that:

$$\mathbf{X}^{-1}\mathbf{A}\mathbf{X} = \mathbf{\Lambda}, \tag{2.118}$$

where $\mathbf{\Lambda}$ is the diagonal matrix with the eigenvalues λ_r as the diagonal elements. Equation 2.118 represents a similarity transformation and in this case it diagonalizes the matrix $\mathbf{A}$.

2.15 Vectors in four dimensions

Most of this chapter has been devoted to three-dimensional vectors, each component of which corresponds to one of the three spatial dimensions. In the last section it was seen that these vectors can be thought of as 3×1 matrices and be manipulated accordingly. However, a 3×1 matrix is just a special case of an $n \times 1$ matrix, and such properties as scalar and vector products can be generalized from three to many dimensions, although it is hard to see the significance of say a vector product in n dimensions. But is there any point in doing this? It is merely noted here that when special relativity is considered in Chapter 4, it will be convenient to think of time as a fourth dimension, and that the theory may be developed in terms of four-dimensional vectors or four-vectors. Further discussion of this topic is postponed till then.

Bibliography

Vectors and Matrices

Arfken, G. (1970), *Mathematical Methods for Physicists*, Academic Press, New York: Chapters 1 and 4.

Margenau, H., and Murphy, G. M. (1956), *The Mathematics of Physics and Chemistry*, McGraw-Hill, New York: Volume I, Chapters 4 and 10.

Vectors

Feynman, Leighton and Sands: Volume I, Chapters 11 and 20; Volume II, Chapters 2 and 3.

The Dirac delta function

Arfken, G. (1970), *Mathematical Methods for Physicists*, Academic Press, New York: pages 69 and 413-417.

CHAPTER THREE

Classical Mechanics

Before considering special relativity, and in particular relativistic mechanics, a brief review of classical mechanics will prove useful. This serves two purposes, to prepare the ground for contrasting special relativity with classical theory and to review such topics as Lagrangians and Hamiltonians.

3.1 Inertial frames and Galileo's relativity principle

In this book it is assumed that the motion of electrons and nuclei is independent of their dimensions, that is, they may be regarded as point masses or particles. The justification for this assumption is that electrons and nuclei are very small compared with the atoms and molecules in which they are found; the radii of electrons, nuclei and atoms are of the order of 10^{-15}, 10^{-14} and 10^{-10} m, respectively. The position of an electron or a nucleus at a particular time can thus be specified by its vector position **r**, that is by three cartesian coordinates. In general a system of N particles requires $3N$ coordinates to specify its instantaneous configuration and the system is said to have $3N$ degrees of freedom.

Although a system can be characterized instantaneously by $3N$ cartesian coordinates, other coordinates may be used. Any set of $3N$ coordinates q_n ($n = 1, \ldots, 3N$), that completely specify the system, is acceptable, and these are called generalized coordinates; for example, one of these generalized coordinates could be the separation between two particles. However, to be able to measure coordinates at all, we must have a frame of reference, that is a system of coordinate axes and, since the evolution of a system with time is of interest, a clock fixed in this frame of reference.

There are an infinite number of possible frames of reference, but the only ones of interest in classical mechanics are those in which space is homogeneous and isotropic and time is homogeneous, in agreement with experience. These frames are called inertial frames and in particular a freely moving particle (one that experiences no external forces) has a constant velocity in such a frame. There are still an arbitrarily large number of inertial frames, but these move uniformly with respect to one another.

Now experiment shows that the laws of physics are the same no matter what inertial frame is used; this is Galileo's principle of relativity. Thus, the same phenomena will be described by similar equations in two different inertial frames. Suppose that the position of a particle is $\mathbf{r}$ in a frame K and $\mathbf{r}'$ in another frame K', that is moving uniformly with a velocity $\mathbf{v}$ relative to the first frame K. If the two frames are coincident at time $t = 0$, then $\mathbf{r}$ and $\mathbf{r}'$ are related by the Galilean transformation:

$$\mathbf{r} = \mathbf{r}' + \mathbf{v}t. \tag{3.1}$$

It will be noted that time is here assumed to be the same in the two frames, $t = t'$. It is this assumption that time is absolute that distinguishes a classical theory from a relativistic one; special relativity will be considered in the next chapter, and for the remainder of this chapter it will be assumed that time is indeed independent of the inertial frame used. With this assumption an equation expressing a physical law must be invariant to the transformation (Equation 3.1); that is, the equation must have the same form in different inertial frames.

3.2 The principle of least action

The state of a mechanical system is completely specified at an instant of time, if all the generalized coordinates q_n are known. If all the generalized velocities $\dot{q}_n$ (= $\mathrm{d}q_n/\mathrm{d}t$)are also known at the same instant of time, then in principle the subsequent motion may be calculated. This is achieved by setting up equations of motion relating the coordinates q_n, the velocities $\dot{q}_n$ and the accelerations $\ddot{q}_n$. Solution of these equations for the q_n as functions of time, enables the evolution of the system to be followed.

One way of obtaining the equations of motion would be to use Newton's laws. However, there are other formulations of classical mechanics and the one given here, the principle of least action or Hamilton's principle, is the most important. This principle may be shown to be consistent with Newton's laws, but neither should be regarded as a consequence of the other; they are alternative fundamental postulates.

The principle of least action states that a system may be characterized by the Lagrangian $\mathcal{L}(q_n, \dot{q}_n, t)$, which is a function of the generalized coordinates, the generalized velocities and also the time. The motion of the system between two times, t_a and t_b, at which the coordinates are known, is such that the action S, which is defined by:

$$S = \int_{t_a}^{t_b} \mathcal{L}(q_n, \dot{q}_n, t)\mathrm{d}t, \qquad (3.2)$$

takes the least possible value. Thus, the action is a line integral along the path taken by the system between t_a and t_b and the path actually followed is such that the action is least. (Strictly speaking this is only true for an infinitesimally short path and the condition for a finite path is that the action should have a stationary value; however, it will be seen that in deriving the equations of motion from the principle this point is of no consequence.)

It is not easy to visualize the physical meaning of the least action principle. However, there are other minimum principles in physics, the most transparent being Fermat's principle of least time, which states that light travels between two points in such a way that the time taken is a minimum. Use of this principle leads immediately to the rectilinear propagation of light in a uniform medium, the laws of reflection at a surface and the laws of refraction for light crossing the boundary between two different media. For detailed discussions of this principle the reader is referred to the bibliography at the end of the chapter.

3.3 Lagrange's equations of motion

By minimizing the action (Equation 3.2) equations of motion may be obtained. For simplicity this will be done initially for a system characterized by only one coordinate q, the extension to many degrees of freedom being made later. The fixed values of q at times t_a and t_b are taken to be q_a and q_b (see Fig. 3.1). Let us suppose that the function of time $q(t)$ represents the path along which the action integral is minimized (or in general has a stationary value). A slight displacement from this optimum path to a new path, $q(t) + \delta q(t)$, can have no first-order effect on the action, that is the corresponding change δS in the action integral is zero:

$$\delta S = \delta \int_{t_a}^{t_b} \mathcal{L}(q, \dot{q}, t)\mathrm{d}t = 0. \qquad (3.3)$$

This may be expanded as follows:

$$\delta S = \int_{t_a}^{t_b} \delta\mathcal{L}(q, \dot{q}, t)\mathrm{d}t$$

$$= \int_{t_a}^{t_b} [(\partial\mathcal{L}/\partial q)\delta q + (\partial\mathcal{L}/\partial\dot{q})\delta\dot{q}]\mathrm{d}t = 0. \tag{3.4}$$

It might be thought that, since the Lagrangian is a function of time, a term involving $(\partial\mathcal{L}/\partial t)$ should appear. However, in varying the path we are interested in the change in the Lagrangian for each line segment of the action integral, that is each change is considered at a fixed time. Since $\delta\dot{q}\, \mathrm{d}t$ is the same as $\mathrm{d}(\delta q)$, the second term in Equation 3.4 may be integrated by parts:

$$\delta S = \left[\left(\frac{\partial\mathcal{L}}{\partial\dot{q}}\right)\delta q\right]_{t_a}^{t_b} + \int_{t_a}^{t_b}\left[\left(\frac{\partial\mathcal{L}}{\partial q}\right) - \frac{\mathrm{d}}{\mathrm{d}t}\left(\frac{\partial\mathcal{L}}{\partial\dot{q}}\right)\right]\delta q\, \mathrm{d}t = 0. \tag{3.5}$$

Now the first term in this equation vanishes, since q is required to have the fixed values q_a and q_b at times t_a and t_b, so that $\delta q(t_a)$ and $\delta q(t_b)$ are both

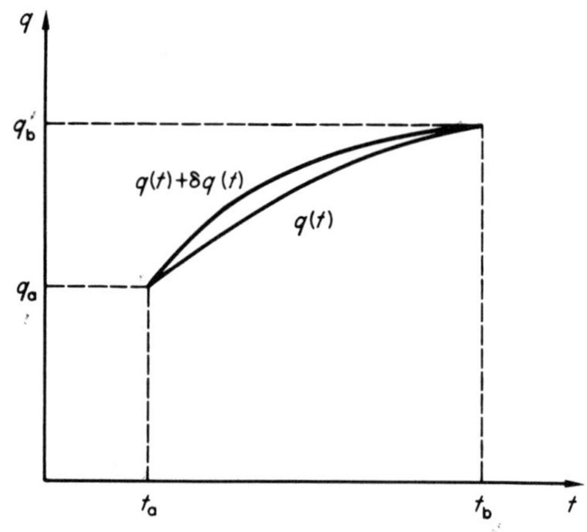

Fig. 3.1. The path taken by a one-coordinate system between (q_a, t_a) and (q_b, t_b); $q(t)$ is the path along which the action is assumed to be least and $q(t) + \delta q(t)$ is a slightly displaced path.

zero. Thus, the integral in Equation 3.5 must vanish and this must be true for any arbitrary infinitesimal change in path δq. The only way that this can happen is for the integrand to vanish also:

$$\frac{\mathrm{d}}{\mathrm{d}t}\left(\frac{\partial\mathcal{L}}{\partial\dot{q}}\right) - \frac{\partial\mathcal{L}}{\partial q} = 0. \tag{3.6}$$

This result may be generalized to a system with many degrees of freedom. Corresponding to each generalized coordinate q_n there can be an arbitrary infinitesimal change in the line integral, so that there will be an equation similar to Equation 3.6 for each coordinate:

$$\frac{\mathrm{d}}{\mathrm{d}t}\left(\frac{\partial \mathcal{L}}{\partial \dot{q}_n}\right) - \left(\frac{\partial \mathcal{L}}{\partial q_n}\right) = 0 \qquad (n = 1, \ldots, 3N). \tag{3.7}$$

These equations are known as Lagrange's equations of motion (or Euler's equations in the general context of the calculus of variations, which is the mathematical technique that has been used here). These partial differential equations may be solved for the particular system under consideration to give the q_n explicitly as functions of time; in general, there will be $6N$ arbitrary constants, which may be fixed by initial conditions.

Of course, the form of the Lagrangian must be known before the equations of motion may be solved. Before considering this in the next section, it is noted that the equations of motion are unaltered by multiplication of the Lagrangian by an arbitrary constant; however, the same constant must be used for all Lagrangians, if isolated systems are to be brought together and treated as one with a combined Lagrangian. In addition, a Lagrangian is undetermined to the extent that addition of a total time derivative of a function $f(q_n, t)$ of coordinates and time will not affect the equations of motion. This may be seen by noting that addition of $(\mathrm{d}f/\mathrm{d}t)$ to the Lagrangian adds to the action integral (Equation 3.2) a term:

$$\int_{t_a}^{t_b} (\mathrm{d}f/\mathrm{d}t)\mathrm{d}t = [f(q_n, t)]_{t_a}^{t_b} = \text{constant}, \tag{3.8}$$

since the q_n are assumed fixed at t_a and t_b. Consequently such a term can have no influence in the variation procedure used in developing the equations of motion.

3.4 The Lagrangian for a system of particles

It has already been stated that the principle of least action is consistent with Newton's laws and indeed they may be derived from the principle via Lagrange's equations of motion. The derivation given here is not completely rigorous, but the consistency of the two approaches should be apparent. It is first necessary to obtain an appropriate Lagrangian and this is done for a free particle before considering a collection of particles.

The Lagrangian of a free particle in an inertial frame cannot depend on the vector position $\mathbf{r}$ of the particle, since this would imply that space was not homogeneous, nor can it depend on the time explicitly, if the homogeneity of time is to be preserved. This means that the Lagrangian can only involve the velocity $\dot{\mathbf{r}}$ of the particle. However, the isotropy of space demands that it must be independent of the direction of the velocity so that it must be concluded that the Lagrangian of a free particle can only be a function of $\dot{r}^2$, that is $\mathcal{L}(\dot{r}^2)$. Substitution of this result into Lagrange's equations of motion (Equation 3.7) gives immediately:

$$\frac{\mathrm{d}}{\mathrm{d}t}\left(\frac{\partial \mathcal{L}}{\partial \dot{\mathbf{r}}}\right) = 0 \quad \text{or} \quad \frac{\partial \mathcal{L}}{\partial \dot{\mathbf{r}}} = \text{constant.} \tag{3.9}$$

(When the derivative of a scalar with respect to a vector is written, we mean the vector whose components are the derivatives of the scalar with respect to the corresponding component of the vector. Thus the second of Equations 3.9 represents three equations, a typical one being $(\partial \mathcal{L}/\partial \dot{r}_x) =$ constant). Since the Lagrangian is a non-linear function of the velocity, the velocity itself must be constant. This gives the first of Newton's laws, that in an inertial frame a free particle moves with constant velocity.

The precise form of the Lagrangian may be determined by appealing to the law of relativity, since on performing a Galilean transformation (Equation 3.1) between two inertial frames the equations of motion must be unchanged so that the Lagrangian can only be changed by at most the addition of a total time derivative of a function of coordinates and time. This can be shown to be possible only if the Lagrangian is directly proportional to $\dot{r}^2$; to agree with other approaches to classical mechanics the proportionality constant is taken to be $\frac{1}{2}m$, where m is the mass of the particle:

$$\mathcal{L} = \tfrac{1}{2}m\dot{r}^2. \tag{3.10}$$

The Lagrangian in the K' inertial frame of Section 3.1, which is related to the present frame by the transformation of Equation 3.1, is then given by:

$$\begin{aligned} \mathcal{L}' &= \tfrac{1}{2}m(\dot{r}')^2 = \tfrac{1}{2}m(\dot{\mathbf{r}} - \mathbf{v})^2 \\ &= \tfrac{1}{2}m\dot{r}^2 - m\dot{\mathbf{r}}.\mathbf{v} + \tfrac{1}{2}mv^2 \\ &= \mathcal{L} - \mathrm{d}(m\mathbf{r}.\mathbf{v} - \tfrac{1}{2}mv^2t)/\mathrm{d}t. \end{aligned} \tag{3.11}$$

This separation of $\mathcal{L}'$ into $\mathcal{L}$ and a total time derivative would not be possible if the Lagrangian were a more complicated function of $\dot{r}^2$ than Equation 3.10.

For a system of particles the Lagrangian is the sum of the Lagrangians for the individual particles together with a term representing the interaction between the particles and this can only be a function of the positions of the particles. If the particles are distinguished by the subscript i, the total Lagrangian may be written:

$$\mathcal{L} = \sum_i \tfrac{1}{2} m_i \dot{r}_i^2 - V(\mathbf{r}_i) = T(\dot{\mathbf{r}}_i) - V(\mathbf{r}_i). \quad (3.12)$$

The first term is the kinetic energy T while V is the potential energy; the reason for these names and the presence of the minus sign will become apparent in the next section. The repeated subscript summation convention is not appropriate for the first term, since the subscript i differentiates particles and not vector components. Substitution of Equation 3.12 into Lagrange's equations of motion (Equation 3.7) gives three equations for each particle:

$$m_i(\mathrm{d}\dot{\mathbf{r}}_i/\mathrm{d}t) = -(\partial V/\partial \mathbf{r}_i) = \mathbf{F}_i. \quad (3.13)$$

By interpreting $-(\partial V/\partial \mathbf{r}_i)$ as the vector $\mathbf{F}_i$, the force on particle i, Newton's second law is obtained, the Equations 3.13 being Newton's equations of motion which equate forces with mass accelerations.

Up till now the discussion has been in terms of cartesian coordinates, but a transformation may be made to any appropriate generalized coordinates. The kinetic energy is in general a quadratic function of generalized velocities, but cross terms can appear and the coefficients may be functions of the coordinates. However, the Lagrangian may still be written in terms of the kinetic energy T and the potential energy V:

$$\mathcal{L} = T(q_n, \dot{q}_n) - V(q_n). \quad (3.14)$$

Since the kinetic energy can also be a function of the generalized coordinates, the generalized force F_n must be defined as:

$$F_n = (\partial \mathcal{L}/\partial q_n). \quad (3.15)$$

Earlier it was stated that the Lagrangian can also be a function of time, and this occurs if the particles are subject to some sort of time-dependent external potential; this possibility will be encountered later.

3.5 Constants of motion

For a closed system, that is one that is not subject to an external potential, there are a number of functions of the coordinates and velocities that remain constant throughout the motion. Of these constants of motion the most important are those that are a direct consequence of the homogeneity and isotropy of time and space. Here they will be derived using cartesian coordinates, although the extension to generalized coordinates will be discussed.

The homogeneity of time implies that for a closed system the Lagrangian does not depend explicitly on time. Thus:

$$(\mathrm{d}\mathcal{L}/\mathrm{d}t) = \sum_i [(\partial\mathcal{L}/\partial\mathbf{r}_i).\dot{\mathbf{r}}_i + (\partial\mathcal{L}/\partial\dot{\mathbf{r}}_i).\ddot{\mathbf{r}}_i]. \tag{3.16}$$

Substitution for $(\partial\mathcal{L}/\partial\mathbf{r}_i)$ using Lagrange's equations of motion (Equation 3.7) gives:

$$\frac{\mathrm{d}\mathcal{L}}{\mathrm{d}t} = \sum_i \left[\dot{\mathbf{r}}_i.\frac{\mathrm{d}}{\mathrm{d}t}\left(\frac{\partial\mathcal{L}}{\partial\dot{\mathbf{r}}_i}\right) + \left(\frac{\partial\mathcal{L}}{\partial\dot{\mathbf{r}}_i}\right).\ddot{\mathbf{r}}_i\right]$$

$$= \sum_i \frac{\mathrm{d}}{\mathrm{d}t}\left[\dot{\mathbf{r}}_i.\left(\frac{\partial\mathcal{L}}{\partial\dot{\mathbf{r}}_i}\right)\right] = \frac{\mathrm{d}}{\mathrm{d}t}\left[\sum_i \dot{\mathbf{r}}_i.\left(\frac{\partial\mathcal{L}}{\partial\dot{\mathbf{r}}_i}\right)\right]. \tag{3.17}$$

Rearrangement of this equation shows that:

$$E = \sum_i \dot{\mathbf{r}}_i.(\partial\mathcal{L}/\partial\dot{\mathbf{r}}_i) - \mathcal{L}, \tag{3.18}$$

does not vary with time and is a constant of motion. Substitution of the explicit Lagrangian (Equation 3.12) for a system of particles now gives:

$$E = T(\dot{\mathbf{r}}_i) + V(\mathbf{r}_i). \tag{3.19}$$

E is the total energy of the system, and is composed of two parts, the kinetic energy T, which is a function of the velocities, and the potential energy V, which is a function of the coordinates only. When generalized coordinates are being used, the equations analogous to Equations 3.18 and 3.19 are:

$$E = \sum_n \dot{q}_n(\partial\mathcal{L}/\partial\dot{q}_n) - \mathcal{L} = T(q_n, \dot{q}_n) + V(q_n). \tag{3.20}$$

Since space is homogeneous an arbitrary infinitesimal displacement $\boldsymbol{\delta}$ of a system of particles must leave the Lagrangian unaltered:

$$\delta\mathcal{L} = \sum_i (\partial\mathcal{L}/\partial\mathbf{r}_i).\boldsymbol{\delta} = \boldsymbol{\delta}.\sum_i (\partial\mathcal{L}/\partial\mathbf{r}_i) = 0. \tag{3.21}$$

Since $\boldsymbol{\delta}$ is arbitrary, this can only be interpreted as showing that:

$$\sum_i (\partial\mathcal{L}/\partial\mathbf{r}_i) = 0, \tag{3.22}$$

or using Lagrange's equation of motion (Equation 3.7):

$$\sum_i \frac{\mathrm{d}}{\mathrm{d}t}\left(\frac{\partial\mathcal{L}}{\partial\dot{\mathbf{r}}_i}\right) = \frac{\mathrm{d}}{\mathrm{d}t}\sum_i\left(\frac{\partial\mathcal{L}}{\partial\dot{\mathbf{r}}_i}\right) = 0. \tag{3.23}$$

Thus $\sum_i (\partial \mathcal{L}/\partial \dot{\mathbf{r}}_i)$ is a constant of motion and is called the total momentum of the system. It is clear that, unlike the energy, momentum is additive and each particle has associated with it a momentum:

$$\mathbf{p}_i = (\partial \mathcal{L}/\partial \dot{\mathbf{r}}_i); \tag{3.24}$$

substitution of the Lagrangian (Equation 3.12) shows that this is equal to $m_i \dot{\mathbf{r}}_i$. In addition, substitution of Equation 3.24 into Lagrange's equations of motion and use of Equation 3.13 shows that:

$$\dot{\mathbf{p}}_i = (\partial \mathcal{L}/\partial \mathbf{r}_i) = -(\partial V/\partial \mathbf{r}_i) = \mathbf{F}_i, \tag{3.25}$$

where $\mathbf{F}_i$ is the force on particle i. From this equation it is apparent that Equation 3.22 shows that $\sum_i \mathbf{F}_i = 0$ and in particular for a system of just two particles we may recover Newton's third law, that to every action there is an equal and opposite reaction.

Equations 3.24 and 3.25 may be generalized:

$$p_n = (\partial \mathcal{L}/\partial \dot{q}_n); \qquad \dot{p}_n = (\partial \mathcal{L}/\partial q_n) = F_n, \tag{3.26}$$

where p_n is a generalized momentum and is said to be conjugate to the coordinate q_n. The energy (Equation 3.20) can now be written in the simpler form:

$$E = \sum_n \dot{q}_n p_n - \mathcal{L}. \tag{3.27}$$

Finally, the isotropy of space implies that the Lagrangian is invariant to infinitesimal rotations of the axis system. Use of this condition leads to a third constant of motion, the angular momentum. The rotation of a system of particles through an angle $\delta\theta$ may be represented by a vector $\delta\boldsymbol{\theta}$ directed along the axis of rotation, the sense of the rotation being given by the right-hand screw rule. If the axis passes through the origin of the coordinates, the rotation changes a position vector $\mathbf{r}$ by an amount $\delta\mathbf{r} = \delta\boldsymbol{\theta} \wedge \mathbf{r}$ and the change in the time derivative of $\mathbf{r}$ is given by $\delta\dot{\mathbf{r}} = \delta\boldsymbol{\theta} \wedge \dot{\mathbf{r}}$. The corresponding change in the Lagrangian is thus:

$$\begin{aligned} \delta\mathcal{L} &= \sum_i [(\partial \mathcal{L}/\partial \mathbf{r}_i).\delta\mathbf{r}_i + (\partial \mathcal{L}/\partial \dot{\mathbf{r}}_i).\delta\dot{\mathbf{r}}_i] \\ &= \sum_i [\dot{\mathbf{p}}_i.\delta\boldsymbol{\theta} \wedge \mathbf{r}_i + \mathbf{p}_i.\delta\boldsymbol{\theta} \wedge \dot{\mathbf{r}}_i] \\ &= \delta\boldsymbol{\theta}.[\mathrm{d}(\sum_i \mathbf{r}_i \wedge \mathbf{p}_i)/\mathrm{d}t], \end{aligned} \tag{3.28}$$

where Equations 3.7 and 3.24 have been used for $(\partial \mathcal{L}/\partial \mathbf{r}_i)$ and $(\partial \mathcal{L}/\partial \dot{\mathbf{r}}_i)$. Since $\mathcal{L}$ is invariant to rotation, $\delta\mathcal{L}$ must vanish and, because $\delta\boldsymbol{\theta}$ is arbitrary, Equation 3.28 shows that:

$$\mathbf{L} = \sum_i \mathbf{r}_i \wedge \mathbf{p}_i \tag{3.29}$$

is a constant of motion. This is the angular momentum and as for the linear momentum it is an additive property of a system of particles.

3.6 The Hamiltonian

The Lagrangian is a function of coordinates and velocities, but it is often more appropriate to describe a mechanical system in terms of coordinates and momenta; for example, quantum mechanics is usually formulated in terms of momenta rather than velocities, although in that case the momenta are, of course, operators. The equations of motion may be rewritten in terms of momenta in the following way. The total differential of the Lagrangian expressed in terms of generalized coordinates is:

$$\mathrm{d}\mathcal{L} = \sum_n (\partial\mathcal{L}/\partial q_n)\mathrm{d}q_n + \sum_n (\partial\mathcal{L}/\partial\dot{q}_n)\mathrm{d}\dot{q}_n, \tag{3.30}$$

where again a closed system is assumed so that the Lagrangian does not depend explicitly on time. Substitution of Equations 3.26 into Equation 3.30 gives:

$$\begin{aligned}\mathrm{d}\mathcal{L} &= \sum_n (\dot{p}_n\mathrm{d}q_n + p_n\mathrm{d}\dot{q}_n)\\ &= \sum_n [\dot{p}_n\mathrm{d}q_n + \mathrm{d}(\dot{q}_n p_n) - \dot{q}_n\mathrm{d}p_n],\end{aligned} \tag{3.31}$$

or on rearrangement:

$$\mathrm{d}\mathcal{H} = \mathrm{d}(\sum_n \dot{q}_n p_n - \mathcal{L}) = \sum_n [-\dot{p}_n\mathrm{d}q_n + \dot{q}_n\mathrm{d}p_n]. \tag{3.32}$$

The new symbol $\mathcal{H}$ represents the Hamiltonian:

$$\mathcal{H}(q_n, p_n) = \sum_n \dot{q}_n p_n - \mathcal{L}, \tag{3.33}$$

and reference to Equation 3.27 shows that it is just the energy of the system. Hamilton's equations of motion may now be obtained directly from Equation 3.32:

$$(\partial\mathcal{H}/\partial q_n) = -\dot{p}_n; \qquad (\partial\mathcal{H}/\partial p_n) = \dot{q}_n. \tag{3.34}$$

If the Lagrangian of a system is known as a function of coordinates and velocities, the corresponding Hamiltonian may be found as a function of coordinates, velocities and momenta from Equation 3.33. However, the Hamiltonian is a function of coordinates and momenta alone and the dependence on velocities must be removed for it to be useful. This may be done by using the first of Equations 3.26 to obtain the momenta as functions of the velocities and coordinates, solving this set of $3N$ simultaneous equations for the velocities as a function of coordinates and momenta and substituting in Equation 3.33.

As developed here, the Hamiltonian is a function of coordinates and momenta, but like the Lagrangian it may also depend explicitly on time if the system is not closed. However, it will be seen in Chapter 5 that external potentials may be included without modification of the above equations.

Bibliography

Classical mechanics

Goldstein, H. (1950), *Classical Mechanics*, McGraw-Hill, New York.
Landau, L. D., and Lifshitz, E. M. (1960), *Mechanics*, Pergamon Press, Oxford.

The principle of least action and Fermat's principle

Atkins, P. W. (1970), *Molecular Quantum Mechanics*, Clarendon Press, Oxford: pages 27–31.
Feynman, Leighton and Sands: Volume I, Chapter 26; Volume II, Chapter 19.

CHAPTER FOUR

Special Relativity

The theory of classical mechanics, which was discussed briefly in the previous chapter, is classical in two senses. First, it is not a quantum mechanical theory and is thus not appropriate on a molecular scale. Secondly, it takes no account of special relativity. The use of the word 'classical' is thus ambiguous, since in different contexts it may mean non-quantal or non-relativistic or both. It is the purpose of this chapter to remove one of the restrictions and develop a relativistic but non-quantum-mechanical theory for a free particle.

4.1 Einstein's principle of relativity

The relativity principle, that the laws of physics are the same in all inertial frames, has been met already in Section 3.1. However, the Galilean transformation (Equation 3.1) used there to relate inertial frames, is based on the assumption that time is absolute, that is the time interval between two events will be the same no matter what inertial frame is being used. We will see that this assumption is not in fact correct.

Experiment shows that, if two particles interact in some way, a change in one does not have an instantaneous effect on the other, but that there is a finite time lapse between the cause and the effect. This in itself points out a difficulty with classical (non-relativistic) mechanics, since the interaction is described by a potential energy that is taken to be a function of the particle positions alone, an assumption that implies the instantaneous propagation of interactions. If there is a finite delay between cause and effect, there must be a maximum velocity of propagation of interaction that determines the minimum delay before a change in one part of a system begins to affect another part. The existence of a maximum velocity of propagation means that a particle cannot move faster than this velocity. In addition, the principle of relativity tells us that this maximum velocity of propagation must be the same in all inertial frames and is thus a universal constant. In fact it is the velocity of light in a vacuum, $c = 2.997925 \times 10^8$ m s^{-1}. By linking the principle of relativity with a finite maximum velocity of propagation of interaction, Einstein's special principle of relativity is obtained and the theory based on it is said to be relativistic in contrast to the non-relativistic theory based on an infinite velocity of propagation.

The non-relativistic theory of Chapter 3, and in particular the Galilean transformation (Equation 3.1), may now be seen to be inconsistent with Einstein's theory of relativity. Differentiation of Equation 3.1 with respect to time shows that in non-relativistic theory velocity vectors may be summed and in particular the velocity of light will be different in different inertial frames in contradiction to the principle of relativity. These assertions are in agreement with the experiments of Michelson and Morley, who showed that the velocity of light does not depend on its direction of propagation, despite the fact that the earth is moving. Of course, the fact that non-relativistic mechanics is wrong does not invalidate Newton's laws and the mechanics associated with them. Because the velocity of light is so large, deviations from a non-relativistic theory are usually negligible and the classical theory is a very good approximation in most cases. The relativistic equations that are derived in this chapter may be used to calculate the discrepancies between the two theories and, if we pretend that the velocity of light is infinite, that is if we revert to a non-relativistic theory, these discrepancies vanish.

If the Galilean transformation (Equation 3.1) is not correct, time cannot be absolute and must evolve differently in different inertial frames. The time lapse between two events will depend on the inertial frame used and this must be specified. In addition, two events occuring in different parts of space may be simultaneous when observed from one inertial frame, but they will not be simultaneous in other frames. The fact that it must now be thought of as being relative, puts time on the same footing as spatial coordinates and it is often convenient to think of a four-dimensional space in which an event is described by three space coordinates and one time coordinate. Points in this space are called world points to distinguish them from points in three-dimensional space and the history of a particle is described by a world line. For a particle that is moving with a constant velocity in a straight line the corresponding world line is straight.

4.2 The interval

Although time is not absolute and the time lapse between two events depends on the inertial frame used, there is another quantity, the interval, that is independent of the inertial frame.

To demonstrate this the invariance of the velocity of light is first expressed mathematically. Suppose that in an inertial frame K a signal leaves the point x_1, y_1, z_1 at time t_1 with velocity c and arrives at another point x_2, y_2, z_2 at time t_2. The distance between the two points may be related to the time taken by the signal to travel between them:

$$[(x_2-x_1)^2+(y_2-y_1)^2+(z_2-z_1)^2]^{\frac{1}{2}}-c(t_2-t_1) = 0. \tag{4.1}$$

In another frame K' the same signal travels between the world points x'_1, y'_1, z'_1, t'_1 and x'_2, y'_2, z'_2, t'_2 and a similar equation may be written:

$$[(x'_2-x'_1)^2+(y'_2-y'_1)^2+(z'_2-z'_1)^2]^{\frac{1}{2}}-c(t'_2-t'_1) = 0. \tag{4.2}$$

This suggests that for any two events with world points x_1, y_1, z_1, t_1 and x_2, y_2, z_2, t_2 in K the quantity:

$$s_{12} = [c^2(t_2-t_1)^2-(x_2-x_1)^2-(y_2-y_1)^2-(z_2-z_1)^2]^{\frac{1}{2}} \tag{4.3}$$

is of importance; s_{12} is called the interval between the two events. Since the velocity of light is the same in all inertial frames, Equations 4.1 and 4.2 show that, if the interval s_{12} between two events in K is zero, the interval s'_{12} in K' for the same two events also vanishes.

If the two events occur at two world points that are infinitesimally close to one another and whose coordinates are separated by $\mathrm{d}x$, $\mathrm{d}y$, $\mathrm{d}z$, $\mathrm{d}t$ in K, then the interval $\mathrm{d}s$ is given by:

$$\mathrm{d}s^2 = c^2\mathrm{d}t^2-\mathrm{d}x^2-\mathrm{d}y^2-\mathrm{d}z^2; \tag{4.4}$$

the corresponding interval $\mathrm{d}s'$ in K' is given by a similar equation and as before if $\mathrm{d}s = 0$ then so does $\mathrm{d}s'$. In addition, the principle of relativity requires that the laws of physics are the same in the two frames so that the transformation from $\mathrm{d}x'$, $\mathrm{d}y'$, $\mathrm{d}z'$, $\mathrm{d}t'$ to $\mathrm{d}x$, $\mathrm{d}y$, $\mathrm{d}z$, $\mathrm{d}t$ must be linear and homogeneous (no additive constants). Consequently, $\mathrm{d}s'^2$ must be a homogeneous quadratic function of $\mathrm{d}x$, $\mathrm{d}y$, $\mathrm{d}z$, $\mathrm{d}t$ just as $\mathrm{d}s^2$ is. It can now be seen that, since $\mathrm{d}s'$ vanishes whenever $\mathrm{d}s$ does, no matter what the values $\mathrm{d}x$, $\mathrm{d}y$, $\mathrm{d}z$ and $\mathrm{d}t$ take, $\mathrm{d}s'^2$ can differ from $\mathrm{d}s^2$ by at most a multiplicative factor k:

$$\mathrm{d}s'^2 = k\mathrm{d}s^2. \tag{4.5}$$

The coefficient k cannot depend on space or time coordinates, since this would imply that different points in space and time are not equivalent, in contradiction to the homogeneity of space and time. Thus it can only depend on the relative velocity **v** of the two inertial frames K and K'. However, the isotropy of space demands that k cannot depend on the direction of the relative motion of the two frames and k can only be a function of the magnitude of **v**. If ds^2 had been expressed in terms of ds'^2 the corresponding equation to Equation 4.5 would have been:

$$ds^2 = k'ds'^2, \tag{4.6}$$

where k' is the same function of the magnitude of $-\mathbf{v}$ as k is of the magnitude of $+\mathbf{v}$. Consequently $k = k'$ and comparison of Equations 4.5 and 4.6 shows that k must be ± 1. The negative solution can be dismissed immediately, since ds'^2 must tend to ds^2 as the relative velocities of the two frames tends to zero, so that:

$$ds^2 = ds'^2. \tag{4.7}$$

Similarly ds' must tend to ds as **v** tends to zero and $ds = ds'$. Integration to give finite intervals then gives the final result that the interval between two events is invariant to the inertial frame used:

$$s_{12} = s'_{12}, \tag{4.8}$$

and this invariance may be attributed to the fact that the velocity of light is the same in all inertial frames.

4.3 The Lorentz transformation

Now that the interval between two events has been shown to be invariant to the inertial frame used, the explicit transformation between space and time coordinates in different frames can be found. Two inertial frames K and K' are again considered and for simplicity it is assumed that the x and x' axes coincide, and the y and z axes are parallel to the y' and z' axes, respectively; the velocity of the frame K' relative to K is v along the x axis (see Fig. 4.1).

It has already been noted that time is not absolute and that it is convenient to think of a four-dimensional space in which one of the coordinates is related to time. The time and space coordinates may be made equivalent by introducing the new time coordinate:

$$\tau = ict, \tag{4.9}$$

since the interval (Equation 4.3) between two events is then given by:

$$s_{12}^2 = -[(\tau_2 - \tau_1)^2 + (x_2 - x_1)^2 + (y_2 - y_1)^2 + (z_2 - z_1)^2], \tag{4.10}$$

so that, if τ is the coordinate associated with the fourth dimension, the interval is related to the distance between two points in four-dimensional space. The transformation from one inertial frame K with coordinates x, y, z, τ to another K' with coordinates x', y', z', τ' can thus be thought of as a transformation in four-dimensional space. Since the interval is independent of the inertial frame, the length of a line must be preserved in such a transformation, that is only orthogonal transformations are of interest.

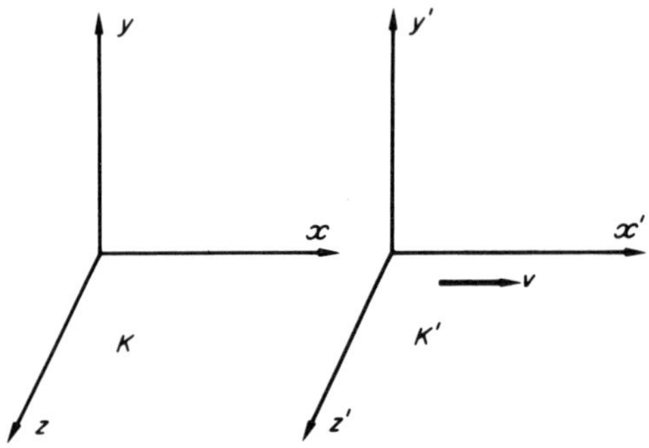

Fig. 4.1. The relationship between the inertial frames K and K'.

All orthogonal transformations may be broken down into rotations and parallel displacements of the coordinate system. The latter are of no interest since they only change the origins of the space and time coordinates and a 'mixing' of space and time is required. Any rotation can be resolved into six rotations parallel to the six orthogonal planes $xy, yz, zx, x\tau, y\tau, z\tau$ in the four-dimensional space. The first three of these account for the usual space rotations and of the last three, which mix space and time, a rotation parallel to the $x\tau$ plane is chosen for specific consideration, since the y and z coordinates will not change, as in the model that has been chosen for consideration (Fig. 4.1).

For an orthogonal rotation through an angle θ in the $x\tau$ plane:

$$\begin{aligned} x &= x' \cos\theta - \tau' \sin\theta; \\ \tau &= x' \sin\theta + \tau' \cos\theta. \end{aligned} \tag{4.11}$$

To relate θ to the relative velocity v of K and K', the motion of the origin of the K' frame is considered, that is x' is set equal to zero. Substitution in Equations 4.11 gives:

$$x = -\tau' \sin\theta; \qquad \tau = \tau' \cos\theta, \tag{4.12}$$

or combining these:

$$\tan\theta = -x/\tau. \tag{4.13}$$

Now, if the origins of the two frames are assumed to coincide at $\tau = t = 0$, the origin of the K' frame has the coordinate $x = vt = -iv\tau/c$ in the K frame, where Equation 4.9 has been used to relate t and τ. Substitution in Equation 4.13 now gives:

$$\tan\theta = iv/c. \tag{4.14}$$

Using familiar trigonometric relations, $\sin\theta$ and $\cos\theta$ may now be recovered:

$$\begin{aligned} \sin\theta &= \tan\theta(1+\tan^2\theta)^{-\frac{1}{2}} = iv\gamma/c; \\ \cos\theta &= (1+\tan^2\theta)^{-\frac{1}{2}} = \gamma, \end{aligned} \tag{4.15}$$

where γ is defined as:

$$\gamma = (1 - v^2/c^2)^{-\frac{1}{2}}. \tag{4.16}$$

The transformation (Equations 4.11) now become:

$$\begin{aligned} x &= \gamma(x' - iv\tau'/c); \\ \tau &= \gamma(ivx'/c + \tau'). \end{aligned} \tag{4.17}$$

Finally the time coordinate t may be reintroduced using Equation 4.9 to give:

$$\begin{aligned} x &= \gamma(x' + vt'); \qquad y = y'; \\ z &= z'; \qquad t = \gamma(vx'/c^2 + t'), \end{aligned} \tag{4.18}$$

as the Lorentz transformation for the model under discussion, in which K' moves with velocity v along the x axis relative to K.

The transformation (Equations 4.18) may be inverted:

$$\begin{aligned} x' &= \gamma(x - vt); \qquad y' = y; \\ z' &= z; \qquad t' = \gamma(-vx/c^2 + t), \end{aligned} \tag{4.19}$$

where the definition (Equation 4.16) of γ is unaltered. This inversion is most easily performed by noting that, if the primed and unprimed coordinates are interchanged in Equations 4.18, the roles of the two frames are also reversed and v must be replaced by the velocity of K relative to K', that is by $-v$.

One of the checks that can be applied to relativistic equations is that, if the velocity of light c is permitted to become infinite, the correct non-relativistic equations should be recovered. In this case Equations 4.18 become:

$$x = x' + vt'; \quad y = y'; \quad z = z'; \quad t = t', \tag{4.20}$$

which are appropriate Galilean transformation equations for the situation under consideration (cf. Equation 3.1).

Of course, the Lorentz transformation of Equations 4.18 applies to the special case when the space axes of the two frames are parallel and their relative motion is along the x axis. Any problem may be reduced to this situation by an appropriate transformation of the space coordinates in the two frames. However, in the more general case where the spatial axes of the two frames are parallel, but the relative motion is characterized by a general vector $\mathbf{v}$, the transformation may be written as:

$$\mathbf{r} = \mathbf{r}' + \mathbf{v}[(\mathbf{v}.\mathbf{r}')(\gamma - 1)/v^2 + \gamma t']; \qquad t = \gamma[(\mathbf{v}.\mathbf{r}')/c^2 + t']; \tag{4.21}$$

the proof is left as an exercise.

Finally, we note that all relativistic equations expressing physical laws must be invariant to the Lorentz transformation, just as all non-relativistic equations must be invariant to the Galilean transformation (Equation 3.1); this is the principle of relativity. Indeed, the reason that Lorentz's name rather than Einstein's is linked with the transformation is that Lorentz had found earlier that Maxwell's equations of electromagnetism are invariant to this transformation rather than the Galilean transformation. It is for this reason that it is appropriate that electromagnetic theory is developed relativistically as in the next chapter. Invariance to transformations is generally expressed in the theory of groups and Lorentz transformations can be shown to constitute a group, though this fact is not pursued here.

4.4 Contraction, dilation and paradoxes

In this section some of the more entertaining of the consequences of relativity are briefly discussed, although a result which will be used later is obtained in the process.

The length of a rod, which is at rest in the inertial frame K and lies parallel to the x axis, is $\Delta x = x_2 - x_1$, where x_2 and x_1 are the coordinates of the ends of the rod. In the frame K', which is moving along the x axis with velocity v relative to K, the length of the rod, $\Delta x' = x_2' - x_1'$, will be different. This length may be calculated by using the first of Equations 4.18 to determine x_1' and x_2' at time t' in terms of x_1 and x_2:

$$x_1 = \gamma(x_1' + vt'); \qquad x_2 = \gamma(x_2' + vt'). \tag{4.22}$$

Elimination of t' and use of Equation 4.16 gives:

$$\Delta x' = \Delta x/\gamma = \Delta x(1 - v^2/c^2)^{\frac{1}{2}}. \tag{4.23}$$

If the first of Equations 4.19 had been used instead, the positions x_1' and x_2' would have been determined at two different times t_1' and t_2', as can be seen from the last of Equations 4.19. In the method used here the times at which x_1 and x_2 are determined is not important, since the rod is at rest in K. From Equation 4.23 it can now be seen that the length of the rod in K', the frame in which it is moving, is less than the length in its rest frame K. This shortening is known as the Lorentz (or Fitzgerald) contraction.

A similar phenomenon occurs when time lapses are considered. For a clock that is at rest in the K frame and is consequently moving at velocity $-v$ in the K' frame, the time between two events that occur at x, y, z at t_1 and t_2 is $\Delta t = t_2 - t_1$. The corresponding time interval, $\Delta t' = t_2' - t_1'$, in K' may be obtained from Equations 4.19:

$$t_1' = \gamma(t_1 + vx/c^2); \qquad t_2' = \gamma(t_2 + vx/c^2) \tag{4.24}$$

from which:

$$\Delta t' = \gamma \Delta t = \Delta t(1 - v^2/c^2)^{-\frac{1}{2}}. \tag{4.25}$$

Here the fact that x has the same value for each event has been used; if Equations 4.18 were used, allowance would have to be made for the fact that the value of x' is different for the two events. Equation 4.25 shows that the time (Δt) measured by a moving clock is always less than the time $(\Delta t')$ measured in the rest frame. This phenomenon is known as time dilation.

The infinitesimal interval ds is given by Equation 4.4. Now a clock measures the time in its own rest frame so that the infinitesimals dx, dy and dz are all zero and $c\mathrm{d}t = \mathrm{d}s$. Thus the time lapse measured by a clock is:

$$\Delta t = t_b - t_a = \int_{t_a}^{t_b} \mathrm{d}t = c^{-1} \int_a^b \mathrm{d}s, \tag{4.26}$$

where the second integral is a line integral along the world line of the clock. If the clock is at rest, its world line will be straight and parallel to the t axis and, as we have seen, it will indicate a greater time than if it had been subjected to a journey which returned to its starting point. Thus the line integral $\int_a^b \mathrm{d}s$ is a maximum for a straight world line joining the world points a and b. Although this fact has been established for a world line parallel to the t axis, that is for a clock at rest, the integral will be a maximum for any straight world line, since the interval s is independent of the inertial frame used and a straight world line in one frame will be straight in all other inertial frames. This result may be interpreted as showing that the length of a line between two world points is greatest when the line is straight! However, this is not as ridiculous as it seems, since one of the coordinates, $\tau = ict$, in the four-dimensional space is really imaginary. The fact that $\int_a^b \mathrm{d}s$ is a maximum for a straight world line will be used in the next section.

Both the Lorentz contraction and time dilation are real phenomena, though after more than half a century the paradoxes associated with them are still discussed frequently and there are those who remain unconvinced. We will confine our discussion to one experimental manifestation of time dilation and one of the paradoxes associated with it.

Mu-mesons, which are created by cosmic rays in the upper atmosphere at a height of approximately 10^4 m, are observed on the ground. Their observed lifetime in the laboratory is only 2×10^{-6} s suggesting that those reaching the ground must travel at velocities of the order of $10c$! However, the lifetime measured in the laboratory is appropriate for the rest frame of a mu-meson and, as a result of time dilation, the lifetime will be greater by a factor of $(1 - v^2/c^2)^{-\frac{1}{2}}$ in a frame in which the mu-meson is moving with velocity v. Mu-mesons with velocities close to that of light can thus reach the ground before they decay.

One of the more famous paradoxes is the so-called clock paradox. If two clocks are synchronized in the rest frame and one then undergoes a round-trip at high velocity, on its return the travelled clock will be found to be running slow relative to the other as a result of time dilation. The paradox arises when it is argued that the two clocks should run at the same rate, since there is a symmetry between them in that their roles could be reversed by choosing a different rest frame. However, although this symmetry between the clocks exists when they are moving uniformly with respect to one another, it is not present during the acceleration and deceleration that the travelling clock must experience during its journey.

4.5 The transformation of velocities

In a relativistic theory velocities cannot be added vectorially, since it would be possible to have a resultant velocity greater than that of light. To see how velocities should be combined, the way in which velocities transform from one inertial frame to another must be considered.

The two inertial frames K and K', that are illustrated in Fig. 4.1, are used again. The Lorentz transformation, Equations 4.18, may be written in terms of infinitesimals:

$$\begin{aligned} dx &= \gamma(dx' + v\,dt'); \qquad dy = dy'; \\ dz &= dz'; \qquad dt = \gamma(v\,dx'/c^2 + dt'). \end{aligned} \tag{4.27}$$

Division of the first of these equations by the fourth gives the x component of a velocity $\mathbf{u}$ in the K frame in terms of the corresponding component of $\mathbf{u}'$ the velocity in the K' frame:

$$u_x = \frac{dx}{dt} = \frac{dx' + v dt'}{v dx'/c^2 + dt'} = \frac{u'_x + v}{1 + u'_x v/c^2}; \tag{4.28}$$

similarly:

$$u_y = u'_y/\gamma(1 + u'_x v/c^2); \qquad u_z = u'_z/\gamma(1 + u'_x v/c^2). \tag{4.29}$$

Unlike the y and z components of a position vector the corresponding components of a velocity are changed in this transformation, since time intervals are changed. Equations 4.28 and 4.29 both give the classical equations, if c is made infinite.

For a particle moving with velocity u' parallel to the x' axis in the frame K', Equations 4.28 and 4.29 show that in the K frame it moves parallel to the x axis with a velocity u given by:

$$u = (u' + v)/(1 + u'v/c^2). \tag{4.30}$$

This equation expresses the result of adding two velocities, and can be seen to reduce to the classical result, $u = u' + v$, for low velocities. In addition, it is a simple exercise to show from Equation 4.30 that the relativistic sum of two velocities can never exceed the velocity of light.

4.6 The relativistic mechanics of a free particle

In the classical (non-relativistic) mechanics of the last chapter, Newton's second law implies that a particle can be accelerated to any velocity by the continued application of a force. This is relativistically impossible, and there must be an increased resistance to acceleration as the particle's velocity approaches the velocity of light; indeed, this resistance must become infinite in the limit that the particle has velocity c. The reason that the classical theory is unsatisfactory is that the equations of motion are not invariant to a Lorentz transformation, which is necessary for relativistic validity. Of course, this does not mean that Newton's laws are wrong; they work very well when low velocities are involved and the corresponding relativistic equations approximate to them for such situations. However, the theory of Chapter 3 needs modification when high velocities are considered. In fact, the only change necessary is in the Lagrangian used, since invariance to a Galilean transformation was only invoked to determine the explicit form of the non-relativistic Lagrangian of a free particle (Section 3.4).

The principle of least action (Section 3.2) requires that the action integral S has a minimum for actual motion between two situations, a and b, specified by coordinates and times. Thus S can be thought of as an integral along a world line between two world points. Now S must be independent of the inertial frame used and must therefore be a scalar, since Lorentz transformations do not preserve the magnitudes or directions of vectors. This means that the integrand of the action integral must be a scalar as well as being a first-order differential. The only scalar that satisfies these conditions is the infinitesimal interval ds, which is defined in Equation 4.4. Thus:

$$S = k \int_a^b \mathrm{d}s, \tag{4.31}$$

where k is a constant and a and b are two world points. The sign of k is negative, since the action is required to be a minimum for actual motion and it was seen in Section 4.4 that $\int_a^b \mathrm{d}s$ is maximum for a straight world line and could be made arbitrarily small by a suitable choice of a curved world line.

From Equation 4.4 ds is given by:

$$\begin{aligned} \mathrm{d}s &= (c^2\mathrm{d}t^2 - \mathrm{d}x^2 - \mathrm{d}y^2 - \mathrm{d}z^2)^{\frac{1}{2}} \\ &= c\mathrm{d}t(1 - u^2/c^2)^{\frac{1}{2}}, \end{aligned} \tag{4.32}$$

where $\mathbf{u}$ is the velocity of the particle. Substitution in Equation 4.31 gives:

$$S = k \int_{t_a}^{t_b} c(1 - u^2/c^2)^{\frac{1}{2}}\, \mathrm{d}t, \tag{4.33}$$

which may be compared with the definition (Equation 3.2) of the action integral in terms of the Lagrangian:

$$S = \int_{t_a}^{t_b} \mathcal{L}\, \mathrm{d}t, \tag{4.34}$$

showing that relativistically:

$$\mathcal{L} = kc(1 - u^2/c^2)^{\frac{1}{2}}. \tag{4.35}$$

The constant k characterizes the particle and may be determined by comparing this Lagrangian in the classical limit with the non-relativistic Lagrangian of Chapter 3, which in the current notation is just $\frac{1}{2}mu^2$. Expansion of Equation 4.35 as a power series in (u/c) gives:

$$\mathcal{L} = kc - \tfrac{1}{2}kc(u/c)^2 + \ldots; \tag{4.36}$$

the first term may be ignored, since it is a constant and makes a contribution to the action integral (Equation 4.34) that is independent of the path of the integral. If k is taken to be $-mc$ the second term reduces to $\frac{1}{2}mu^2$ and the remaining terms all vanish in the classical limit. The relativistic Lagrangian for a free particle is thus:

$$\mathcal{L} = -mc^2(1 - u^2/c^2)^{\frac{1}{2}}. \tag{4.37}$$

From this Lagrangian the momentum and Hamiltonian of the particle may be determined from the definitions of the previous chapter. Using Equation 3.24 the momentum of the particle is:

$$\mathbf{p} = (\partial\mathcal{L}/\partial\mathbf{u}) = m\mathbf{u}(1 - u^2/c^2)^{-\frac{1}{2}}. \tag{4.38}$$

The Hamiltonian is given by Equation 3.33 to be:

$$\mathcal{H} = \mathbf{p}.\mathbf{u} - \mathcal{L} = mc^2(1 - u^2/c^2)^{-\frac{1}{2}}. \tag{4.39}$$

Of course, to be useful the Hamiltonian must be expressed in terms of the momentum rather than the velocity, but Equation 4.38 may be used to eliminate $\mathbf{u}$ from Equation 4.39. This is achieved most easily by noting that:

$$\begin{aligned}\mathcal{H}^2 &= m^2c^4/(1 - u^2/c^2) = m^2c^4 + [m^2c^2u^2/(1 - u^2/c^2)] \\ &= m^2c^4 + c^2p^2.\end{aligned} \tag{4.40}$$

The required result is thus:

$$\mathcal{H} = c(m^2c^2 + p^2)^{\frac{1}{2}}, \tag{4.41}$$

where the positive square root has been taken to agree with the non-relativistic limit.

When c is allowed to become infinite, the momentum (Equation 4.38) goes over to the classical expression $m\mathbf{u}$ as expected, but the Hamiltonian (Equation 4.39) becomes infinite! This difficulty may be avoided by expanding $\mathcal{H}$ as a power series in (u/c) and then neglecting terms in $(u/c)^4$ and higher. This gives:

$$\mathcal{H} = mc^2 + \tfrac{1}{2}mu^2; \tag{4.42}$$

similarly, in the non-relativistic limit Equation 4.41 reduces to:

$$\mathcal{H} = mc^2 + p^2/2m. \tag{4.43}$$

Both of these expressions are the same as the non-relativistic ones except for the presence of an extra term mc^2. This term is present even when the particle is at rest and is called the rest energy.

In this book the mass m of a particle is taken to be a constant independent of velocity, but an equation such as 4.39 may be viewed in a different way by rewriting it as $\mathcal{H} = Mc^2$, where M, which is still thought of as the mass, varies with velocity and is related to the rest mass m by $M = m(1 - u^2/c^2)^{-\frac{1}{2}}$. Thus, when a particle is accelerated the energy it gains may be associated with its extra mass, suggesting that all mass is a manifestation of energy, even the rest mass. These ideas led to the famous equation $E = Mc^2$ and are supported by experiments involving nuclei and fundamental particles in which mass and energy are interconverted.

4.7 Four-vectors

It has already been noted that time may be thought of as a fourth dimension and that the location and time of an event may be represented by a world point in four-dimensional space. The position of this world point may be specified by a vector with four components r_α ($\alpha = 1, 2, 3, 4$), where r_1, r_2 and r_3 are the spatial coordinates x, y and z respectively and r_4 is related to the time coordinate. To distinguish this type of vector from those in three-dimensional space, they are called four-vectors and here Greek subscripts are used to label their components.

By analogy with three-dimensional vectors the scalar product of the position four-vector with itself is defined as:

$$r_\alpha r_\alpha = r_1^2 + r_2^2 + r_3^2 + r_4^2, \tag{4.44}$$

where the repeated subscript convention is again used, but since a four-vector is involved, the summation is over the four possible values of the subscript α. If r_4 is to chosen to be $\tau = ict$ (Equation 4.9), then Equation 4.44 becomes:

$$r_\alpha r_\alpha = -c^2t^2 + x^2 + y^2 + z^2, \tag{4.45}$$

and the scalar product $r_\alpha r_\alpha$ is invariant to Lorentz transformations, since it is then directly proportional to the square of the interval s (see Section 4.2). With this choice of r_4 the Lorentz transformation (Equations 4.18) is:

$$\begin{aligned} r_1 &= \gamma(r_1' - ivr_4'/c); \qquad r_2 = r_2'; \qquad r_3 = r_3'; \\ r_4 &= \gamma(ivr_1'/c + r_4'), \end{aligned} \tag{4.46}$$

where γ retains its original meaning (Equation 4.16).

A general four-vector may now be defined as one having components that transform in precisely the same way as the r_α under a Lorentz transformation. With this definition the scalar product of any two four-vectors, for example $A_\alpha B_\alpha$, is invariant to such transformations and is independent of the inertial frame used, as can be seen by expansion of the product.

Although it was not at all obvious, there were four quantities in the previous section that constitute the components of a four-vector. The way in which the components of the momentum $\mathbf{p}$ (Equation 4.38) of a particle behave under a Lorentz transformation may be established using Equations 4.28 and 4.29, which describe the transformation of velocities. The algebra is simple but tedious, and will not be given here, the result being that the components of $\mathbf{p}$ are the spatial components, p_1, p_2 and p_3, of a four-vector the corresponding time component being $p_4 = i\mathcal{H}/c$, where $\mathcal{H}$ is the particle's Hamiltonian or energy (Equation 4.39). It follows from the invariance of the scalar product of two four-vectors that $p_\alpha p_\alpha$ should be invariant:

$$p_\alpha p_\alpha = p^2 - \mathcal{H}^2/c^2; \tag{4.47}$$

from Equation 4.41 this is just equal to $-m^2c^2$, where m is the rest mass of the particle, confirming that it is indeed invariant.

The Lorentz transformation can be thought of as an orthogonal transformation in four-dimensional space. This means that, in addition to scalar products being invariant, any expression or equation that can be written entirely in terms of four-vectors will be unchanged by a Lorentz transformation. Of course the components of the four-vector will be changed, but in the new frame the expression or equation will have the same form as in the old and the physical content will be unaltered. This behaviour is analagous to the three-dimensional situation, where the meaning of a vector equation is independent of the orthogonal axis system used. Thus, by writing an equation in a four-vector notation its Lorentz invariance is automatically ensured.

The Lorentz transformation of the position four-vector can be written generally as:

$$r_\alpha = a_{\alpha\beta} r'_\beta. \tag{4.48}$$

The $a_{\alpha\beta}$ are the components of a four-by-four matrix, and, since the transformation is orthogonal, this matrix is orthogonal and the components satisfy the relations (cf. Equation 2.104):

$$a_{\alpha\beta} a_{\alpha\gamma} = \delta_{\beta\gamma}; \tag{4.49}$$

the inverse relation to Equation 4.48 is thus:

$$r'_\beta = a_{\alpha\beta} r_\alpha. \tag{4.50}$$

Now under the transformation the operator $(\partial/\partial r_\alpha)$ transforms as:

$$\begin{aligned}(\partial/\partial r_\alpha) &= (\partial r'_\beta/\partial r_\alpha)(\partial/\partial r'_\beta)\\ &= a_{\alpha\beta}(\partial/\partial r'_\beta),\end{aligned} \tag{4.51}$$

so that $(\partial/\partial r_\alpha)$ is the component of a four-vector operator. Other four-vectors will be encountered in the next chapter.

The concept of four-vectors, the components of which transform in the same way as the r_α, can be extended to four-tensors, the components of a four-tensor of rank n transforming in the same way as products of n components of the position four-vector. Thus a four-tensor of rank one is just a four-vector, while a four-tensor of rank two has sixteen components $F_{\alpha\beta}$ $(\alpha, \beta = 1, 2, 3, 4)$

which transform in the same way as the sixteen quadratic products $r_\alpha r_\beta$, that is:

$$F_{\alpha\beta} = a_{\alpha\gamma} a_{\beta\delta} F'_{\gamma\delta}. \tag{4.52}$$

Four-tensors of higher rank are defined in an analogous fashion and examples of four-tensors will be encountered in the next chapter. The advantage of using a four-tensor notation is that any equation that can be written entirely in terms of four-tensors will automatically be Lorentz invariant.

The theory of four-vectors may be developed further along the same lines as for three-vectors; for example the permutation symbol $\epsilon_{\alpha\beta\gamma\delta}$ can be introduced. However, any further discussion will be postponed till it is needed. Finally, it should be noted that many texts use superscripts as well as subscripts to distinguish the covariant and contravariant components of vectors. This distinction is not necessary in the special theory of relativity; indeed, we would gain nothing from this sophistication and the price would be greater complexity.

Bibliography

Special Relativity

Feynman, Leighton and Sands: Volume I, Chapters 15–17.
Goldstein, H. (1950), *Classical Mechanics*, McGraw-Hill, New York: Chapter 6.
Landau and Lifschitz: Chapters 1 and 2.
Rindler, W. (1966), *Special Relativity*, Oliver and Boyd, Edinburgh.

CHAPTER FIVE

The Interaction of Charged Particles with Electromagnetic Fields

So far we have been concerned with the mechanics of a free particle, and the possibility of a particle interacting with an electromagnetic field must now be considered. This will lead naturally to a discussion of the electromagnetic field itself. It has already been noted that Maxwell's equations, which describe electromagnetic fields, were found to be invariant to Lorentz transformations rather than Galilean transformations and that this discovery preceded Einstein's theory of special relativity. Consequently it is appropriate that electromagnetic theory should be developed from a relativistic stand point, although this was not the case historically. The approach used here is a logical one, but of course the final justification of the theory is that it agrees with experiment.

5.1 Units

It is at this point that a decision must be made about units. In mechanics the transition between different systems of units is trivial and usually involves only factors of 10. However, in electromagnetic theory the appearance or non-appearance of the factors 4π, c, ϵ_0 and μ_0 depends on the system of units adopted.

In the past the system of units used by chemists and chemical physicists has usually been the mixed or Gaussian system, which is based on centimetre-gramme-second (CGS) units, but in which magnetic quantities are measured in electromagnetic units (e.m.u.) and electrostatic quantities in electrostatic units (e.s.u.). Because of the way these units are defined, equations connecting the two sorts of quantity contain factors of c. This appearance of c is not a consequence of special relativity, but merely reflects the separate initial development of electrostatics and magnetostatics; in a relativistic theory these scaling factors involving c are confusing. In addition, since there is no separate dimension for electromagnetic properties in the CGS system, the dimensions of electromagnetic quantities involve square roots. Finally, in equations using the mixed system of units factors of 4π appear when they are not expected geometrically, and do not appear when they are.

It has now been agreed internationally that the Systême International (SI) should be used for units. Although the transition from the old system to the SI system is not trivial, it will be adopted here, since it has definite advantages. First, the ampere is employed as a fundamental unit in addition to the metre, kilogramme and second, that is the units are based on an MKSA system. This means that square roots no longer appear in dimension expressions. Secondly, all electromagnetic quantities are expressed in units coherent with these four, so that the scaling factors involving c no longer appear; when c does appear in an equation it is usually a consequence of special relativity. Thirdly, the system is rationalized in that factors of 4π only appear where they are expected on geometrical grounds.

There are other consequences of the adoption of SI units. In the mixed system the permeability and permittivity of free space, μ_0 and ϵ_0, are dimensionless constants and are frequently omitted from equations, since they are both equal to unity. However, in the SI system they have dimensions, are not equal to unity and must be included in equations. In addition, the magnetic field strength **H** is often used instead of the magnetic flux density **B** in the mixed system, since they are equal in free space, although observable physical properties depend on **B**. In the SI system **B** and **H** differ by a factor of $\mu_0 = 4\pi \times 10^{-7}$ kg m s^{-2} A^{-2} and equations must be written in terms of **B**.

The details of the SI system as far as electromagnetic phenomena are concerned will appear at appropriate points in the text. These details are summarized in Appendix A together with a few comments on the conversion from the mixed CGS system to the SI system.

5.2 The electromagnetic potentials

The relativistic mechanics of a particle was developed in the last chapter by using the principle of least action. The appropriate action integral was seen to be (Section 4.6):

$$S = \int_a^b (-mc\, ds), \tag{5.1}$$

where s is the interval and the integration is performed between the world points a and b. If the particle interacts with an electromagnetic field, further terms have to be added to this action integral, and these terms will involve quantities characteristic of the particle and the field. The extra terms must, of course, be Lorentz invariant, but beyond that little can be said about them and an appeal to experiment is necessary. That is, the justification of the terms that will be added is that the equations that they lead to are in agreement with experimental observations.

Experiments indicate that only one new parameter, the charge q, is needed to characterize a particle as far as its interaction with an electromagnetic field is concerned, but that four quantities A_α ($\alpha = 1, 2, 3, 4$), which are functions of time and the position of the particle, are needed for the field. The A_α are the components of a four-vector and thus under a Lorentz transformation they behave in a similar manner to the r_α, the components of the position four-vector. The spatial components, A_1, A_2 and A_3, are the components of a three-dimensional vector $\mathbf{A}$, which is called the (magnetic) vector potential, while the time component A_4 is related to the electric or scalar potential ϕ by:

$$A_4 = i\phi/c. \tag{5.2}$$

A theory that agrees with experiment may be developed by taking the interaction terms in the action integral to be:

$$\int_a^b qA_\alpha \mathrm{d}r_\alpha; \tag{5.3}$$

the integrand is proportional to the invariant formed by the scalar product of the potential four-vector and the infinitesimal position four-vector, which has components $\mathrm{d}r_\alpha$.

The total action integral thus appears as:

$$S = \int_a^b (-mc\mathrm{d}s + q\mathbf{A}.\mathrm{d}\mathbf{r} - q\phi \mathrm{d}t), \tag{5.4}$$

where the scalar product in Equation 5.3 has been expanded in terms of scalars and three-dimensional vectors. The corresponding Lagrangian may be extracted from the action by recalling that (Equation 4.32):

$$\mathrm{d}s = c(1 - u^2/c^2)^{\frac{1}{2}} \mathrm{d}t, \tag{5.5}$$

where $\mathbf{u}$ is the velocity of the particle:

$$\mathbf{u} = \mathrm{d}\mathbf{r}/\mathrm{d}t. \tag{5.6}$$

Substitution for $\mathrm{d}\mathbf{r}$ and $\mathrm{d}s$ in Equation 5.4 gives:

$$S = \int_{t_a}^{t_b} [-mc^2(1 - u^2/c^2)^{\frac{1}{2}} + q\mathbf{A}.\mathbf{u} - q\phi]\,\mathrm{d}t, \tag{5.7}$$

and comparison with the definition (Equation 4.34) of the action integral in terms of the Lagrangian $\mathcal{L}$ shows that:

$$\mathcal{L} = -mc^2(1-u^2/c^2)^{\frac{1}{2}} + q\mathbf{A}.\mathbf{u} - q\phi. \tag{5.8}$$

The introduction of an electromagnetic field can thus be accounted for by the addition of two terms to the Lagrangian.

The momentum of the particle is given by Equation 3.24:

$$\begin{aligned}\mathbf{p} &= \partial\mathcal{L}/\partial\mathbf{u} = m\mathbf{u}(1-u^2/c^2)^{-\frac{1}{2}} + q\mathbf{A} \\ &= \boldsymbol{\pi} + q\mathbf{A},\end{aligned} \tag{5.9}$$

where $\boldsymbol{\pi}$ is called the mechanical momentum, that is the momentum in the absence of the field; $\mathbf{p}$ is the total momentum conjugate to the velocity $\mathbf{u}$ and is distinguished from $\boldsymbol{\pi}$ by being referred to as the canonical momentum. The momentum of a particle is thus modified by the addition of a term involving the vector potential $\mathbf{A}$.

Equation 3.33 gives the Hamiltonian of the particle:

$$\mathcal{H} = \mathbf{p}.\mathbf{u} - \mathcal{L} = mc^2(1-u^2/c^2)^{-\frac{1}{2}} + q\phi. \tag{5.10}$$

This may be expressed more appropriately in terms of the canonical momentum $\mathbf{p}$ in the same way that Equation 4.41 was obtained for the case of no field:

$$\begin{aligned}\mathcal{H} &= c\,[m^2c^2 + (\mathbf{p}-q\mathbf{A})^2]^{\frac{1}{2}} + q\phi \\ &= c(m^2c^2 + \pi^2)^{\frac{1}{2}} + q\phi.\end{aligned} \tag{5.11}$$

Comparison of Equations 5.9 and 5.11 with Equations 4.38 and 4.41 shows that to generalize from a free particle to a particle in a field, all that is necessary is to replace $\mathbf{p}$ by $\boldsymbol{\pi} = (\mathbf{p} - q\mathbf{A})$ and $\mathcal{H}$ by $(\mathcal{H} - q\phi)$. For example, the non-relativistic limit of the Hamiltonian (Equation 5.11) is, by comparison with the non-relativistic field-free Hamiltonian (Equation 4.43):

$$\mathcal{H} = mc^2 + q\phi + \pi^2/2m. \tag{5.12}$$

In the SI system of units the charge q has dimensions TA, while the vector and scalar potentials, $\mathbf{A}$ and ϕ, have dimensions $MLT^{-2}A^{-1}$ and $ML^2T^{-3}A^{-1}$, respectively. For the mixed system of units Equations 5.9 to 5.12 are only modified in that the mechanical momentum must be rewritten as:

$$\boldsymbol{\pi} = \mathbf{p} - (q/c)\mathbf{A}, \tag{5.13}$$

where the factor c^{-1} has no relativistic significance; however, there are other hidden differences, since the dimensions of q are modified to $M^{1/2}L^{3/2}T^{-1}$ and $\mathbf{A}$ and ϕ have the same dimensions $M^{1/2}L^{1/2}T^{-1}$.

5.3 The field vectors

Now that the Lagrangian for a charged particle in an electromagnetic field is known, the particle's equations of motion may be determined from the Lagrange equations 3.7, which for a single particle may be written as:

$$\frac{\mathrm{d}}{\mathrm{d}t}\left(\frac{\partial \mathcal{L}}{\partial \mathbf{u}}\right) - \left(\frac{\partial \mathcal{L}}{\partial \mathbf{r}}\right) = 0. \tag{5.14}$$

It is first noted that $(\partial\mathcal{L}/\partial\mathbf{u})$ is by definition the canonical momentum $\mathbf{p} = (\boldsymbol{\pi} + q\mathbf{A})$, while the operator $(\partial/\partial\mathbf{r})$ is just the operator $\boldsymbol{\nabla}$, so that Equation 5.14 can be rewritten as:

$$(\mathrm{d}\boldsymbol{\pi}/\mathrm{d}t) + q(\mathrm{d}\mathbf{A}/\mathrm{d}t) - \boldsymbol{\nabla}\mathcal{L} = 0. \tag{5.15}$$

Now, since $\mathbf{A}$ is a function of space and time coordinates:

$$\begin{aligned}(\mathrm{d}\mathbf{A}/\mathrm{d}t) &= (\partial\mathbf{A}/\partial t) + (\mathrm{d}r_i/\mathrm{d}t)(\partial\mathbf{A}/\partial r_i)\\ &= (\partial\mathbf{A}/\partial t) + (\mathbf{u}.\boldsymbol{\nabla})\mathbf{A}.\end{aligned} \tag{5.16}$$

In addition, the gradient of the Lagrangian (Equation 5.8) is:

$$\begin{aligned}\boldsymbol{\nabla}\mathcal{L} &= q\boldsymbol{\nabla}(\mathbf{A}.\mathbf{u}) - q\boldsymbol{\nabla}\phi\\ &= q[(\mathbf{u}.\boldsymbol{\nabla})\mathbf{A} + \mathbf{u}_\wedge(\boldsymbol{\nabla}_\wedge\mathbf{A})] - q\boldsymbol{\nabla}\phi,\end{aligned} \tag{5.17}$$

where use has been made of the vector relation, Equation 2.66, and the fact that $\mathbf{u}$ does not depend explicitly on $\mathbf{r}$. Substitution of Equations 5.16 and 5.17 into Equation 5.15 followed by rearrangement gives finally:

$$(\mathrm{d}\boldsymbol{\pi}/\mathrm{d}t) = q\mathbf{u}_\wedge(\boldsymbol{\nabla}_\wedge\mathbf{A}) + q[-\boldsymbol{\nabla}\phi - (\partial\mathbf{A}/\partial t)]. \tag{5.18}$$

The term $(\mathrm{d}\boldsymbol{\pi}/\mathrm{d}t)$ is just the force on the particle, since in the classical limit this reduces to $m(\mathrm{d}\mathbf{u}/\mathrm{d}t)$.

The right-hand side of Equation 5.18 has been separated into a velocity dependent part and a velocity independent part, so that this equation may be rewritten as:

$$(\mathrm{d}\boldsymbol{\pi}/\mathrm{d}t) = q\mathbf{u}_\wedge\mathbf{B} + q\mathbf{E}; \tag{5.19}$$

in this form it is known as Lorentz's equation. The vector $\mathbf{B}$ is called the magnetic flux density and is defined by:

$$\mathbf{B} = \boldsymbol{\nabla}_\wedge\mathbf{A}, \tag{5.20}$$

while the electric field strength $\mathbf{E}$ is given by:

$$\mathbf{E} = -\boldsymbol{\nabla}\phi - (\partial\mathbf{A}/\partial t). \tag{5.21}$$

In a vacuum **B** and **E** together completely specify the electromagnetic field, but in media other than free space two additional vectors are needed, the magnetic field strength **H** and the electric displacement **D**. In the mixed system of units **B** and **E** are respectively the same as **H** and **D** for the special case of a vacuum and in particular **H** is frequently used (incorrectly) in place of **B**. However, this equality does not occur when the SI system is used and it is necessary to distinguish **B** and **H**. From now on we will only be concerned with **B** and **E**, and when we refer (loosely) to a magnetic or electric field, it is to be understood that these are characterized by the magnetic flux density **B** and the electric field strength **E**.

From Equations 5.20 and 5.21 **B** and **E** may be seen to have the dimensions $MT^{-2}A^{-1}$ and $MLT^{-3}A^{-1}$, respectively, in the SI system. In the mixed unit system Equation 5.20 is unaltered, but the term involving **A** in Equation 5.21 is modified by a factor of c^{-1}; again the dimensions in the mixed unit system are more complicated than this implies.

5.4 The Lorentz transformation of electric and magnetic fields

The components of the magnetic flux density **B** and the electric field strength **E** are related to the components of the four-potential A_α and the position four-vector r_α by Equations 5.20 and 5.21. We might enquire how these components behave under a Lorentz transformation. It will be seen that they transform in the same way as quadratic products of the components of two four-vectors, that is as the components of a four-tensor of second rank.

The sixteen quantities $F_{\alpha\beta}$ $(\alpha, \beta = 1, 2, 3, 4)$, which are defined as:

$$F_{\alpha\beta} = (\partial A_\beta/\partial r_\alpha) - (\partial A_\alpha/\partial r_\beta), \tag{5.22}$$

are components of a second rank four-tensor. Since the A_α are components of the four-potential, they behave in the same way as the components of the position four-vector (Equation 4.48) under a Lorentz transformation, that is:

$$A_\alpha = a_{\alpha\beta}A'_\beta. \tag{5.23}$$

In addition the operators $(\partial/\partial r_\alpha)$ were shown to be the components of a four-vector (Equation 4.51):

$$(\partial/\partial r_\alpha) = a_{\alpha\beta}(\partial/\partial r'_\beta), \tag{5.24}$$

so that:

$$F_{\alpha\beta} = a_{\alpha\gamma}a_{\beta\delta}F'_{\gamma\delta}; \tag{5.25}$$

comparison with Equation 4.52 shows that the $F_{\alpha\beta}$ are indeed the components of a four-tensor. However, explicit consideration of these components (Equation 5.22), and use of Equations 5.20 and 5.21 shows that they are related to the components of **B** and **E**; this relationship may be indicated by a matrix-like array:

$$(F_{\alpha\beta}) = \begin{pmatrix} 0 & B_z & -B_y & -iE_x/c \\ -B_z & 0 & B_x & -iE_y/c \\ B_y & -B_x & 0 & -iE_z/c \\ iE_x/c & iE_y/c & iE_z/c & 0 \end{pmatrix} \tag{5.26}$$

Thus the components of **B** and **E** are related to the components of a four-tensor, which is called the electromagnetic field tensor and which is antisymmetric, since $F_{\alpha\beta} = -F_{\beta\alpha}$.

The transformation properties of the $F_{\alpha\beta}$ and hence the components of **B** and **E** may now be determined. As usual the special case depicted in Fig. 4.1 is considered, that is the $\alpha = 2$ and 3 components of a four-vector are unaltered by the transformation. Thus F'_{22}, F'_{23} and F'_{33} are unchanged, while F'_{12} and F'_{42} transform as the $\alpha = 1$ and $\alpha = 4$ components of the same four-vector; the same remark applies to F'_{13} and F'_{43}. The remaining components are related to these because of their antisymmetry, except for F'_{11}, F'_{14} and F'_{44}, which can easily be shown to be unchanged for an antisymmetric four-tensor. We conclude that $F_{\alpha\beta} = F'_{\alpha\beta}$ except for:

$$\begin{aligned} F_{12} &= \gamma(F'_{12} - ivF'_{42}/c); \qquad & F_{42} &= \gamma(ivF'_{12}/c + F'_{42}); \\ F_{13} &= \gamma(F'_{13} - ivF'_{43}/c); \qquad & F_{43} &= \gamma(ivF'_{13}/c + F'_{43}). \end{aligned} \tag{5.27}$$

By identifying the $F_{\alpha\beta}$ with the components of **B** and **E** (Equation 5.26), we obtain:

$$\begin{aligned} B_x &= B'_x; \quad B_y = \gamma(B'_y - vE'_z/c^2); \quad B_z = \gamma(B'_z + vE'_y/c^2); \\ E_x &= E'_x; \quad E_y = \gamma(E'_y + vB'_z); \quad E_z = \gamma(E'_z - vB'_y). \end{aligned} \tag{5.28}$$

Finally, these equations may be generalized to the case where the primed frame moves with a general velocity **v** relative to the unprimed frame; for the non-relativistic limit $v \ll c$, they take the particularly simple form:

$$\mathbf{B} = \mathbf{B}' + \mathbf{v}_\wedge \mathbf{E}'/c^2; \qquad \mathbf{E} = \mathbf{E}' - \mathbf{v}_\wedge \mathbf{B}'. \tag{5.29}$$

The magnetic flux density and the electric field strength are thus interrelated, so that if one vanishes in one frame of reference it will not necessarily be zero in another. For example, a particle that is moving in an inertial frame in which there is a magnetic field, but no electric field, will nevertheless experience an electric field in its own rest frame.

For a four-vector, with components F_α say, one can construct a quantity that is invariant to Lorentz transformations; this quantity is just the scalar product $F_\alpha F_\alpha$. One would expect to be able to construct invariants from the components $F_{\alpha\beta}$ of the electromagnetic field tensor. Since the $F_{\alpha\beta}$ transform as products of pairs of components of four-vectors, one such invariant will be $F_{\alpha\beta}F_{\alpha\beta}$, where summation over both α and β is implied. Expanding this in terms of **B** and **E** gives:

$$F_{\alpha\beta}F_{\alpha\beta} = 2(B^2 - E^2/c^2) = \text{invariant}; \tag{5.30}$$

this may be checked by explicit use of Equations 5.28.

There is another invariant of second degree this being **E.B**. However, it is not a true scalar in that it changes sign on inversion of the coordinate system, that is when **r** is replaced by $-\mathbf{r}$. This may be seen by noting that ordinary vectors such as **A** and **∇** also change sign and are called polar vectors; Equation 5.21 shows that **E** is also a polar vector, but a vector such as **B**, which may be written as the product of two polar vectors, is unchanged on inversion and is called an axial vector. Thus the scalar product **E.B** is antisymmetric with respect to inversion and is called a pseudoscalar. The expression 5.30 is a true scalar and will be employed in Chapter 6, where the electromagnetic field is considered in more detail.

5.5 Gauge transformations

The electromagnetic field vectors **B** and **E** may be determined from the potentials **A** and ϕ by Equations 5.20 and 5.21. However, since it is **B** and **E** that determine physically observable phenomena such as the force on a charged particle (Equations 5.19), it is pertinent to ask to what extent **A** and ϕ are determined by **B** and **E**. It will be seen that there is some freedom in the choice of these potentials and this may be attributed to the fact that integration of Equation 5.20 or Equation 5.21 introduces an arbitrary integration constant.

Let **A** and ϕ be replaced by some new potentials given by:

$$\mathbf{A}' = \mathbf{A} - \nabla f; \qquad \phi' = \phi + (\partial f/\partial t), \tag{5.31}$$

where f is any scalar function of position and time coordinates. Substitution into Equation 5.20 gives the magnetic field associated with $\mathbf{A}'$:

$$\mathbf{B}' = \nabla_\wedge \mathbf{A}' = \nabla_\wedge \mathbf{A} - \nabla_\wedge(\nabla f) = \nabla_\wedge \mathbf{A} = \mathbf{B}, \tag{5.32}$$

where the fact (Equation 2.70) that the curl of a gradient vanishes has been used. The electric field corresponding to ϕ' and $\mathbf{A}'$ may be obtained from Equation 5.21:

$$\mathbf{E}' = -\nabla\phi' - (\partial\mathbf{A}'/\partial t)$$

$$= -\nabla\phi - \nabla(\partial f/\partial t) - (\partial\mathbf{A}/\partial t) + (\partial\nabla f/\partial t) = \mathbf{E}. \tag{5.33}$$

Thus the potentials are not uniquely defined, since the electric and magnetic fields are unaltered by the transformation of the potentials described by Equations 5.31; such a transformation is called a gauge transformation.

It is clear that a change of gauge must not affect the equations describing physical phenomena and that these must be gauge invariant. For instance, we might enquire what happens to the Lagrangian (Equation 5.8) under the gauge transformation (Equations 5.31). Substitution shows immediately that the Lagrangian is modified by the addition of:

$$-q\mathbf{u}.(\nabla f) - q(\partial f/\partial t) = -q(\mathrm{d}f/\mathrm{d}t), \tag{5.34}$$

since f is a function of $\mathbf{r}$ and t. However, these extra terms do not affect the predictions made using the Lagrangian, since it was seen in Section 3.3 that the equations of motion are unaltered by the addition of a total time derivative to the Lagrangian.

The choice of the potentials is arbitrary to the extent that the scalar f in Equations 5.31 may be chosen, and this freedom can be used to simplify the equations that arise. Two different gauges are commonly used, the first being the Coulomb or radiation gauge in which f is chosen so that the divergence of the vector potential vanishes:

$$\nabla.\mathbf{A} = 0. \tag{5.35}$$

In the Lorentz gauge, however, the condition imposed is that:

$$\nabla.\mathbf{A} + (\partial\phi/\partial t)/c^2 = 0. \tag{5.36}$$

The latter has the advantage of being a relativistically invariant condition, as can be seen by noting that the left-hand side of Equation 5.36 is the scalar product of two four-vectors with components $(\partial/\partial r_\alpha)$ and A_α. In steady state problems the two gauges are identical and, in addition, there is no advantage in using the Lorentz condition, if a non-relativistic approximation is being considered. Even the application of a condition such as Equation 5.35 or Equation 5.36 does not remove the arbitrariness of f completely, but we would gain nothing by restricting it further at present.

In the special case of uniform fields, that is fields that are the same at all points in space and time, particularly simple expressions for the potentials may be used. If they are written as:

$$\mathbf{A} = \tfrac{1}{2}\mathbf{B}_\wedge\mathbf{r}; \qquad \phi = -\mathbf{E}.\mathbf{r}, \tag{5.37}$$

where **B** and **E** are both constant vectors, substitution in Equations 5.20 and 5.21 and the use of vector relations from Chapter 2 shows that the corresponding fields are just **B** and **E**:

$$-\nabla\phi - (\partial \mathbf{A}/\partial t) = \nabla(\mathbf{E}.\mathbf{r}) = \mathbf{E} \tag{5.38}$$

and

$$\nabla_\wedge \mathbf{A} = \tfrac{1}{2}\nabla_\wedge(\mathbf{B}_\wedge \mathbf{r}) = \mathbf{B}. \tag{5.39}$$

These potentials satisfy both the Coulomb and the Lorentz conditions.

Equation 5.12 gives the energy of a slow-moving charged particle in an electromagnetic field. However, it is expressed in terms of the potentials rather than the fields. For uniform fields the potentials (Equations 5.37) may be substituted. Here we confine our attention to the term involving the magnetic potential which may be expanded as follows:

$$\begin{aligned}(\mathbf{p} - q\mathbf{A})^2/2m &= (p^2 - 2q\,\mathbf{p}.\mathbf{A} + q^2 A^2)/2m \\ &= (p^2 - q\,\mathbf{p}.\mathbf{B}_\wedge \mathbf{r} + q^2 A^2)/2m.\end{aligned} \tag{5.40}$$

The second term in parentheses may be written as:

$$-(q/2m)\mathbf{B}.\mathbf{r}_\wedge \mathbf{p} = -(q/2m)\mathbf{B}.\mathbf{l} = -\boldsymbol{\mu}.\mathbf{B}, \tag{5.41}$$

where $\mathbf{l}$ is the angular momentum of the particle and $\boldsymbol{\mu}$ is the associated magnetic moment. This term represents the energy of the interaction between the magnetic field and a magnetic moment $\boldsymbol{\mu} = (q/2m)\mathbf{l}$.

5.6 Maxwell's equations

Equations 5.20 and 5.21 relate the electromagnetic fields to the potentials:

$$\mathbf{B} = \nabla_\wedge \mathbf{A}; \qquad \mathbf{E} = -\nabla\phi - (\partial \mathbf{A}/\partial t). \tag{5.42}$$

However, the potentials may readily be eliminated from these expressions to give equations involving only the fields.

If we operate on both sides of the first of Equations 5.42 with the divergence operator and use the fact (Equation 2.71) that the divergence of a curl vanishes, the result is:

$$\nabla.\mathbf{B} = \nabla.(\nabla_\wedge \mathbf{A}) = 0. \tag{5.43}$$

In Section 2.9 it was seen that the divergence of a vector represents a loss per unit volume. Indeed, if $\nabla.\mathbf{B}$ is integrated over a finite volume, we have from Equation 5.43 and Gauss' theorem (Equation 2.57):

$$\int \nabla.\mathbf{B}\,d\tau = \int \mathbf{B}.d\mathbf{s} = 0, \tag{5.44}$$

where the second integral is over the surface bounding the volume. Thus the flux of the vector **B** through a closed surface must vanish. This result is related to the non-existence of magnetic monopoles; the divergence of the electric vector **E** will be seen to be non-zero, since isolated electric charges do exist.

The potentials are eliminated from the second of Equations 5.42 by taking the curl of both sides, noting that the curl of a gradient is zero (Equation 2.70) and using the fact that the operators ∇ and $(\partial/\partial t)$ commute:

$$\nabla \wedge \mathbf{E} = -\nabla \wedge (\nabla\phi) - \partial(\nabla \wedge \mathbf{A})/\partial t = -\partial\mathbf{B}/\partial t. \qquad (5.45)$$

Integrating over a surface gives:

$$\int \nabla \wedge \mathbf{E}.\mathbf{ds} = \oint \mathbf{E}.\mathbf{dr} = -\partial(\int \mathbf{B}.\mathbf{ds})/\partial t, \qquad (5.46)$$

where Stokes' theorem (Equation 2.63) has been used. This equation shows that the circulation of the electric vector **E** round a closed contour is minus the partial time derivative of the magnetic flux through the contour. Equation 5.45 is thus the induction law that relates the electromotive force in a circuit to the change in the magnetic flux through the circuit.

The equations:

$$\nabla.\mathbf{B} = 0 \qquad (5.47)$$

and

$$\nabla \wedge \mathbf{E} = -\partial\mathbf{B}/\partial t \qquad (5.48)$$

may be written as one equation by using four-tensor notation:

$$\frac{\partial F_{\alpha\beta}}{\partial r_\gamma} + \frac{\partial F_{\beta\gamma}}{\partial r_\alpha} + \frac{\partial F_{\gamma\alpha}}{\partial r_\beta} = 0, \qquad (5.49)$$

where the $F_{\alpha\beta}$ are components of the electromagnetic field tensor (Equation 5.26) and r_α is a component of the position four-vector. Since the four-tensor is antisymmetric, the left-hand side of Equation 5.49 vanishes if any pair of the subscripts $\alpha\beta\gamma$ are the same. When $\alpha\beta\gamma$ correspond to the three space components, Equation 5.49 reduces to Equation 5.47, while Equation 5.48 arises when one of the $\alpha\beta\gamma$ corresponds to the time component. Writing Equations 5.47 and 5.48 in terms of a four-tensor $F_{\alpha\beta}$ and a four-vector $(\partial/\partial r_\alpha)$ demonstrates immediately that they are both Lorentz invariant, since the left-hand side of Equation 5.49 is just a third rank four-tensor. This can be checked explicitly for the transformation between the frames illustrated in Fig. 4.1 by using Equations 4.46 and 5.27; each of Equations 5.47 and 5.48 transforms into a different linear combination of the corresponding equations in the new frame.

Equations 5.47 and 5.48 cannot completely specify an electromagnetic field, since for example they do not indicate how **E** depends on time. In fact, these are just two of the four Maxwell's equations, which are required for complete specification of a field. The other pair of equations cannot be derived from the theory that has been developed so far and it is necessary to go back to the action integral and add terms for the field itself; up till now the action integral has only included terms for a charged particle and for its interaction with an electromagnetic field. To avoid a major diversion, the derivation of these equations is postponed until the next chapter, but they are quoted here for the sake of completeness, and because one of the equations is needed in the next section.

The second pair of Maxwell's equations is:

$$\nabla . \mathbf{E} = \rho/\epsilon_0, \tag{5.50}$$

and

$$\nabla \wedge \mathbf{B} = c^{-2}[\mathbf{j}/\epsilon_0 + (\partial \mathbf{E}/\partial t)], \tag{5.51}$$

where ϵ_0 is the permittivity of free space and has the dimensions $M^{-1}L^{-3}T^4A^2$ in the SI system. The other new symbols ρ, the charge density, and **j**, the current density, are functions of space and time, and have dimensions $L^{-3}TA$ and $L^{-2}A$, respectively. Since ρ is the charge per unit volume and **j** is the flow of charge per unit time per unit area across a surface, it is not surprising that a four-vector can be constructed with the components of **j** as the spatial components and $ic\rho$ as the time component; this current four-vector with components j_α is discussed in more detail in Chapter 6. The pair of equations can thus be written as one using four-tensor notation:

$$(\partial F_{\alpha\beta}/\partial r_\beta) = j_\alpha/c^2\epsilon_0, \tag{5.52}$$

as may easily be checked by substitution of Equation 5.26. Again we note that Equation 5.52 must be invariant to Lorentz transformations, since it is written in terms of four-tensors alone.

The integral form of Equation 5.50 may be obtained using Gauss' theorem (Equation 2.57):

$$\int \mathbf{E}.\mathrm{d}\mathbf{s} = \int (\rho/\epsilon_0)\mathrm{d}\tau. \tag{5.53}$$

This equation is called Gauss' law and shows that the flux of **E** through a closed surface is proportional to the total charge enclosed by the surface; unlike the divergence of **B**, the divergence of **E** does not vanish, since isolated electric charges do exist.

By using Stokes' theorem (Equation 2.62) the integral form of Equation 5.51 is:

$$\oint \mathbf{B}.\mathrm{d}\mathbf{r} = \int c^{-2}[\mathbf{j}/\epsilon_0 + (\partial \mathbf{E}/\partial t)].\mathrm{d}\mathbf{s}, \tag{5.54}$$

showing that the circulation of **B** round a closed contour is related to a current through any surface bounded by the contour, this current being made up of the true current and the so-called displacement current, which involves $(\partial\mathbf{E}/\partial t)$ and which must be included when time-dependent fields are considered.

The four Maxwell's equations are thus:

$$\left.\begin{aligned}\nabla.\mathbf{B} &= 0\\ \nabla_\wedge \mathbf{E} &= -(\partial\mathbf{B}/\partial t)\end{aligned}\right\} \quad \frac{\partial F_{\alpha\beta}}{\partial r_\gamma} + \frac{\partial F_{\beta\gamma}}{\partial r_\alpha} + \frac{\partial F_{\gamma\alpha}}{\partial r_\beta} = 0;$$

$$\left.\begin{aligned}\nabla.\mathbf{E} &= \rho/\epsilon_0\\ \nabla_\wedge \mathbf{B} &= c^{-2}\,[\mathbf{j}/\epsilon_0 + (\partial\mathbf{E}/\partial t)]\end{aligned}\right\} \quad \frac{\partial F_{\alpha\beta}}{\partial r_\beta} = \frac{j_\alpha}{\epsilon_0 c^2}. \tag{5.55}$$

In the mixed system of units these equations are modified by the replacement of **B** by $\mathbf{B}/c$ and ϵ_0 by $1/4\pi$, but of course the dimensions of all the quantities involved are also altered.

Finally, we should note that the permeability of free space μ_0 often appears in electromagnetic equations, but it is related to the permittivity of free space ϵ_0 by $\mu_0\epsilon_0 = c^{-2}$, so that equations may be written in terms of μ_0, ϵ_0 or both. The quantity μ_0 has the fixed value of $4\pi \times 10^{-7}$ kg m s^{-2}A^{-2} in the SI system, so that ϵ_0 is uncertain to the extent that there is an experimental uncertainty associated with the value of the velocity of light c. Of course use of μ_0 instead of ϵ_0 or vice versa can introduce factors of c, for which special relativity is not responsible.

5.7 The potentials and fields due to a stationary charge

The space surrounding an isolated point charge q has spherical symmetry. This means that at a point **r** relative to the charge as origin the vectors **B** and **E** must be along the radius vector **r**. In addition, the magnitude of the fields can only depend on the distance from the charge.

Although Gauss' law (Equation 5.50) has not yet been derived, it should be familiar to the reader and may be used in its integral form (Equation 5.53) to obtain the magnitude E of the electric field at a distance r from the charge. By considering a sphere of radius r centred at the charge, the surface integral on the left-hand side of Equation 5.53 becomes:

$$\int \mathbf{E}.\mathrm{d}\mathbf{s} = 4\pi r^2 E, \tag{5.56}$$

since **E** is everywhere perpendicular to the surface of the sphere; the volume integral on the right-hand side of Equation 5.53 is just:

$$\int (\rho/\epsilon_0)\mathrm{d}\tau = q/\epsilon_0. \tag{5.57}$$

Equating these two expressions gives:

$$E = q/4\pi\epsilon_0 r^2, \tag{5.58}$$

and, since the direction of the field is along $\mathbf{r}$:

$$\mathbf{E} = q\mathbf{r}/4\pi\epsilon_0 r^3. \tag{5.59}$$

The electric field is given in terms of the potentials by Equation 5.21:

$$\mathbf{E} = -\nabla\phi - (\partial \mathbf{A}/\partial t) = -\nabla\phi, \tag{5.60}$$

since the charge is stationary and the magnetic vector potential $\mathbf{A}$ must be independent of time. When it is recalled that $\nabla(1/r)$ is equal to $-\mathbf{r}/r^3$ (Equation 2.78), it is seen that an appropriate scalar potential is:

$$\phi = q/4\pi\epsilon_0 r. \tag{5.61}$$

To obtain the magnetic field $\mathbf{B}$ the same arguments may be employed except that Equation 5.47 is used instead of Gauss' law. It is immediately apparent that the magnetic field is zero, and that the magnetic vector potential $\mathbf{A}$ may be taken to be zero also.

5.8 The potentials due to a moving charge

From the previous section we know that, in the rest frame K' of a charge q, the potentials at a distance r' from the charge are given by:

$$\mathbf{A}' = 0; \qquad \phi' = q/4\pi\epsilon_0 r'. \tag{5.62}$$

The potentials in a frame K, in which the charge is moving, may be obtained by performing a Lorentz transformation, since A'_x, A'_y, A'_z and $i\phi'/c$ are the components of the potential four-vector. If the velocity of the particle in K is u in the x direction, Equation 4.46 gives:

$$\begin{aligned} A_x &= \gamma(A'_x + u\phi'/c^2); \qquad A_y = A'_y; \\ A_z &= A'_z; \qquad \phi = \gamma(uA'_x + \phi'), \end{aligned} \tag{5.63}$$

where $\gamma = (1 - u^2/c^2)^{-\frac{1}{2}}$, or substituting Equations 5.62 and generalizing to a particle moving with velocity $\mathbf{u}$:

$$\mathbf{A} = \gamma q\mathbf{u}/4\pi\epsilon_0 c^2 r' = \mathbf{u}\phi/c^2; \qquad \phi = \gamma q/4\pi\epsilon_0 r'. \tag{5.64}$$

In their present form these expressions are unacceptable, since they involve r', the distance between the charge and the observer in the K' frame. Two methods will be given to express the potentials in terms of quantities measured in the K frame.

For a potential to be experienced by an observer a signal must be emitted by the charge at some earlier time and this signal is propagated with velocity c. Let us assume that the frames coincide at $t = t' = 0$ and that the observer is situated at the origin of the K frame. The charge is taken to be at position $\mathbf{r}'$ in its own rest frame and at $\mathbf{r}$ in the K frame; $\mathbf{r}$ is clearly time dependent. If the potentials are observed at $t = t' = 0$, the signal must have left the charge at times:

$$t = -r/c; \qquad t' = -r'/c, \tag{5.65}$$

where r is measured at the time the signal is emitted. Now, for the emission of the signal, t' may be related to $\mathbf{r}$ and t by the inverse of the Lorentz transformation (Equations 4.21):

$$t' = \gamma[t - (\mathbf{u}.\mathbf{r})/c^2], \tag{5.66}$$

so that:

$$r' = \gamma[r + (\mathbf{u}.\mathbf{r})/c]. \tag{5.67}$$

Finally, substitution in Equations 5.64 gives:

$$\mathbf{A} = \mathbf{u}\phi/c^2; \qquad \phi = q/4\pi\epsilon_0[r + (\mathbf{u}.\mathbf{r})/c]. \tag{5.68}$$

Although these potentials have been established for a specific situation, they apply generally if $\mathbf{r}$ is interpreted as the position of the charge relative to the observer (in the previous section $\mathbf{r}$ was the position of the observer relative to the charge). The important thing to note is that the expressions 5.68, which are known as the Liénard-Wiechert potentials, must be evaluated at the time the signal leaves the charge. The potentials are thus expressed in terms of the so-called retarded values of $\mathbf{r}$ and $\mathbf{u}$. Clearly the observed potentials only depend on the instantaneous velocity and position of the particle at the retarded time, the time at which the signal is emitted by the charge, so that Equations 5.68 apply to a charge moving in an arbitrary manner. The retarded potentials may be expressed in terms of non-retarded quantities provided that there is sufficient information about the motion of the particle. This is fairly simple for a uniformly moving charge, but rather than do this here another method is presented, in which the retardation effect is automatically accounted for.

Retarded potentials arose because the event to which the Lorentz transformation 5.66 applies is the emission of the signal from the charge and the relative velocity of the two frames must be taken as the charge's velocity at this event. However, if the charge is moving with constant velocity, the same rest frame K' for the charge is appropriate at both the emission of the signal and the observation of the potentials. The position $\mathbf{r}'$ of the charge in K' may be evaluated in terms of $\mathbf{r}$ at any time, and here the time of observation $t = t' = 0$ is chosen. The inverse of the transformation (Equations 4.21) then gives:

$$\mathbf{r}' = \mathbf{r} + (\gamma - 1)(\mathbf{u}.\mathbf{r})\mathbf{u}/u^2, \tag{5.69}$$

from which it is easy to show that:

$$r' = r[1 + (\mathbf{u}.\mathbf{r})^2\gamma^2/c^2r^2]^{\frac{1}{2}}. \tag{5.70}$$

Substitution in Equations 5.64 gives finally:

$$\mathbf{A} = \mathbf{u}\phi/c^2; \qquad \phi = (\gamma q/4\pi\epsilon_0 r)[1 + (\mathbf{u}.\mathbf{r})^2\gamma^2/c^2r^2]^{-\frac{1}{2}}. \tag{5.71}$$

Again these are general expressions if **r** is taken to be the position of the charge relative to the observer. It must be emphasized that Equations 5.71 only apply to a charged particle moving at constant velocity and that the potentials are to be evaluated at the time of observation, since the Lorentz transformation 5.69 applies to the event of observation.

For charges moving with velocities much less than the speed of light it is convenient to expand the potentials (Equations 5.71) as power series in (u/c) and to neglect terms of order higher than quadratic. When this is done, we obtain:

$$\begin{aligned} \mathbf{A} &= (q/4\pi\epsilon_0 cr)[\mathbf{u}/c + \mathrm{O}((u/c)^3)]; \\ \phi &= (q/4\pi\epsilon_0 r)[1 + u^2/2c^2 - (\mathbf{u}.\mathbf{r})^2/2c^2r^2 + \mathrm{O}((u/c)^4)]. \end{aligned} \tag{5.72}$$

In their present form these potentials satisfy the Lorentz gauge condition (Equation 5.36) to the required order. (In working out the divergence of **A** it should be remembered that **r** is the position of the charge relative to the observer and that it is the divergence at the observer that is required; in addition, the position of the observer is fixed and only the charge's position depends explicitly on time.) However, when a non-relativistic approximation is being considered, it is more convenient to work in the Coulomb gauge, in which the divergence of **A** vanishes (Equation 5.35). This may be achieved by performing a gauge transformation (Equations 5.31) with $q(\mathbf{u}.\mathbf{r})/2c^2r$ as the scalar f. The transformed potentials are then:

$$\begin{aligned} \mathbf{A} &= (q/8\pi\epsilon_0 cr)[\mathbf{u}/c + (\mathbf{u}.\mathbf{r})\mathbf{r}/r^2c + \mathrm{O}((u/c)^3)]; \\ \phi &= (q/4\pi\epsilon_0 r)[1 + \mathrm{O}((u/c)^4)], \end{aligned} \tag{5.73}$$

and these may readily be shown to satisfy the Coulomb gauge condition.

These potentials are used in the next section, where the Hamiltonian for two charged particles is obtained. Since we will eventually be considering the motion of electrons in molecules, it might be thought that the result, Equations 5.73, is not appropriate, since the charge is assumed to have constant velocity. However, to order $(u/c)^2$ the same result is obtained if the Liénard-Wiechert potentials for an arbitrarily moving charge are used and these retarded expressions are rewritten in terms of quantities measured at the time of observation. To do this a knowledge of the particle's acceleration is necessary, but as first shown by Darwin the acceleration does not make a contribution to the potentials in the order considered here.

5.9 The interaction of two charged particles

The Lagrangian for an isolated particle with charge q_1, mass m_1 and velocity $\mathbf{u}_1$ in the presence of electromagnetic potentials $\mathbf{A}$ and ϕ is given by Equation 5.8 as:

$$\mathcal{L} = -m_1c^2(1-u_1^2/c^2)^{\frac{1}{2}} + q_1\mathbf{A}.\mathbf{u}_1 - q_1\phi. \tag{5.74}$$

For the remainder of this section it will be assumed that $u_1 \ll c$. If the potentials at $\mathbf{r}_1$, the position of this particle, are entirely due to a second slow-moving particle distinguished from the first by the subscript 2, then they are given by Equations 5.73:

$$\begin{aligned} \mathbf{A} &= (q_2/8\pi\epsilon_0 cr)[\mathbf{u}_2/c + (\mathbf{u}_2.\mathbf{r})\mathbf{r}/r^2c + \mathrm{O}((u_2/c)^3)]; \\ \phi &= (q_2/4\pi\epsilon_0 r)[1 + \mathrm{O}((u_2/c)^4)], \end{aligned} \tag{5.75}$$

where $\mathbf{r} = (\mathbf{r}_2 - \mathbf{r}_1)$ is the position of electron 2 relative to electron 1. Substitution in Equation 5.74 gives:

$$\begin{aligned} \mathcal{L} = &-m_1c^2(1-u_1^2/c^2)^{\frac{1}{2}} + (q_1q_2/8\pi\epsilon_0 r)[(\mathbf{u}_1.\mathbf{u}_2)/c^2 \\ &+ (\mathbf{u}_1.\mathbf{r})(\mathbf{u}_2.\mathbf{r})/c^2r^2 - 2 + \mathrm{O}((u/c)^4)] \end{aligned} \tag{5.76}$$

as the Lagrangian for particle 1.

Now addition of a term such as $-m_2c^2(1-u_2^2/c^2)^{\frac{1}{2}}$ to this Lagrangian cannot affect the equations of motion for particle 1, since it is independent of $\mathbf{r}_1$ and $\mathbf{u}_1$. The Lagrangian for particle 1 subject to potentials due to particle 2, the motion of which is assumed to be known, can thus be written:

$$\begin{aligned} \mathcal{L} = &-m_1c^2(1-u_1^2/c^2)^{\frac{1}{2}} - m_2c^2(1-u_2^2/c^2)^{\frac{1}{2}} \\ &+ (q_1q_2/8\pi\epsilon_0 r)[(\mathbf{u}_1.\mathbf{u}_2)/c^2 + (\mathbf{u}_1.\mathbf{r})(\mathbf{u}_2.\mathbf{r})/c^2r^2 \\ &-2 + \mathrm{O}((u/c)^4)]. \end{aligned} \tag{5.77}$$

However, this Lagrangian is symmetric to interchange of the subscripts 1 and 2, so it must also be an appropriate Lagrangian for particle 2 moving under the influence of potentials due to particle 1. It must be concluded that Equation 5.77 is the Lagrangian for the whole system, particle 1 plus particle 2. For consistency the first two terms should also be expanded as power series in (u/c), but it is appropriate to retain terms of order $mc^2(u/c)^4$; this is because these results are to be applied to molecular problems and in the ground state of the hydrogen atom, for example, $(q_1q_2/8\pi\epsilon_0 r)$ is of the order of $mc^2(u/c)^2$ (see Section 8.7). Thus:

$$\begin{aligned}\mathcal{L} = &\sum_{i=1,2}[-m_ic^2 + \tfrac{1}{2}m_iu_i^2 + m_iu_i^4/8c^2] + (q_1q_2/8\pi\epsilon_0 r)[(\mathbf{u}_1.\mathbf{u}_2)/c^2 \\ &+ (\mathbf{u}_1.\mathbf{r})(\mathbf{u}_2.\mathbf{r})/c^2r^2 - 2] + O(mc^2(u/c)^6).\end{aligned} \tag{5.78}$$

Now that we have a Lagrangian for the system the corresponding Hamiltonian may be found by the usual prescription. From Equation 3.24 the momentum of particle 1 is:

$$\begin{aligned}\mathbf{p}_1 = \partial\mathcal{L}/\partial\mathbf{u}_1 = &\ m_1\mathbf{u}_1 + m_1\mathbf{u}_1u_1^2/2c^2 + (q_1q_2/8\pi\epsilon_0 cr)[\mathbf{u}_2/c \\ &+ \mathbf{r}(\mathbf{u}_2.\mathbf{r})/cr^2] + O(mc(u/c)^5)\end{aligned} \tag{5.79}$$

with a similar expression for $\mathbf{p}_2$, the momentum of particle 2. These equations may be inverted to give the velocities of the particles in terms of their momenta; for particle 1 successive approximation gives:

$$\begin{aligned}\mathbf{u}_1 = &\ \mathbf{p}_1/m_1 - \mathbf{p}_1p_1^2/2m_1^3c^2 - (q_1q_2/8\pi\epsilon_0 m_1m_2c^2)[\mathbf{p}_1/r \\ &+ \mathbf{r}(\mathbf{p}_2.\mathbf{r})/r^3] + O(c(p/mc)^5)\end{aligned} \tag{5.80}$$

and there is a similar equation for $\mathbf{u}_2$. Finally, substitution of these and Equation 5.78 into Equation 3.31 for the Hamiltonian gives:

$$\begin{aligned}\mathcal{H} &= \sum_{i=1,2}\mathbf{u}_i.\mathbf{p}_i - \mathcal{L} \\ &= \sum_{i=1,2}[m_ic^2 + p_i^2/2m_i - p_i^4/8m_i^3c^2] + q_1q_2/4\pi\epsilon_0 r \\ &\quad - (q_1q_2/8\pi\epsilon_0 m_1m_2c^2)[(\mathbf{p}_1.\mathbf{r})/r + (\mathbf{p}_1.\mathbf{r})(\mathbf{r}.\mathbf{p}_2)/r^3] \\ &\quad + O(mc^2(p/mc)^6).\end{aligned} \tag{5.81}$$

This then is the Hamiltonian for two charged particles and, as we saw in Section 5.2, it may be generalized to include external electromagnetic fields by the following replacements:

$$\begin{aligned}&\mathbf{p}_i \longrightarrow \boldsymbol{\pi}_i = \mathbf{p}_i - q_i\mathbf{A}_i; \\ &\mathcal{H} \longrightarrow \mathcal{H} - \sum_{i=1,2} q_i\phi_i,\end{aligned} \tag{5.82}$$

where $\mathbf{A}_i$ and ϕ_i are the external potentials at particle i. This result will be used in Chapter 9 when the quantum mechanical theory for two electrons is considered.

5.10 The Thomas precession

The concept of electron spin was first introduced to account for fine structure in atomic spectra. It was postulated that the electron has an intrinsic angular momentum $\hbar\mathbf{s}$ and an associated magnetic moment $\boldsymbol{\mu} = -(e\hbar/m)\mathbf{s}$, where $-e$ and m are the charge and mass of the electron; in a magnetic field the energy associated with this moment is $-\boldsymbol{\mu}.\mathbf{B}$. However, a discrepancy soon became apparent, since an electron moving with velocity $\mathbf{u}$ ($\ll c$) under the influence of an electric field $\mathbf{E}$, should according to Equations 5.29 experience a magnetic field:

$$\mathbf{B} = (\mathbf{E} \wedge \mathbf{u})/c^2 \tag{5.83}$$

in its own rest frame. The energy of interaction between the spin magnetic moment of the electron and this magnetic field should thus be $-\boldsymbol{\mu}.(\mathbf{E} \wedge \mathbf{u})/c^2$, but in atoms where the electric field is provided by the nucleus experiment showed that this expression was a factor of two too large. This discrepancy can be explained in terms of a relativistic effect, if the concept of spin is accepted, and this may be done without considering quantum mechanics.

To do this the concept of precession must be introduced. In a magnetic field $\mathbf{B}$ an electron experiences a torque $\boldsymbol{\mu} \wedge \mathbf{B}$ by virtue of its intrinsic magnetic moment $\boldsymbol{\mu}$. Now this torque will change the angular momentum of the electron and the rate of change of the angular momentum is just equal to the torque:

$$\mathrm{d}(\hbar\mathbf{s})/\mathrm{d}t = \boldsymbol{\mu} \wedge \mathbf{B} = -(e\hbar/m)(\mathbf{s} \wedge \mathbf{B}). \tag{5.84}$$

Thus the change in the angular momentum is perpendicular to itself and the magnetic field, so that the magnitude of the angular momentum vector and the angle θ that it makes with the magnetic field vector do not change, but instead the angular momentum precesses about the direction of $\mathbf{B}$ (Fig. 5.1). The angular velocity $\boldsymbol{\omega}_L$ of this so-called Larmor precession is in the same direction as $\mathbf{B}$ and its magnitude may be found by considering a short time $\mathrm{d}t$, in which the precession angle $\mathrm{d}\phi$ is $\omega_L \mathrm{d}t$. This angle is given also by:

$$\mathrm{d}\phi = \mathrm{d}(\hbar s)/\hbar s \sin\theta = eB\mathrm{d}t/m, \tag{5.85}$$

where Equation 5.84 has been used, so that:

$$\boldsymbol{\omega}_L = e\mathbf{B}/m. \tag{5.86}$$

All this is analogous to the precession of a gyroscope in a gravitational field. The important thing to note is that the Larmor frequency $\omega_L/2\pi$ is directly proportional to the energy of the interaction of the electron's spin magnetic moment with the magnetic field and that the two may be thought to be interrelated.

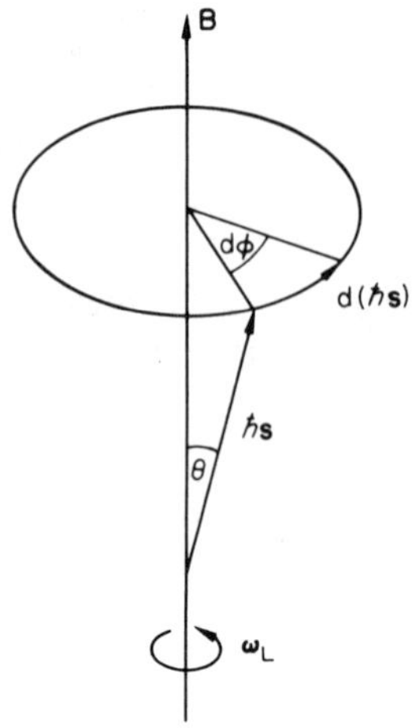

Fig. 5.1. The precession of the angular momentum of the electron about the direction of the magnetic field.

Let us now consider an electron moving with velocity u along the x axis of an inertial frame K in which there is an electric field E directed along the y axis. In the rest frame of the electron the precession will have an angular velocity about the z axis of:

$$\omega_L = -euE/mc^2, \tag{5.87}$$

where Equations 5.83 and 5.86 have been used. Now by virtue of its charge the electron will experience a force due to the electric field and will suffer a resulting acceleration:

$$a = -eE/m, \tag{5.88}$$

in the y direction. It will be seen that as a result of this acceleration the axes of the rest frame of the electron appear to be themselves precessing when viewed from the frame K.

After a short time dt the situation will be as in Fig. 5.2. The rest frame K'' of the electron is moving with velocity $u' = a dt$ in the y' direction relative to a frame K', which is itself moving with velocity u in the x direction relative to the K frame; the frames are assumed to be coincident at $t = 0$ and $u' \ll u \ll c$. Use of Equations 4.28 and 4.29 shows that the velocity $\mathbf{v}$ of the K'' frame relative to the K frame has components:

$$v_x = u; \qquad v_y = u'/\gamma(u), \tag{5.89}$$

where $\gamma(u) = (1 - u^2/c^2)^{-\frac{1}{2}}$. On the other hand the velocity $\mathbf{v}''$ of the K frame relative to the rest frame of the electron has components:

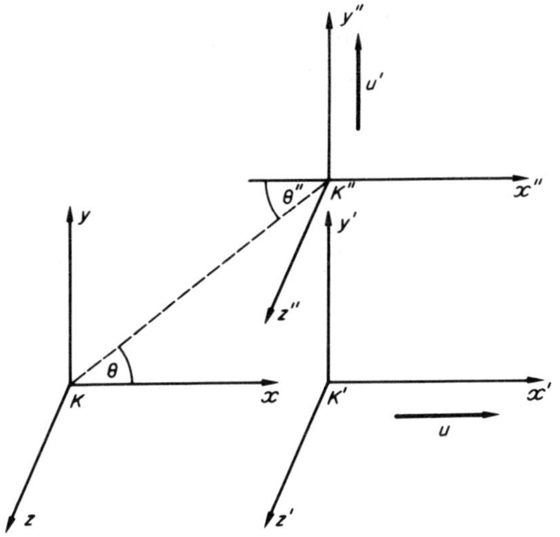

Fig. 5.2. The relationship between the inertial frames K, K' and K''.

$$v''_x = -u/\gamma(u'); \qquad v''_y = -u'. \tag{5.90}$$

Now the principle of relativity tells us that **v** should be equal and opposite to **v**$''$. Equations 5.89 and 5.90 show that their magnitudes are equal but that their directions are not, since:

$$\begin{aligned} \theta &\simeq \tan\theta = v_y/v_x = u'/u\gamma(u); \\ \theta'' &\simeq \tan\theta'' = v''_y/v''_x = u'\gamma(u')/u. \end{aligned} \tag{5.91}$$

This means that for the principle of relativity to be obeyed the K'' axis system must be rotated through an angle:

$$\mathrm{d}\theta = \theta - \theta'' \simeq (u'/u)[1/\gamma(u) - \gamma(u')] \simeq -uu'/2c^2, \tag{5.92}$$

where the assumption that $u' \ll u \ll c$ has been used. Substitution of $u' = a\mathrm{d}t$ and Equation 5.88 shows that the K'' axes precess with angular velocity:

$$\omega_T = +euE/2mc^2, \tag{5.93}$$

when observed from the K frame; this is the Thomas precession. Thus the precession rate of the spin angular momentum in the K frame is:

$$\omega = \omega_L + \omega_T = -euE/2mc^2, \tag{5.94}$$

just half its magnitude in the rest frame of the electron (Equation 5.87).

Since the energy of interaction between a magnetic field and the spin magnetic moment of the electron is directly proportional to the angular velocity of precession, we conclude that, when the magnetic field is due to the motion of an electron relative to a frame in which there is an electric field, special relativity introduces a correction factor of $\frac{1}{2}$; the energy of interaction is thus $-\boldsymbol{\mu}.(\mathbf{E}_\wedge \mathbf{u})/2c^2$ and this is just the spin-orbit interaction term when the electric field arises from a nucleus. Later it will be seen that this correction factor arises naturally from the Dirac equation, which is a relativistic quantum mechanical equation for the electron.

Bibliography

Electromagnetism

Coulson, C. A. (1948), *Electricity*, Oliver and Boyd, Edinburgh.
Feynman, Leighton and Sands: Volume II.
Landau and Lifschitz: Chapters 3, 4 and 5.
Jackson, J. D. (1962), *Classical Electrodynamics*, Wiley, New York.
Panofsky, W. K. H., and Phillips, M. (1962), *Classical Electricity and Magnetism*, Addison-Wesley, Reading, Mass.
Rindler, W. (1966), *Special Relativity*, Oliver and Boyd, Edinburgh: Chapter 6.

The interaction of charged particles

Darwin, C. G. (1920), 'The dynamical motions of charged particles', *Phil. Mag.*, **39**, 537: a relativistic treatment of the motion of two interacting, charged particles.
Landau and Lifschitz: Chapter 8.

The Thomas precession

Thomas, L. H. (1926), 'The motion of the spinning electron', *Nature*, **117**, 514: the original explanation of the discrepancy between theory and experiment for the spin-orbit coupling in atoms.
Eisberg, R. M. (1961), *Fundamentals of Modern Physics*, Wiley, New York: pages 340-344.
Hameka: pages 45-49.

CHAPTER SIX

The Classical Theory of Electromagnetic Fields

The classical field theory that is developed here is not used until Chapter 12, where its quantization is considered. Consequently, a reading of this chapter may be delayed until then, provided that the validity of Gauss' law (Equation 5.50) is accepted; it is recalled that the proof of this equation, which is used in Section 5.7, was postponed until this chapter, together with the proof of Equation 5.51.

The reason for this postponement was that, although the Lagrangian for a charged particle subject to an electromagnetic field had been obtained, the Lagrangian for the field itself had not. There is clearly a fundamental difference between a particle, which is concentrated at a point, and a field, which is spread out through space. The Lagrangian for a particle is a function of its position and velocity, while for the Lagrangian of a field some new variables, which are themselves functions of position $\mathbf{r}$ and time t, are necessary. This means that our ideas about Lagrangians, Hamiltonians and equations of motion will need some modification, when a continuous, rather than a discrete, system is under consideration. To see how the theory needs to be modified we first discuss the mechanics of a continuous system.

6.1 Continuous mechanical systems

Let us consider a one-dimensional system consisting of an infinite chain of identical particles. A continuous system may be obtained by making the separation between neighbouring particles infinitesimally small; the generalization to a three-dimensional system is trivial.

It is assumed that the particles are of mass m and are joined to their neighbours by 'springs' of equal length a and with equal 'force constants' k. The Lagrangian for the system is the difference of the total kinetic energy and the total potential energy (Section 3.4):

$$\mathcal{L} = \sum_i \tfrac{1}{2}[m\dot{\eta}_i^2 - k(\eta_{i+1} - \eta_i)^2], \tag{6.1}$$

where η_i is the displacement of the ith particle from its equilibrium position, $\dot{\eta}_i$ is its velocity and end effects have been neglected. This is just the Lagrangian for coupled oscillators and may be rewritten as:

$$\mathcal{L} = \sum_i \tfrac{1}{2}a[(m/a)\dot{\eta}_i^2 - ka((\eta_{i+1} - \eta_i)/a)^2]. \tag{6.2}$$

To go from this discrete situation to a continuous system the separation a between the particles is replaced by the infinitesimal dx and the summation becomes an integration. At the same time (m/a) becomes the linear mass density μ, ka is replaced by the Young's modulus Y and $(\eta_{i+1} - \eta_i)/a$ is interpreted as $(\partial\eta/\partial x)$. The Lagrangian for the continuous system is thus:

$$\mathcal{L} = \int L(\eta, \dot{\eta}, \partial\eta/\partial x, t)\mathrm{d}x, \tag{6.3}$$

where L is the Lagrangian density, which in the present example is:

$$L = \tfrac{1}{2}[\mu\dot{\eta}^2 - Y(\partial\eta/\partial x)^2]. \tag{6.4}$$

The variable $\eta(x, t)$ may be thought of as a generalized coordinate similar to q in Chapter 3, except that it is a function of position as well as time. The quantity $\dot{\eta}$ is just the partial derivative of η with respect to time.

To obtain the equation of motion from this Lagrangian the method of Section 3.3 is used, that is η is varied so that the action integral between times t_a and t_b is minimized:

$$\delta S = \delta \int_{t_a}^{t_b} \mathcal{L}(\eta, \dot{\eta}, \partial\eta/\partial x)\mathrm{d}t = 0. \tag{6.5}$$

Here it is assumed that the Lagrangian does not depend explicitly on time, and that $\delta\eta$, the change in η, is zero at t_a and t_b and also at the limits of the space integration in Equation 6.3. Substitution of Equation 6.3 in Equation 6.5, expansion and integration by parts gives:

$$\delta S = \delta \int_{t_a}^{t_b} \int L(\eta, \dot{\eta}, \partial\eta/\partial x)\mathrm{d}x\mathrm{d}t$$

$$= \int_{t_a}^{t_b}\int \left[\frac{\partial L}{\partial \eta}\,\delta\eta + \frac{\partial L}{\partial \dot{\eta}}\,\delta\dot{\eta} + \frac{\partial L}{\partial(\partial\eta/\partial x)}\,\delta\left(\frac{\partial\eta}{\partial x}\right)\right]\mathrm{d}x\mathrm{d}t$$

$$= \int_{t_a}^{t_b} \int \left[\frac{\partial L}{\partial \eta} - \frac{\partial}{\partial t}\left(\frac{\partial L}{\partial \dot{\eta}}\right) - \frac{\partial}{\partial x}\left(\frac{\partial L}{\partial(\partial \eta/\partial x)}\right)\right] \delta\eta \mathrm{d}x\mathrm{d}t, \tag{6.6}$$

where the integrated parts have been omitted, since they vanish as a result of the assumed boundary conditions. Now δS must vanish for an arbitrary choice of $\delta\eta$, so that:

$$\frac{\partial L}{\partial \eta} - \frac{\partial}{\partial t}\left(\frac{\partial L}{\partial \dot{\eta}}\right) - \frac{\partial}{\partial x}\left(\frac{\partial L}{\partial(\partial \eta/\partial x)}\right) = 0. \tag{6.7}$$

This is the equation of motion, which is to be compared with Equation 3.6 for a discrete system.

For the example used here, substitution of the Lagrangian density (Equation 6.4) in Equation 6.7 gives:

$$Y(\partial^2\eta/\partial x^2) - \mu(\partial^2\eta/\partial t^2) = 0, \tag{6.8}$$

which is just the one-dimensional wave equation for compressional waves travelling with velocity $(Y/\mu)^{\frac{1}{2}}$.

The definitions of the momentum (Equation 3.24) and the Hamiltonian (Equation 3.33) for a discrete system may readily be extended to the continuous case. The canonical momentum density conjugate to η is thus:

$$p = (\partial L/\partial\dot{\eta}), \tag{6.9}$$

while the Hamiltonian density is:

$$H = \dot{\eta}p - L. \tag{6.10}$$

In our example the momentum density is just $\mu\dot{\eta}$ and the Hamiltonian density, which is given by:

$$H = \tfrac{1}{2}[\mu\dot{\eta}^2 + Y(\partial\eta/\partial x)^2], \tag{6.11}$$

is clearly the sum of kinetic energy and potential energy densities.

So far attention has been confined to a one-dimensional system, but the extension to three dimensions is trivial. Remembering that the subscript i can be x, y or z and that the repeated subscript summation convention is being used, Equation 6.7 generalizes to:

$$\frac{\partial L}{\partial \eta} - \frac{\partial}{\partial t}\left(\frac{\partial L}{\partial \dot{\eta}}\right) - \frac{\partial}{\partial r_i}\left(\frac{\partial L}{\partial(\partial \eta/\partial r_i)}\right) = 0, \tag{6.12}$$

where η is now a function of the position vector $\mathbf{r}$ and the time t; by using four-vector notation this equation may be written in a more abbreviated form:

$$\frac{\partial L}{\partial \eta} - \frac{\partial}{\partial r_\alpha}\left(\frac{\partial L}{\partial(\partial \eta/\partial r_\alpha)}\right) = 0. \tag{6.13}$$

This equation is Lorentz invariant provided L is a scalar density, that is if L is itself Lorentz invariant; this restriction is quite severe and is useful in choosing Lagrangian densities. The definitions of the momentum and Hamiltonian densities are unaltered in the three-dimensional case.

In Section 3.5 constants of motion were found for a system of particles by considering the derivative of the Lagrangian with respect to space and time coordinates. A similar approach is possible for continuous systems. Here we consider $\partial L/\partial r_\alpha$, the derivative of the Lagrangian density with respect to a component of the position four-vector. Since L is a function of η and its derivatives $\partial \eta/\partial r_\alpha$, we may write:

$$\frac{\partial L}{\partial r_\alpha} = \left(\frac{\partial L}{\partial \eta}\right)\left(\frac{\partial \eta}{\partial r_\alpha}\right) + \left(\frac{\partial L}{\partial(\partial \eta/\partial r_\beta)}\right)\left(\frac{\partial(\partial \eta/\partial r_\beta)}{\partial r_\alpha}\right). \tag{6.14}$$

Use of the equation of motion (Equation 6.13) and the fact that $\partial(\partial \eta/\partial r_\beta)/\partial r_\alpha$ is the same as $\partial(\partial \eta/\partial r_\alpha)/\partial r_\beta$ now shows that:

$$\begin{aligned}\frac{\partial L}{\partial r_\alpha} &= \frac{\partial}{\partial r_\beta}\left(\frac{\partial L}{\partial(\partial \eta/\partial r_\beta)}\right)\left(\frac{\partial \eta}{\partial r_\alpha}\right) + \left(\frac{\partial L}{\partial(\partial \eta/\partial r_\beta)}\right)\left(\frac{\partial(\partial \eta/\partial r_\alpha)}{\partial r_\beta}\right) \\ &= \frac{\partial}{\partial r_\beta}\left[\left(\frac{\partial \eta}{\partial r_\alpha}\right)\left(\frac{\partial L}{\partial(\partial \eta/\partial r_\beta)}\right)\right].\end{aligned} \tag{6.15}$$

Since $(\partial L/\partial r_\alpha)$ can be expressed as $\delta_{\alpha\beta}(\partial L/\partial r_\beta)$, this equation may be rewritten as:

$$(\partial T_{\alpha\beta}/\partial r_\beta) = 0, \tag{6.16}$$

where the components of the four-tensor are given by:

$$T_{\alpha\beta} = \delta_{\alpha\beta}L - \left(\frac{\partial \eta}{\partial r_\alpha}\right)\left(\frac{\partial L}{\partial(\partial \eta/\partial r_\beta)}\right) \tag{6.17}$$

It should be noted that the $T_{\alpha\beta}$ are not uniquely determined by Equation 6.16, since its validity is unaffected if $\partial U_{\alpha\beta\gamma}/\partial r_\gamma$ is added to $T_{\alpha\beta}$, where $U_{\alpha\beta\gamma}$ is any tensor antisymmetric in the subscripts β and γ.

Equation 6.16 may be put in a more useful form by interpreting the left-hand side as the four-divergence of a four-vector T_α with components $T_{\alpha\beta}$. It can then be rewritten as:

$$(\partial T_{\alpha 4}/\partial t) = -ic\nabla.\mathbf{T}_\alpha, \tag{6.18}$$

so that the time derivative of the integral of $T_{\alpha 4}$ over all space is:

$$\begin{aligned}\frac{\partial}{\partial t}\int T_{\alpha 4}\mathrm{d}\tau &= \int (\partial T_{\alpha 4}/\partial t)\mathrm{d}\tau \\ &= -ic\int \nabla.\mathbf{T}_{\alpha}\mathrm{d}\tau \\ &= -ic\int \mathbf{T}_{\alpha}.\mathrm{ds}, \end{aligned} \tag{6.19}$$

where Gauss' theorem (Equation 2.57) has been used. If it is assumed that η and its derivatives vanish at infinity, then so do the components of $\mathbf{T}_{\alpha}$ and the surface integral on the right-hand side of Equation 6.19 vanishes. Thus $\int T_{\alpha 4}\mathrm{d}\tau$, the integral of $T_{\alpha 4}$ over all space, is a constant in time and we may define the four-vector:

$$G_{\alpha} = k\int T_{\alpha 4}\mathrm{d}\tau = \text{constant}, \tag{6.20}$$

k being an arbitrary constant.

The time component of the four-vector (Equation 6.20) may be expanded using Equation 6.17:

$$G_4 = k\int T_{44}\mathrm{d}\tau = k\int [L-(\partial\eta/\partial t)(\partial L/\partial(\partial\eta/\partial t))]\mathrm{d}\tau. \tag{6.21}$$

Equations 6.9 and 6.10 now show that, if k has the value $-i/c$:

$$G_4 = (i/c)\int H\mathrm{d}\tau = (i/c)\mathcal{H}, \tag{6.22}$$

where $\mathcal{H}$ is the Hamiltonian. Thus G_4 is just the time component of the 4-momentum, as may be seen by reference to Section 4.7. Consequently the four-vector:

$$G_{\alpha} = -(i/c)\int T_{\alpha 4}\mathrm{d}\tau \tag{6.23}$$

may be interpreted as the four-momentum of the continuous system; the spatial components represent the momentum and describe the way in which the energy moves about in the system. The total energy and momentum in all space are thus conserved, since we have just seen that the four-vector G_{α} is a constant of motion. These conclusions are not affected by the fact that the $T_{\alpha\beta}$ are not uniquely defined by Equation 6.17, since the integral of $\partial U_{\alpha 4\gamma}/\partial r_{\gamma}$ over all space vanishes as may easily be demonstrated using Gauss' theorem (Equation 2.57).

Finally, it should be noted that, although the example considered here is for a continuous mechanical system, the theory applies to any continuous system, and in particular it is appropriate to fields. The Equations 6.9, 6.10 and 6.13 are appropriate to a field that is characterized by a scalar variable η, which is, of course, a function of space and time coordinates; such a field is referred to as a scalar field. However, there is no reason why a field should not be characterized by more than one variable and, for example, these could be the components of a vector. For such a vector field there is an equation of motion similar to Equation 6.13 for each component of the vector. In addition, there is a momentum density conjugate to each field variable and the Hamiltonian density generalizes to:

$$H = \sum_n \dot{\eta}_n(\partial L/\partial\dot{\eta}_n) - L, \tag{6.24}$$

where the subscript n distinguishes the field variables. Similar remarks apply to the tensor $T_{\alpha\beta}$ and the four-momentum G_α.

6.2 The Lagrangian density for an electromagnetic field

In the last chapter it was seen that an electromagnetic field may be characterized by the components A_α of the potential four-vector. These four components are functions of position **r** and time t, and they might be expected to be appropriate field variables, so that there are four equations of motion (Equation 6.13) for the field. However, before we can proceed any further an appropriate Lagrangian must be found and to do this the action integral is again considered.

For a system consisting of charged particles and an electromagnetic field the action integral consists of three parts, one for the particles, one for the field and one for the interaction between the two. When the equations of motion of a particle were obtained, only those terms in the Lagrangian which depend on the position and velocity of the particle were needed; that is the electromagnetic potentials were regarded as known and an explicit term for the field was unnecessary. However, in finding the equations of motion for a field, the potentials are the variables and the motion of the charged particles is assumed to be known. In this case the action for the particles alone is unimportant, since it is not a function of the field variables. The action due to the field itself has yet to be derived, but the interaction term is already known (Equation 5.3), although it will be necessary to devote the next section to rewriting it in terms of the current four-vector.

A number of restrictions may be made on the action integral for the field. In the last section it was seen that the Lagrangian density is a function of the field variables and their partial derivatives with respect to space and time coordinates (if it were to depend explicitly on time, this would imply that time is not homogeneous). However, explicit dependence on the potentials themselves is not possible, since they are not uniquely determined and the Lagrangian density must be invariant to gauge transformations. Thus, only quantities of the type $(\partial A_\alpha/\partial r_\beta)$ are involved, and these are related to the $F_{\alpha\beta}$, the components of the electromagnetic field tensor (Equation 5.22).

It has also been noted that the Lagrangian density must be an invariant, if the Lorentz invariance of the equations of motion is to be ensured. An invariant (Equation 5.30) has already been constructed from the $F_{\alpha\beta}$ and this suggests that for an electromagnetic field alone an appropriate action integral is:

$$a \int_{t_a}^{t_b} \int F_{\alpha\beta} F_{\alpha\beta} \mathrm{d}\tau \mathrm{d}t = 2a \int_{t_a}^{t_b} \int (B^2 - E^2/c^2) \mathrm{d}\tau \mathrm{d}t. \qquad (6.25)$$

Before considering the value of the constant a two comments need to be made about this choice of action integral.

First, the invariant has been chosen to be quadratic in the A_α, although it is clearly possible to construct invariants that are say quartic in the A_α. The justification for this choice depends on the experimental fact that electromagnetic fields may be superimposed, so that the field due to a collection of charges is just the sum of the fields produced by the charges individually. Thus, for an equation of motion for the field to be valid it must be possible to construct new solutions by adding together other solutions. This is only possible if the equation of motion for the field is a differential equation that is linear in the field variables A_α. Reference to the equation of motion (Equation 6.13) indicates that for this to be true the Lagrangian density must be quadratic in the field variables.

Secondly, it was noted in Section 5.4 that **E.B** is also an invariant quadratic in the A_α. However, it was pointed out that this is not a true scalar, since it changes sign on inversion of the coordinate system. In addition, a term proportional to **E.B** in the Lagrangian does not, in fact, affect the equations of motion. In Section 3.3 it was seen that the addition of a total time derivative to a Lagrangian does not affect the equations of motion. The corresponding result for a Lagrangian density, for which the action is an integral over space as well as time coordinates, is that the equations of motion are unaffected by the addition of a term $\partial F_\alpha / \partial r_\alpha$, where F_α is a four-vector. It may be shown that **E.B** is a term of this type, since it may be written as:

$$\mathbf{E.B} = (ic/8) \epsilon_{\alpha\beta\gamma\delta} F_{\alpha\beta} F_{\gamma\delta} = (ic/2) \frac{\partial}{\partial r_\alpha} \left(\epsilon_{\alpha\beta\gamma\delta} A_\beta \frac{\partial A_\delta}{\partial r_\gamma} \right), \qquad (6.26)$$

where $\epsilon_{\alpha\beta\gamma\delta}$ is the unit antisymmetric tensor of rank four analogous to ϵ_{ijk} for three dimensions; the details are left to the reader.

It remains to choose a value for the constant a in Equation 6.25 and clearly this will depend on the system of units employed. For the SI system agreement with experiment is obtained by taking a to be $- c^2 \epsilon_0 / 4$ $(= - 1/4\mu_0)$, where ϵ_0 and μ_0 are respectively the permittivity and permeability of free space; it may easily be checked that this choice makes the action integral dimensionally correct. That a must be negative is due to the fact that the quantity $(\partial^2\mathbf{A}/\partial t^2)$ appears with a negative sign in $F_{\alpha\beta}F_{\alpha\beta}$. However, it must appear with a positive sign in the action, since it may be made arbitrarily large by varying the potentials sufficiently rapidly, and if it had a negative sign the action could not be minimized as required by the principle of least action. The Lagrangian density for a field is thus:

$$L = -c^2\epsilon_0 F_{\alpha\beta}F_{\alpha\beta}/4 = -\tfrac{1}{2}\epsilon_0(c^2B^2 - E^2). \tag{6.27}$$

By substitution of Equation 6.27 in the equations of motion (Equation 6.13) the field equations for free space could now be obtained, but the more general situation in which electric charges are present is considered after the Lagrangian density for the interaction of charges with a field has been determined.

6.3 The current four-vector

The action integral for the interaction of a point charge q with an electromagnetic field is (Equations 5.3):

$$\int_b^a qA_\alpha \mathrm{d}r_\alpha. \tag{6.28}$$

Now that the field itself is under consideration, it is desirable to be able to treat any distribution of charge and not just a single point charge. To do this the charge density ρ is introduced. The total charge in a region of space is then the appropriate volume integral of ρ, that is $\int \rho \mathrm{d}\tau$. A point charge q may be included in this description, since its contribution to the charge density may be written in terms of a Dirac delta function, $q\delta(\mathbf{r}-\mathbf{r}_q)$, where $\mathbf{r}_q$ is the vector position of the charge.

For an infinitesimal region of space $\mathrm{d}\tau$ the included charge is $\mathrm{d}q = \rho \mathrm{d}\tau$. Thus we can write:

$$\mathrm{d}q\mathrm{d}r_\alpha = \rho \mathrm{d}\tau \mathrm{d}r_\alpha = \rho \mathrm{d}\tau \mathrm{d}t(\mathrm{d}r_\alpha/\mathrm{d}t) = j_\alpha \mathrm{d}\tau \mathrm{d}t. \tag{6.29}$$

Now $\mathrm{d}q$ is a scalar. so that the left-hand side of this equation is the component of a four-vector. The same must be true of the right-hand side and, since $\mathrm{d}\tau\mathrm{d}t$ is a scalar and unaltered by a Lorentz transformation:

$$j_\alpha = \rho(\mathrm{d}r_\alpha/\mathrm{d}t), \tag{6.30}$$

is a component of the so-called current four-vector. The spatial components of this four-vector can be written as $\mathbf{j} = \rho\mathbf{u}$, where $\mathbf{u}$ is the charge velocity at a particular point, so that **j** is the flow of charge per unit time per unit area across a surface; the time component $ic\rho$ is proportional to the charge density.

If the charge q is replaced by the volume integral of the charge density, the action (Equation 6.28) becomes:

$$\int_b^a \int \rho \mathrm{d}\tau \, A_\alpha \mathrm{d}r_\alpha = \int_{t_b}^{t_a} \int j_\alpha A_\alpha \mathrm{d}\tau \mathrm{d}t, \tag{6.31}$$

where Equation 6.29 has been used. The Lagrangian density for the interaction between a charge distribution and an electromagnetic field is thus $j_\alpha A_\alpha$.

6.4 The second pair of Maxwell's equations

The results of the two preceding sections show that the Lagrangian density for an electromagnetic field in the presence of a charge distribution is:

$$L = j_\alpha A_\alpha - c^2\epsilon_0 F_{\alpha\beta}F_{\alpha\beta}/4, \tag{6.32}$$

where the components of the electromagnetic field tensor are given by:

$$F_{\alpha\beta} = (\partial A_\beta/\partial r_\alpha) - (\partial A_\alpha/\partial r_\beta). \tag{6.33}$$

Strictly speaking there should also be a term for the charges themselves, but this is independent of the field variables A_α, and does not affect the field equations to be derived.

There are four equations of motion for the field, one for each of the components A_α, and reference to Equation 6.13 shows them to be given by:

$$\frac{\partial L}{\partial A_\alpha} - \frac{\partial}{\partial r_\beta}\left(\frac{\partial L}{\partial(\partial A_\alpha/\partial r_\beta)}\right) = 0. \tag{6.34}$$

Substitution of Equations 6.32 and 6.33 gives:

$$j_\alpha + (c^2\epsilon_0/4)\frac{\partial}{\partial r_\beta}\left\{\frac{\partial}{\partial(\partial A_\alpha/\partial r_\beta)}\left[\left(\frac{\partial A_\delta}{\partial r_\gamma} - \frac{\partial A_\gamma}{\partial r_\delta}\right)\left(\frac{\partial A_\delta}{\partial r_\gamma} - \frac{\partial A_\gamma}{\partial r_\delta}\right)\right]\right\}$$

$$= j_\alpha + \tfrac{1}{2}c^2\epsilon_0\frac{\partial}{\partial r_\beta}\left\{\frac{\partial}{\partial(\partial A_\alpha/\partial r_\beta)}\left[\frac{\partial A_\delta}{\partial r_\gamma}\frac{\partial A_\delta}{\partial r_\gamma} - \frac{\partial A_\delta}{\partial r_\gamma}\frac{\partial A_\gamma}{\partial r_\delta}\right]\right\}$$

$$= j_\alpha + c^2\epsilon_0\frac{\partial}{\partial r_\beta}\left[\frac{\partial A_\alpha}{\partial r_\beta} - \frac{\partial A_\beta}{\partial r_\alpha}\right] = j_\alpha + c^2\epsilon_0\frac{\partial F_{\beta\alpha}}{\partial r_\beta} = 0. \tag{6.35}$$

Finally, we note that $F_{\alpha\beta} = -F_{\beta\alpha}$, so that:

$$(\partial F_{\alpha\beta}/\partial r_\beta) = j_\alpha/c^2\epsilon_0. \tag{6.36}$$

This is just the second pair of Maxwell's equations written in four-tensor notation; it was quoted (Equation 5.52) and interpreted in the last chapter.

For convenience Maxwell's equations are repeated here in full and abbreviated form:

$$\left.\begin{aligned} \nabla.\mathbf{B} &= 0 \\ \nabla_\wedge \mathbf{E} &= -(\partial \mathbf{B}/\partial t) \end{aligned}\right\} \quad \frac{\partial F_{\alpha\beta}}{\partial r_\gamma} + \frac{\partial F_{\beta\gamma}}{\partial r_\alpha} + \frac{\partial F_{\gamma\alpha}}{\partial r_\beta} = 0; \qquad \begin{aligned} &(6.37\text{a}) \\ &(6.37\text{b}) \end{aligned}$$

$$\left.\begin{aligned} \nabla.\mathbf{E} &= \rho/\epsilon_0 \\ \nabla_\wedge \mathbf{B} &= [\mathbf{j}/\epsilon_0 + (\partial \mathbf{E}/\partial t)]/c^2 \end{aligned}\right\} \quad \frac{\partial F_{\alpha\beta}}{\partial r_\beta} = \frac{j_\alpha}{\epsilon_0 c^2}. \qquad \begin{aligned} &(6.37\text{c}) \\ &(6.37\text{d}) \end{aligned}$$

It is recalled that the first pair of equations arise from the expressions 5.42 for the fields in terms of the potentials:

$$\mathbf{B} = \nabla_\wedge \mathbf{A}; \qquad \mathbf{E} = -\nabla\phi - (\partial \mathbf{A}/\partial t). \tag{6.38}$$

It can be shown immediately that the second pair of Maxwell's equations is consistent with charge conservation by taking the divergence of Equation 6.37d and substituting Equation 6.37c:

$$\begin{aligned} \nabla.\nabla_\wedge \mathbf{B} &= \nabla.\mathbf{j}/c^2\epsilon_0 + [\partial(\nabla.\mathbf{E})/\partial t]/c^2 \\ &= (\nabla.\mathbf{j} + \partial\rho/\partial t)/c^2\epsilon_0. \end{aligned} \tag{6.39}$$

Now the left-hand side of this equation vanishes, since it is the divergence of a curl (Equation 2.71), so that:

$$\nabla.\mathbf{j} + \partial\rho/\partial t = 0; \tag{6.40}$$

this is just the charge conservation equation 2.55. This result could have been obtained using four-tensor notation by operating on the combined equation for Equations 6.37c and 6.37d with the operator $\partial/\partial r_\alpha$ and summing over the subscript α (that is taking the four-divergence of Equation 6.36):

$$(\partial j_\alpha/\partial r_\alpha)/c^2\epsilon_0 = (\partial^2 F_{\alpha\beta}/\partial r_\alpha \partial r_\beta); \tag{6.41}$$

the right-hand side of this equation vanishes, since the components $F_{\alpha\beta}$ of the electromagnetic field tensor are antisymmetric, giving:

$$\partial j_\alpha/\partial r_\alpha = 0, \tag{6.42}$$

which is just Equation 6.40 written in terms of four-vectors.

We started this section with the Lagrangian density (Equation 6.32) for an electromagnetic field in the presence of a charge distribution. The results of Section 6.1 can be used to obtain the momentum and Hamiltonian of the field, and for this it is assumed that no charges are present, that is $j_\alpha = 0$. The Lagrangian density is then $-c^2\epsilon_0 F_{\alpha\beta}F_{\alpha\beta}/4$ and its derivative with respect to $\partial A_\gamma/\partial r_\beta$ is:

$$\partial L/\partial(\partial A_\gamma/\partial r_\beta) = -\tfrac{1}{2}c^2\epsilon_0 (F_{\beta\gamma} - F_{\gamma\beta}) = -c^2\epsilon_0 F_{\beta\gamma}, \qquad (6.43)$$

where Equation 6.33 has been used. Thus, the components of the tensor $T_{\alpha\beta}$ (Equation 6.17) may be written:

$$\begin{aligned} T_{\alpha\beta} &= \delta_{\alpha\beta}L - (\partial A_\gamma/\partial r_\alpha)(\partial L/\partial(\partial A_\gamma/\partial r_\beta)) \\ &= -c^2\epsilon_0 [\delta_{\alpha\beta}F_{\gamma\delta}F_{\gamma\delta}/4 - (\partial A_\gamma/\partial r_\alpha)F_{\beta\gamma}]. \end{aligned} \qquad (6.44)$$

This expression may be made symmetric in the subscripts α and β by adding $-c^2\epsilon_0(\partial A_\alpha/\partial r_\gamma)F_{\beta\gamma}$ to the right-hand side of this equation:

$$T_{\alpha\beta} = -c^2\epsilon_0 [\delta_{\alpha\beta}F_{\gamma\delta}F_{\gamma\delta}/4 - F_{\alpha\gamma}F_{\beta\gamma}]. \qquad (6.45)$$

This ruse is permissible, since we saw in Section 6.1 that $T_{\alpha\beta}$ is undetermined to the extent that a term $\partial U_{\alpha\beta\gamma}/\partial r_\gamma$ may be added to it, where $U_{\alpha\beta\gamma}$ is antisymmetric in β and γ; furthermore, this modification of $T_{\alpha\beta}$ does not affect the total Hamiltonian or momentum of the field.

The Hamiltonian of the field may now be obtained from Equations 6.22 and 6.23:

$$\begin{aligned} \mathcal{H} &= -\int T_{44}\mathrm{d}\tau = c^2\epsilon_0 \int (F_{\gamma\delta}F_{\gamma\delta}/4 - F_{4\gamma}F_{4\gamma})\mathrm{d}\tau \\ &= \tfrac{1}{2}\epsilon_0 \int (c^2B^2 + E^2)\mathrm{d}\tau, \end{aligned} \qquad (6.46)$$

where explicit substitution for the electromagnetic field tensor (Equation 5.26) has been made. The components of the momentum $\mathbf{G}$ are (Equation 6.23):

$$G_i = -(i/c)\int T_{i4}\mathrm{d}\tau = -ic\epsilon_0 \int F_{i\gamma}F_{4\gamma}\mathrm{d}\tau, \qquad (6.47)$$

where explicit substitution for the electromagnetic field tensor (Equation 5.26) **B** to give:

$$\mathbf{G} = \epsilon_0 \int \mathbf{E}_\wedge \mathbf{B}\mathrm{d}\tau = c^{-2}\int \mathbf{S}\mathrm{d}\tau, \qquad (6.48)$$

where $\mathbf{S} = \epsilon_0 c^2 \mathbf{E}_\wedge \mathbf{B}$ is the so-called Poynting vector.

6.5 Electromagnetic waves

Maxwell's equations may easily be shown to be consistent with the existence of electromagnetic waves, and to do this **B** and **E** are each separately eliminated from them.

Taking the curl of Equation 6.37d gives:

$$\nabla_\wedge(\nabla_\wedge \mathbf{B}) = (\nabla_\wedge \mathbf{j})/\epsilon_0 c^2 + (\partial\nabla_\wedge \mathbf{E}/\partial t)/c^2. \qquad (6.49)$$

Now the left-hand side is just equal to $-\nabla^2\mathbf{B}$ from Equation 2.72 and the fact that $\nabla.\mathbf{B}$ vanishes (Equation 6.37a). Substitution of Equation 6.37b for $\nabla_\wedge \mathbf{E}$ followed by rearrangement then gives:

$$\nabla^2\mathbf{B} - (\partial^2\mathbf{B}/\partial t^2)/c^2 = -(\nabla_\wedge \mathbf{j})/\epsilon_0 c^2. \tag{6.50}$$

An equation in which only $\mathbf{E}$ appears may be obtained in a similar fashion by starting with Equation 6.37b:

$$\nabla^2\mathbf{E} - (\partial^2\mathbf{E}/\partial t^2)/c^2 = \nabla\rho/\epsilon_0 + (\partial\mathbf{j}/\partial t)/\epsilon_0 c^2. \tag{6.51}$$

In the special case of a vacuum $\mathbf{j}$ and ρ both vanish and Equations 6.50 and 6.51 reduce to:

$$\nabla^2\mathbf{B} - (\partial^2\mathbf{B}/\partial t^2)/c^2 = 0; \qquad \nabla^2\mathbf{E} - (\partial^2\mathbf{E}/\partial t^2)/c^2 = 0. \tag{6.52}$$

These represent six wave equations, one for each of the components of $\mathbf{B}$ and $\mathbf{E}$. In each case the velocity of the wave is just c, the velocity of light. When charges are present, Equations 6.50 and 6.51 hold and the terms in $\mathbf{j}$ and ρ on the right-hand sides of these equations are damping terms which represent energy dissipation.

The wave equations may also be derived using the electromagnetic potentials; for simplicity we confine ourselves to the case of a field in free space. In Section 5.5 it was seen that there is some freedom in the choice of potentials, and a gauge transformation (Equation 5.31) can always be applied. Here it is supposed that the transformation has been chosen so that the scalar potential ϕ vanishes everywhere. The electromagnetic fields (Equation 6.38) are then given by:

$$\mathbf{B} = \nabla_\wedge \mathbf{A}; \qquad \mathbf{E} = -(\partial\mathbf{A}/\partial t). \tag{6.53}$$

When these expressions are substituted into Equation 6.37d with $\mathbf{j}$ set equal to zero, we obtain:

$$\nabla_\wedge(\nabla_\wedge \mathbf{A}) = -(\partial^2\mathbf{A}/\partial t^2)/c^2, \tag{6.54}$$

and after use of Equation 2.72 and rearrangement, this becomes

$$\nabla^2\mathbf{A} - \nabla(\nabla.\mathbf{A}) - (\partial^2\mathbf{A}/\partial t^2)/c^2 = 0. \tag{6.55}$$

Now, although one condition has already been imposed on the potentials by insisting that ϕ is zero everywhere, the vector potential $\mathbf{A}$ is not completely determined and a further condition may be imposed. It is recalled that under a gauge transformation (Equation 5.31):

$$\mathbf{A}' = \mathbf{A} - \nabla f; \qquad \phi' = \phi + (\partial f/\partial t), \tag{6.56}$$

so that if the scalar f is chosen to be time-independent the scalar potential remains zero. Consequently, the gradient of any function of position may be subtracted from the vector potential and, by choosing f so that $\nabla^2 f = \mathbf{\nabla}.\mathbf{A}$ everywhere, the vector potential may be chosen to satisfy the Coulomb gauge condition (Equation 5.35), $\mathbf{\nabla}.\mathbf{A}' = 0$. Of course, f must be time-independent, so that $\mathbf{\nabla}.\mathbf{A}$ must be also for this choice to be possible. However, Equations 6.37c and 6.53 show that in free space ($\rho = 0$):

$$\mathbf{\nabla}.\mathbf{E} = 0 = -\partial(\mathbf{\nabla}.\mathbf{A})/\partial t, \tag{6.57}$$

showing that $\mathbf{\nabla}.\mathbf{A}$ is indeed time-independent. Now that the potentials are fixed the primes may be dropped, so that the conditions satisfied by the potentials are:

$$\mathbf{\nabla}.\mathbf{A} = 0; \qquad \phi = 0. \tag{6.58}$$

It is noted that these potentials also satisfy the Lorentz gauge condition (Equation 5.36).

Since the vector potential has been chosen to be divergenceless, Equation 6.55 reduces to:

$$\nabla^2\mathbf{A} - (\partial^2\mathbf{A}/\partial t^2)/c^2 = 0. \tag{6.59}$$

Again we have a wave equation, this time with the vector potential as variable. The wave equations 6.52 for $\mathbf{B}$ and $\mathbf{E}$ in free space may be recovered from Equation 6.59 by operating on it with $\mathbf{\nabla}\wedge$ and $\partial/\partial t$, respectively, and using Equations 6.53. However, it is the wave equation for $\mathbf{A}$ that is used in the next section.

6.6 Solution of the wave equation for free space

The wave equation 6.59 represents three equations, one for each component of the vector potential $\mathbf{A}$. The general solution of one of these equations is:

$$\mathrm{A}_i = f(\mathbf{k}.\mathbf{r} - \omega t), \tag{6.60}$$

where f is any function of $(\mathbf{k}.\mathbf{r} - \omega t)$ and ω is an arbitrary constant, while the arbitrary wave vector $\mathbf{k}$ is restricted in that its magnitude has the value ω/c; these assertions may be checked by direct substitution of Equation 6.60 in Equation 6.59. Of course, boundary and other conditions may be used to restrict the solution 6.60, but first it is convenient to write it in an alternative form and this is done by solving Equation 6.59 by a different method.

Equation 6.59 is solved by the separation of variables method by substituting:

$$\mathbf{A}(\mathbf{r}, t) = q(t)\mathbf{A}(\mathbf{r}), \tag{6.61}$$

to give two differential equations, one for $q(t)$ and one for the components of $\mathbf{A}(\mathbf{r})$:

$$d^2q/dt^2 + \omega^2 q = 0; \qquad \nabla^2\mathbf{A} + (\omega^2/c^2)\mathbf{A} = 0, \tag{6.62}$$

where ω is a constant. The solutions of these equations are:

$$q(t) = q(0) \exp(-i\omega t); \qquad \mathbf{A}(\mathbf{r}) = A\,\boldsymbol{\varepsilon} \exp(i\mathbf{k}.\mathbf{r}), \tag{6.63}$$

where $q(0)$, A, ω and $\mathbf{k}$ are arbitrary except that the magnitude of $\mathbf{k}$ is restricted to ω/c; in addition, if $\mathbf{A}(\mathbf{r}, t)$ is to be finite for all values of $\mathbf{r}$ and t, both $\mathbf{k}$ and ω must be real.

In Equations 6.63 $\boldsymbol{\varepsilon}$ is a unit vector, which is called the polarization vector and which can only be perpendicular to the wave vector $\mathbf{k}$. This restriction arises because $\mathbf{A}(\mathbf{r}, t)$ is to satisfy the Coulomb gauge condition (Equation 6.58), so that $\mathbf{A}(\mathbf{r})$ must be divergenceless. Application of this condition then gives:

$$\boldsymbol{\varepsilon}.\mathbf{k} = 0, \tag{6.64}$$

which shows that $\boldsymbol{\varepsilon}$ and $\mathbf{k}$ must be orthogonal. For a given $\boldsymbol{\varepsilon}$ and $\mathbf{k}$ there is another unit vector that is orthogonal to both, so that there are two linearly independent choices of $\boldsymbol{\varepsilon}$ and these are distinguished by the subscript $\lambda = 1, 2$; the vectors $\boldsymbol{\varepsilon}_1, \boldsymbol{\varepsilon}_2$ and $\mathbf{k}$ are mutually orthogonal.

Clearly any pair of functions of the form 6.63 is also a solution of the wave equation 6.59, so that the general solution is a sum over all possible values for $\mathbf{k}$ and all polarizations λ. However, in writing down this general solution we insist that $\mathbf{A}(\mathbf{r}, t)$ be real, since the fields $\mathbf{B}$ and $\mathbf{E}$ derived from it must be real. This may be accomplished by taking:

$$\mathbf{A}(\mathbf{r}, t) = \sum_{\mathbf{k},\lambda} [q_{\mathbf{k}\lambda}(t)\mathbf{A}_{\mathbf{k}\lambda}(\mathbf{r}) + q^*_{\mathbf{k}\lambda}(t)\mathbf{A}^*_{\mathbf{k}\lambda}(\mathbf{r})], \tag{6.65}$$

where the asterisks indicate complex conjugation,

$$q_{\mathbf{k}\lambda}(t) = q_{\mathbf{k}\lambda}(0) \exp(-i\omega_{\mathbf{k}} t) \tag{6.66}$$

and

$$\mathbf{A}_{\mathbf{k}\lambda}(\mathbf{r}) = A_{\mathbf{k}\lambda}\boldsymbol{\varepsilon}_{\mathbf{k}\lambda} \exp(i\mathbf{k}.\mathbf{r}). \tag{6.67}$$

The general solution (Equation 6.60) can be written in this form by a suitable choice of the coefficients and the wave vectors $\mathbf{k}$; Equation 6.65 is just a Fourier transform of Equation 6.60.

We will shortly need an orthonormalization condition for the vectors $\mathbf{A}_{\mathbf{k}\lambda}(\mathbf{r})$, but they cannot be normalized, since they are defined for all space. This problem may be circumvented by considering the special case of a large cubical volume V of side L centred on the origin. In calculations V can be made arbitrarily large and in the case of observable properties it will be seen that V disappears by cancellation. In addition, a periodic boundary condition is imposed by requiring that $\mathbf{A}_{\mathbf{k}\lambda}(-\frac{1}{2}L, y, z) = \mathbf{A}_{\mathbf{k}\lambda}(+\frac{1}{2}L, y, z)$ with similar conditions for the other pairs of faces of the cube. The solutions of the second of Equations 6.62 are still given by Equation 6.67, but with the additional restriction that:

$$\mathbf{k} = (2\pi n_{\mathbf{k}x}/L, 2\pi n_{\mathbf{k}y}/L, 2\pi n_{\mathbf{k}z}/L), \qquad (6.68)$$

where $n_{\mathbf{k}x}$, $n_{\mathbf{k}y}$ and $n_{\mathbf{k}z}$ are integers; these solutions are similar to the wave functions for a particle in a box.

It may now be demonstrated that the $\mathbf{A}_{\mathbf{k}\lambda}$ are orthogonal to one another:

$$\int \mathbf{A}^*_{\mathbf{k}\lambda}.\mathbf{A}_{\mathbf{k}'\lambda'}\mathrm{d}\tau = (\boldsymbol{\varepsilon}_{\mathbf{k}\lambda}.\boldsymbol{\varepsilon}_{\mathbf{k}'\lambda'})A^*_{\mathbf{k}\lambda}A_{\mathbf{k}'\lambda'} \int \exp\,[i(\mathbf{k}-\mathbf{k}').\mathbf{r}]\mathrm{d}\tau$$

$$= \delta_{\mathbf{k}\mathbf{k}'}\delta_{\lambda\lambda'}|A_{\mathbf{k}\lambda}|^2 V, \qquad (6.69)$$

where the periodic boundary conditions have been used to evaluate the volume integral over the cube; use has also been made of the fact that the two possible polarization vectors for a given $\mathbf{k}$ are orthogonal to one another. It will be seen later that it is convenient to normalize the $\mathbf{A}_{\mathbf{k}\lambda}$ to ϵ_0^{-1}, so that the choice:

$$A_{\mathbf{k}\lambda} = (\epsilon_0 V)^{-\frac{1}{2}} \qquad (6.70)$$

is appropriate and the orthonormalization condition (Equation 6.69) becomes:

$$\int \mathbf{A}^*_{\mathbf{k}\lambda}.\mathbf{A}_{\mathbf{k}'\lambda'}\mathrm{d}\tau = \delta_{\mathbf{k}\mathbf{k}'}\delta_{\lambda\lambda'}\epsilon_0^{-1}. \qquad (6.71)$$

Although $\mathbf{k}$ can only take on discrete values (Equation 6.68), the side of the cube may be made arbitrarily large and a sum over $\mathbf{k}$ may be replaced by an integral:

$$\sum_{\mathbf{k}} f(\mathbf{k}) \longrightarrow \int f(\mathbf{k})\rho(\mathbf{k})\mathrm{d}\mathbf{k}. \qquad (6.72)$$

The function $\rho(\mathbf{k})$ is a number density and is defined so that $\rho(\mathbf{k})\mathrm{d}\mathbf{k}$ is the number of discrete values that $\mathbf{k}$ may adopt between $\mathbf{k}$ and $\mathbf{k} + \mathrm{d}\mathbf{k}$. From Equation 6.68 it is apparent that:

$$\rho(\mathbf{k}) = (L/2\pi)^3 = V/(2\pi)^3. \qquad (6.73)$$

Alternatively, it may be convenient to replace $\mathbf{k}$ by the polar coordinates k, θ, ϕ. Since all directions of $\mathbf{k}$ are equally likely, we have:

$$\rho(\mathbf{k})\mathrm{d}\mathbf{k} = \rho(k)k^2\mathrm{d}k\mathrm{d}\Omega = (Vk^2/(2\pi)^3)\mathrm{d}k\mathrm{d}\Omega, \qquad (6.74)$$

where $\mathrm{d}\Omega = \sin\theta\mathrm{d}\theta\mathrm{d}\phi$ is an element of solid angle. Finally, it is noted that $\omega(= kc)$ could be used as a variable instead of k, so that:

$$\rho(\mathbf{k})\mathrm{d}\mathbf{k} = \rho(\omega)\mathrm{d}\omega\mathrm{d}\Omega = (V\omega^2/(2\pi c)^3)\mathrm{d}\omega\mathrm{d}\Omega. \qquad (6.75)$$

This result will not be used until Chapter 12.

6.7 The characteristic vibrations of an electromagnetic field

The Hamiltonian for an electromagnetic field is given by Equation 6.46:

$$\mathcal{H} = \int \tfrac{1}{2}\epsilon_0(c^2B^2 + E^2)\mathrm{d}\tau. \tag{6.76}$$

If the discussion is again confined to a cubical volume V, this may be evaluated using the solutions of the wave equation that have just been found, since from Equations 6.53 and 6.65 to 6.67:

$$\mathbf{B} = \nabla_\wedge \mathbf{A} = \sum_{\mathbf{k},\lambda} i[q_{\mathbf{k}\lambda}(t)\mathbf{k}_\wedge \mathbf{A}_{\mathbf{k}\lambda} - q^*_{\mathbf{k}\lambda}(t)\mathbf{k}_\wedge \mathbf{A}^*_{\mathbf{k}\lambda}] \tag{6.77}$$

and

$$\mathbf{E} = -(\partial\mathbf{A}/\partial t) = \sum_{\mathbf{k},\lambda} i\omega_{\mathbf{k}}[q_{\mathbf{k}\lambda}(t)\mathbf{A}_{\mathbf{k}\lambda} - q^*_{\mathbf{k}\lambda}(t)\mathbf{A}^*_{\mathbf{k}\lambda}]\,; \tag{6.78}$$

it may easily be seen that a given wave vector and its contributions to **B** and **E** are mutually orthogonal. From now on the dependence of the $q_{\mathbf{k}\lambda}$ on time will be omitted for brevity.

The contribution of the electric field (Equation 6.78) to the energy (Equation 6.76) is:

$$\begin{aligned}\int \tfrac{1}{2}\epsilon_0 E^2\mathrm{d}\tau &= -\tfrac{1}{2}\epsilon_0 \sum_{\mathbf{k},\lambda}\sum_{\mathbf{k}',\lambda'} \omega_{\mathbf{k}}\omega_{\mathbf{k}'}[q_{\mathbf{k}\lambda}q_{\mathbf{k}'\lambda'}\int \mathbf{A}_{\mathbf{k}\lambda}.\mathbf{A}_{\mathbf{k}'\lambda'}\mathrm{d}\tau \\ &\quad - q^*_{\mathbf{k}\lambda}q_{\mathbf{k}'\lambda'}\int \mathbf{A}^*_{\mathbf{k}\lambda}.\mathbf{A}_{\mathbf{k}'\lambda'}\mathrm{d}\tau - q_{\mathbf{k}\lambda}q^*_{\mathbf{k}'\lambda'}\int \mathbf{A}_{\mathbf{k}\lambda}.\mathbf{A}^*_{\mathbf{k}'\lambda'}\mathrm{d}\tau \\ &\quad + q^*_{\mathbf{k}\lambda}q^*_{\mathbf{k}'\lambda'}\int \mathbf{A}^*_{\mathbf{k}\lambda}.\mathbf{A}^*_{\mathbf{k}'\lambda'}\mathrm{d}\tau] \\ &= \frac{1}{2}\sum_{\mathbf{k},\lambda} \omega^2_{\mathbf{k}}[-q_{\mathbf{k}\lambda}q_{-\mathbf{k}\lambda} + q^*_{\mathbf{k}\lambda}q_{\mathbf{k}\lambda} + q_{\mathbf{k}\lambda}q^*_{\mathbf{k}\lambda} - q^*_{\mathbf{k}\lambda}q^*_{-\mathbf{k}\lambda}],\end{aligned} \tag{6.79}$$

where the orthonormalization condition (Equation 6.71) has been used; the reason for normalizing the $\mathbf{A}_{\mathbf{k}\lambda}$ to ϵ_0^{-1} should now be apparent. (It will be noticed that the order of $q_{\mathbf{k}\lambda}$ and $q^*_{\mathbf{k}\lambda}$ has been preserved in Equation 6.79, although this order has no significance in a classical theory. However, in the quantum theory of Chapter 12 the order is important and for this reason it is maintained in Equation 6.79 and subsequent equations.)

Using a modified form of the normalization condition (Equation 6.71) the contribution of the magnetic field **B** (Equation 6.77) to the Hamiltonian is found to be:

$$\begin{aligned}\int \tfrac{1}{2}\epsilon_0 c^2B^2\mathrm{d}\tau &= \tfrac{1}{2}c^2 \sum_{\mathbf{k},\lambda,\lambda'} (\mathbf{k}_\wedge \boldsymbol{\varepsilon}_{\mathbf{k}\lambda}).(\mathbf{k}_\wedge \boldsymbol{\varepsilon}_{\mathbf{k}\lambda'})[q_{\mathbf{k}\lambda}q_{-\mathbf{k}\lambda'} + q^*_{\mathbf{k}\lambda}q_{\mathbf{k}\lambda'} \\ &\quad + q_{\mathbf{k}\lambda}q^*_{\mathbf{k}\lambda'} + q^*_{\mathbf{k}\lambda}q^*_{-\mathbf{k}\lambda'}].\end{aligned} \tag{6.80}$$

The quadruple vector product may be expanded using Equation 2.39:

$$(\mathbf{k}\wedge\boldsymbol{\varepsilon}_{\mathbf{k}\lambda}).(\mathbf{k}\wedge\boldsymbol{\varepsilon}_{\mathbf{k}\lambda'}) = k^2(\boldsymbol{\varepsilon}_{\mathbf{k}\lambda}.\boldsymbol{\varepsilon}_{\mathbf{k}\lambda'}) - (\mathbf{k}.\boldsymbol{\varepsilon}_{\mathbf{k}\lambda'})(\mathbf{k}.\boldsymbol{\varepsilon}_{\mathbf{k}\lambda})$$
$$= k^2\delta_{\lambda\lambda'} = (\omega_{\mathbf{k}}^2/c^2)\delta_{\lambda\lambda'}, \quad (6.81)$$

where we have used the orthogonality of **k** and the two polarization vectors associated with it. Thus:

$$\int \tfrac{1}{2}\epsilon_0 c^2 B^2 \mathrm{d}\tau = \tfrac{1}{2}\sum_{\mathbf{k},\lambda} \omega_{\mathbf{k}}^2(q_{\mathbf{k}\lambda}q_{-\mathbf{k}\lambda} + q_{\mathbf{k}\lambda}^* q_{\mathbf{k}\lambda} + q_{\mathbf{k}\lambda}q_{\mathbf{k}\lambda}^* + q_{\mathbf{k}\lambda}^* q_{-\mathbf{k}\lambda}^*). \quad (6.82)$$

Addition of Equations 6.79 and 6.82 now gives the Hamiltonian for the electromagnetic field:

$$\mathcal{H} = \sum_{\mathbf{k},\lambda} \omega_{\mathbf{k}}^2(q_{\mathbf{k}\lambda}^* q_{\mathbf{k}\lambda} + q_{\mathbf{k}\lambda}q_{\mathbf{k}\lambda}^*). \quad (6.83)$$

It is important to note that the total energy for the field is the sum of the energies of the fields associated with each wave vector **k** and polarization λ. In addition, the Hamiltonian is expressed entirely in terms of the $q_{\mathbf{k}\lambda}(t)$, which are the solutions of the first of the differential equations 6.62.

Using the same techniques the total momentum of the field **G** (Equation 6.48) may also be evaluated:

$$\mathbf{G} = \sum_{\mathbf{k},\lambda} \omega_{\mathbf{k}}\mathbf{k}(q_{\mathbf{k}\lambda}q_{-\mathbf{k}\lambda} + q_{\mathbf{k}\lambda}^* q_{\mathbf{k}\lambda} + q_{\mathbf{k}\lambda}q_{\mathbf{k}\lambda}^* + q_{\mathbf{k}\lambda}^* q_{-\mathbf{k}\lambda}^*). \quad (6.84)$$

Since the summation is over all **k**, both positive and negative, the first and last terms in the parentheses make no contribution to the total sum. (This depends on the order of $q_{\mathbf{k}\lambda}$ and $q_{-\mathbf{k}\lambda}$ as well as that of $q_{\mathbf{k}\lambda}^*$ and $q_{-\mathbf{k}\lambda}^*$ being unimportant; this is of course true classically and it will be seen to be true also in the quantized theory of Chapter 12.) Thus:

$$\mathbf{G} = \sum_{\mathbf{k},\lambda} \omega_{\mathbf{k}}\mathbf{k}(q_{\mathbf{k}\lambda}^* q_{\mathbf{k}\lambda} + q_{\mathbf{k}\lambda}q_{\mathbf{k}\lambda}^*) \quad (6.85)$$

and as for the energy the total momentum may be thought of as the sum of contributions from the individual wave vectors and polarizations.

It is now convenient to define the new real functions of time:

$$Q_{\mathbf{k}\lambda} = q_{\mathbf{k}\lambda} + q_{\mathbf{k}\lambda}^*; \qquad P_{\mathbf{k}\lambda} = -i\omega_{\mathbf{k}}(q_{\mathbf{k}\lambda} - q_{\mathbf{k}\lambda}^*). \quad (6.86)$$

Reference to Equations 6.62 shows that the $Q_{\mathbf{k}\lambda}$ are just the real solutions of the differential equation:

$$\mathrm{d}^2Q/\mathrm{d}t^2 + \omega^2 Q = 0, \quad (6.87)$$

while from Equations 6.63 it is seen that the $P_{\mathbf{k}\lambda}$ are the time derivatives of the $Q_{\mathbf{k}\lambda}$:

$$P_{\mathbf{k}\lambda} = (\mathrm{d}Q_{\mathbf{k}\lambda}/\mathrm{d}t) = \dot{Q}_{\mathbf{k}\lambda}. \quad (6.88)$$

The inverse of the transformation (Equations 6.86) is:

$$q_{\mathbf{k}\lambda} = (\omega_{\mathbf{k}} Q_{\mathbf{k}\lambda} + iP_{\mathbf{k}\lambda})/2\omega_{\mathbf{k}}, \tag{6.89}$$

so that in terms of the $Q_{\mathbf{k}\lambda}$ and $P_{\mathbf{k}\lambda}$ the field Hamiltonian (Equation 6.83) is:

$$\mathcal{H} = \tfrac{1}{2} \sum_{\mathbf{k},\lambda} (P_{\mathbf{k}\lambda}^2 + \omega_{\mathbf{k}}^2 Q_{\mathbf{k}\lambda}^2). \tag{6.90}$$

The advantage of these new variables is that $P_{\mathbf{k}\lambda}$ may be interpreted as the momentum conjugate to $Q_{\mathbf{k}\lambda}$, since Equations 6.87 and 6.88 are then consistent with Hamilton's equations of motion 3.34. In the first place the partial derivative of the Hamiltonian (Equation 6.90) with respect to $P_{\mathbf{k}\lambda}$ is just $P_{\mathbf{k}\lambda}$, which by Equations 3.34 must be equal to $\dot{Q}_{\mathbf{k}\lambda}$, in accord with Equation 6.88:

$$(\partial\mathcal{H}/\partial P_{\mathbf{k}\lambda}) = P_{\mathbf{k}\lambda} = \dot{Q}_{\mathbf{k}\lambda}. \tag{6.91}$$

Secondly, Hamilton's other equation requires that:

$$(\partial\mathcal{H}/\partial Q_{\mathbf{k}\lambda}) = \omega_{\mathbf{k}}^2 Q_{\mathbf{k}\lambda} = -\dot{P}_{\mathbf{k}\lambda}; \tag{6.92}$$

substitution of Equation 6.91 in Equation 6.92 now gives:

$$\ddot{Q}_{\mathbf{k}\lambda} + \omega_{\mathbf{k}}^2 Q_{\mathbf{k}\lambda} = 0, \tag{6.93}$$

which is just Equation 6.87 written in a different notation.

Again we note that the total Hamiltonian for the field may be written as a sum of independent Hamiltonians, one for each wave vector and polarization:

$$\mathcal{H} = \sum_{\mathbf{k},\lambda} \mathcal{H}_{\mathbf{k}\lambda}; \qquad \mathcal{H}_{\mathbf{k}\lambda} = \tfrac{1}{2}(P_{\mathbf{k}\lambda}^2 + \omega_{\mathbf{k}}^2 Q_{\mathbf{k}\lambda}^2). \tag{6.94}$$

Now $\mathcal{H}_{\mathbf{k}\lambda}$ can be seen to be the Hamiltonian for a classical one-dimensional harmonic oscillator, so that the field may be thought of as a collection of independent vibrations or radiation oscillators. In addition, $\mathcal{H}_{\mathbf{k}\lambda}$ is expressed in terms of a coordinate $Q_{\mathbf{k}\lambda}$ and its conjugate momentum $P_{\mathbf{k}\lambda}$; this is the starting point for the quantization of the electromagnetic field, which is discussed in Chapter 12.

Finally, we may note that the field momentum can also be expressed in terms of the $Q_{\mathbf{k}\lambda}$ and $P_{\mathbf{k}\lambda}$ and that the total momentum (Equation 6.85) is the sum of the contributions from the individual radiation osciallators:

$$\mathbf{G} = \sum_{\mathbf{k},\lambda} \mathbf{G}_{\mathbf{k}\lambda}, \tag{6.95}$$

where

$$\begin{aligned}\mathbf{G}_{\mathbf{k}\lambda} &= \tfrac{1}{2}(\mathbf{k}/\omega_{\mathbf{k}})(P_{\mathbf{k}\lambda}^2 + \omega_{\mathbf{k}}^2 Q_{\mathbf{k}\lambda}^2) \\ &= (\mathbf{k}/\omega_{\mathbf{k}})\mathcal{H}_{\mathbf{k}\lambda}.\end{aligned} \tag{6.96}$$

Bibliography

The classical theory of fields

Hameka: Section 6–1.

Landau and Lifschitz: Chapters 4 and 6.

Panofsky, W. K. H., and Phillips, M. (1962), *Classical Electricity and Magnetism*, Addison-Wesley, Reading, Mass.

Sakurai: Chapter 1.

Continuous mechanical systems

Goldstein, H. (1950), *Classical Mechanics*, McGraw-Hill, New York: Chapter 11.

CHAPTER SEVEN

Relativistic Wave Equations

We are now in a position to discuss the quantization of the classical relativistic theory that has been developed in the preceding chapters. In a sense this chapter is abortive, since it will become apparent that, although the wave equations considered here may be appropriate to some particles, they do not describe the electron; their major deficiency is that they cannot account for any of the properties associated with the spin of the electron. However, this discussion outlines the situation that faced Dirac and led him to propose the equation that now bears his name; the Dirac equation is the subject of the next chapter. In addition, the opportunity is taken to consider such topics as the gauge invariance of quantum mechanical equations.

7.1 Quantization of classical equations

The classical expression for the energy (or Hamiltonian) of a particle with mass m and charge q moving under the influence of electromagnetic potentials $\mathbf{A}$ and ϕ is (Equation 5.11):

$$E = c(m^2c^2 + \pi^2)^{\frac{1}{2}} + q\phi, \tag{7.1}$$

where $\boldsymbol{\pi}$ is the mechanical momentum:

$$\boldsymbol{\pi} = \mathbf{p} - q\mathbf{A}. \tag{7.2}$$

The usual prescription for the transition from a classical equation to a quantum mechanical equation is to interpret the energy and the canonical momentum as operators:

$$E = i\hbar(\partial/\partial t); \qquad \mathbf{p} = -i\hbar\boldsymbol{\nabla}, \tag{7.3}$$

and to allow these to operate on a wave function ψ:

$$i\hbar(\partial\psi/\partial t) = \{c[m^2c^2 + (\mathbf{p} - q\mathbf{A})^2]^{\frac{1}{2}} + q\phi\}\psi. \tag{7.4}$$

This wave equation is unsatisfactory for two reasons. First, time and space coordinates do not appear symmetrically, so that it is not clear whether it is Lorentz invariant. Secondly, the presence of a square root term makes the equation practically unusable, except when **A** vanishes or the particle is uncharged; this problem is due to the fact that now **p** is an operator, **p** and **A** do not commute and the meaning of the square root term is ambiguous. The restriction to a chargeless particle or to field-free situations is not acceptable, if a wave equation for an electron is being sought.

Of course, if we are interested in a slow-moving particle, a non-relativistic approximation may be obtained by expanding the square root term in Equation 7.1 as a power series in (π/mc):

$$E = mc^2 + q\phi + \pi^2/2m - \pi^4/8m^3c^2 + O(mc^2(\pi/mc)^6). \tag{7.5}$$

Neglect of terms of order $mc^2(\pi/mc)^4$ and above and use of Equations 7.3 gives for the quantum-mechanical version of this equation:

$$i\hbar(\partial\psi/\partial t) = (mc^2 + q\phi + \pi^2/2m)\psi. \tag{7.6}$$

Except for the mc^2 term this is just the usual time-dependent Schrödinger equation for a charged particle in an electromagnetic field.

The presence of mc^2 is merely due to the use of a different origin for the energy scale. It may be removed formally by introducing a phase factor into the wave function:

$$\psi \longrightarrow \psi' = \psi \exp(-imc^2t/\hbar). \tag{7.7}$$

The left-hand side of Equation 7.6 then becomes:

$$\begin{aligned} i\hbar(\partial\psi'/\partial t) &= i\hbar\frac{\partial}{\partial t}[\psi \exp(-imc^2t/\hbar)] \\ &= \exp(-imc^2t/\hbar)[i\hbar\partial\psi/\partial t + mc^2\psi]. \end{aligned} \tag{7.8}$$

The exponential commutes with the operators on the right-hand side of Equation 7.6, which after rearrangement and premultiplication throughout by $\exp(imc^2t/\hbar)$ becomes:

$$i\hbar(\partial\psi/\partial t) = (q\phi + \pi^2/2m)\psi. \tag{7.9}$$

Thus the energy origin has been shifted by changing the phase factor of the wave function; clearly this can have no effect on the physical properties calculated from the wave function.

Equation 7.9 is a time-dependent Schrödinger equation with the Hamiltonian operator:

$$\mathcal{H} = q\phi + \pi^2/2m, \tag{7.10}$$

and this may be expanded by substituting for the mechanical momentum $\boldsymbol{\pi}$ (Equation 7.2):

$$\mathcal{H} = q\phi + p^2/2m - q(\mathbf{p.A} + \mathbf{A.p})/2m + q^2A^2/2m. \tag{7.11}$$

The cross terms may be simplified by noting that:

$$\mathbf{p.A} = \mathbf{A.p} - i\hbar(\boldsymbol{\nabla}.\mathbf{A}). \tag{7.12}$$

In obtaining this result it must be remembered that we are now dealing with operators and that they must always be thought of as operating on a wave function:

$$\begin{aligned}\mathbf{p.A}\psi &= -i\hbar\boldsymbol{\nabla}.\mathbf{A}\psi = -i\hbar[\mathbf{A}.(\boldsymbol{\nabla}\psi) + (\boldsymbol{\nabla}.\mathbf{A})\psi] \\ &= [\mathbf{A.p} - i\hbar(\boldsymbol{\nabla}.\mathbf{A})]\,\psi,\end{aligned} \tag{7.13}$$

where the vector relation 2.64 has been used. If the Coulomb gauge (Equation 5.35) is being employed, the divergence of **A** vanishes and the Hamiltonian (Equation 7.11) becomes:

$$\mathcal{H} = q\phi + p^2/2m - (q/m)(\mathbf{A.p}) + q^2A^2/2m. \tag{7.14}$$

The first two terms in this Hamiltonian represent respectively the energy of interaction of the charge with the electric field and the kinetic energy of the particle. For a steady magnetic field **B** the vector potential may be written as $\mathbf{A} = \frac{1}{2}\mathbf{B}_\wedge\mathbf{r}$ (Equations 5.37), so that the third term:

$$-(q/m)(\mathbf{A.p}) = -(q/2m)\mathbf{B}.(\mathbf{r}_\wedge\mathbf{p}) \tag{7.15}$$

represents the interaction of the magnetic field with the magnetic moment $(q/2m)(\mathbf{r}_\wedge\mathbf{p})$ associated with the orbital angular momentum of the particle. The final term represents the diamagnetic interaction; this arises because the magnetic field induces some orbital angular momentum and the associated magnetic moment then interacts with the magnetic field.

It is clear that there is nothing in the Hamiltonian (Equation 7.14) that can be interpreted as either a magnetic moment or an angular momentum due to the spin of the particle. If higher terms, such as $-\pi^4/8m^3c^2$, were retained in the power series expansion (Equation 7.5) all the additional terms in the resulting quantum mechanical Hamiltonian could be described as relativistic corrections to the kinetic energy or the magnetic interactions in Equation 7.14. It must be concluded that a wave equation based on Equation 7.5 does not include anything attributable to spin.

7.2 Gauge invariance of quantum mechanical equations

In the last section it was assumed that the gauge of the electromagnetic potentials can be chosen in quantum mechanical equations just as it can be in classical equations. This assumption is now justified.

It is recalled that in a gauge transformation new potentials:

$$\mathbf{A}' = \mathbf{A} - \nabla f; \qquad \phi' = \phi + (\partial f/\partial t), \tag{7.16}$$

are introduced (Equations 5.31), where f is any scalar function of position and time. This transformation does not alter the corresponding magnetic flux density and electric field strength. It will now be shown that, provided we also introduce a phase factor into the wave function, a quantum mechanical equation is essentially unchanged, as it must be if it is to be gauge invariant. The new wave function is chosen to be:

$$\psi' = \psi \exp(-iqf/\hbar); \tag{7.17}$$

although the matrix elements of individual operators may be different for ψ and ψ', the physical results calculated from these wave functions are the same.

The way in which $\boldsymbol{\pi}\psi$ behaves under the transformation is considered first:

$$\begin{aligned}\boldsymbol{\pi}\psi \longrightarrow (\mathbf{p} - q\mathbf{A}')\psi' &= (\mathbf{p} - q\mathbf{A} + q\nabla f)\psi \exp(-iqf/\hbar)\\ &= \exp(-iqf/\hbar)(\mathbf{p} - q\nabla f - q\mathbf{A} + q\nabla f)\psi\\ &= \exp(-iqf/\hbar)\boldsymbol{\pi}\psi.\end{aligned} \tag{7.18}$$

Thus $\boldsymbol{\pi}\psi$ is merely multiplied on the left by the phase factor in Equation 7.17, and the same is true of $\pi^2\psi$, $\pi^4\psi$ and so on. If $(i\hbar\partial/\partial t - q\phi)\psi$ is now examined, it will be seen that the effect of the transformation is the same:

$$\begin{aligned}(i\hbar\partial/\partial t - q\phi)\psi \longrightarrow &(i\hbar\partial/\partial t - q\phi - q\partial f/\partial t)\psi \exp(-iqf/\hbar)\\ &= \exp(-iqf/\hbar)(i\hbar\partial/\partial t + q\partial f/\partial t - q\phi - q\partial f/\partial t)\psi\\ &= \exp(-iqf/\hbar)(i\hbar\partial/\partial t - q\phi)\psi.\end{aligned} \tag{7.19}$$

The wave equation 7.6 can be rearranged so that, besides constants such as mc^2, the only operators that appear are $\boldsymbol{\pi}$ and $(i\hbar\partial/\partial t - q\phi)$. The combined result of the gauge transformation (Equation 7.16) and the change of phase (Equation 7.17) is to premultiply every term in this equation by $\exp(-iqf/\hbar)$, so that it is essentially unchanged. This conclusion is unaffected, if higher terms in the expansion 7.5 are retained.

It is clear that any wave equation that, except for constants, can be written entirely in terms of the operators $\boldsymbol{\pi}$ and $(i\hbar\partial/\partial t - q\phi)$ is also gauge invariant.

7.3 The Klein-Gordon equation

The wave equation 7.4 has the disadvantages that it is not obviously Lorentz invariant, and that it is difficult to interpret and use. The first attempt to avoid these difficulties was to return to the classical Equation 7.1 and rewrite it as:

$$(E-q\phi)^2 = c^2(m^2c^2+\pi^2), \tag{7.20}$$

before attempting to quantize it. Equations 7.20 and 7.1 are not exactly equivalent, since the possibility that:

$$E = q\phi - c(m^2c^2+\pi^2)^{\frac{1}{2}} \tag{7.21}$$

has been introduced. The leading term in an expansion of Equation 7.21 is $-mc^2$ so that Equation 7.20 has some solutions with negative energy, that is solutions which correspond to particles with negative mass.

If Equation 7.20 is quantized in the usual manner (Equations 7.3) the corresponding wave equation is:

$$(i\hbar\partial/\partial t - q\phi)^2\psi = c^2[m^2c^2+(\mathbf{p}-q\mathbf{A})^2]\psi; \tag{7.22}$$

this is the Klein-Gordon equation. Since the square root term is no longer present, it is unambiguous, and in addition the operators $\boldsymbol{\nabla}$ and $\partial/\partial t$ both appear to second order so that the space and time coordinates occur symmetrically and the Lorentz invariance of the equation is at least credible, if not obvious. The invariance may be confirmed by noting that Equation 7.22 may be rewritten as:

$$(m^2c^4 + c^2\pi_\alpha\pi_\alpha)\psi = 0, \tag{7.23}$$

where

$$\pi_\alpha = -i\hbar\partial/\partial r_\alpha - q\mathrm{A}_\alpha, \tag{7.24}$$

is the component of a four-vector, since both $\partial/\partial r_\alpha$ (Equation 4.51) and A_α (Equation 5.2) are. Of course, the wave function will also be changed by a Lorentz transformation, but it will be the same function of position and time no matter what inertial frame is used.

Although the Klein-Gordon equation does not suffer from the disadvantages of the wave equation 7.4, there are a number of objections to it which arise from the way in which it is constructed. It has already been seen that the Klein-Gordon equation has negative energy solutions; in fact, this is also true of the Dirac equation, which is considered in the next chapter, but Dirac was able to use this apparent deficiency to advantage. However, there are more serious objections to the Klein-Gordon equation, one being that for its solutions $\psi^*\psi$ cannot be regarded as a probability density, since its integral over all space is time dependent.

For a time-dependent Schrödinger equation:

$$\mathcal{H}\psi = i\hbar(\partial\psi/\partial t), \tag{7.25}$$

the time and space variables may be separated by the substitution:

$$\psi(\mathbf{r}, t) = \chi(\mathbf{r})\theta(t). \tag{7.26}$$

This leads to two equations, the time-independent Schrödinger equation, $\mathcal{H}\chi = E\chi$, and an equation for $\theta(t)$:

$$i\hbar(\mathrm{d}\theta/\mathrm{d}t) = E\theta, \tag{7.27}$$

the solution of which is:

$$\theta = k \exp(-iEt/\hbar), \tag{7.28}$$

where k is an arbitrary constant. The probability density $\psi^*\psi$ is thus proportional to $\chi^*\chi$ and is time independent. Integration of $\psi^*\psi$ over all space shows that a particle described by the wave Equation 7.25 has a constant probability of being somewhere in space.

This is no longer true for a wave equation, such as the Klein-Gordon equation, in which the second derivative with respect to time appears. To demonstrate this a simpler example is considered:

$$\mathcal{H}\psi = -\hbar^2(\partial^2\psi/\partial t^2). \tag{7.29}$$

The variables may again be separated by the substitution of Equation 7.26, but the time-dependent part $\theta(t)$ now satisfies the equation:

$$(\mathrm{d}^2\theta/\mathrm{d}t^2) = -\alpha^2\theta, \tag{7.30}$$

where α $(= \sqrt{E}/\hbar)$ is a constant. The general solution of this differential equation is:

$$\theta = k_1 \sin \alpha t + k_2 \cos \alpha t, \tag{7.31}$$

so that $\psi^*\psi$ is in general time dependent and so is its integral over all space. Consequently, the probability of a particle described by the Klein-Gordon equation being somewhere in space varies with time.

In actuality many of the problems associated with the Klein-Gordon equation can be overcome, and it has been used to describe certain fundamental particles, such as mesons, which can decay. In this connection it can also be noted that the wave equation 6.59 for the electromagnetic field is just the Klein-Gordon equation for a particle with zero mass and charge. Of course, if a wave equation is to describe a decaying particle, some of the basic ideas of quantum mechanics must be reformulated. For example, although the Klein-Gordon equation is not consistent with the conservation of the number of particles, it has been shown to be consistent with the conservation of electric charge, which can be thought of as a more basic requirement; for an electron, charge conservation and number conservation are equivalent.

Perhaps the most serious objection to the Klein-Gordon equation as a wave equation for an electron is that it does not lead to anything that can be interpreted as spin. It must be concluded that, although the Klein-Gordon equation may be appropriate to certain spinless particles, it is not a satisfactory equation for the electron and that a wave equation linear in the operator $(\partial/\partial t)$ is necessary for the description of an electron.

Bibliography

Gauge invariance of quantum mechanical equations

Davies, D. W. (1967), *The Theory of the Electric and Magnetic Properties of Molecules*, Wiley, New York: Section 2.3.
Hameka: pages 41-42.

The Klein-Gordon equation

Foldy: pages 1-4 and 40-46.

CHAPTER EIGHT

The Dirac Equation

The last chapter was devoted to showing that direct quantization of classical equations does not lead to a wave equation for the electron. We will now discuss the Dirac equation for the electron. Although much of this chapter will be concerned with its justification, it will be assumed from the outset that it is indeed appropriate to the electron to the extent that we will refer to an electron with charge $-e$, where e is positive, rather than a particle with an arbitrary charge q; in addition, m will now be taken to be the mass of the electron.

Let us briefly summarize the problem that faced Dirac. For a particle with mass m and charge $-e$ in the presence of electromagnetic potentials $(\mathbf{A}, \phi)$ the classical relativistic expression for the energy is:

$$E = -e\phi + c[m^2c^2 + \pi^2]^{\frac{1}{2}}, \tag{8.1}$$

where

$$\boldsymbol{\pi} = \mathbf{p} + e\mathbf{A} \tag{8.2}$$

is the mechanical momentum. In the previous chapter this was interpreted as a quantum mechanical equation by replacing the energy and the canonical momentum by operators:

$$E = i\hbar(\partial/\partial t); \qquad \mathbf{p} = -i\hbar\nabla, \tag{8.3}$$

and allowing these to operate on a wave function ψ. However, such an interpretation was seen to be unsatisfactory, since time and space coordinates do not appear symmetrically and, except when $\mathbf{A}$ vanishes, the equation is unusable. These objections were removed at the expense of introducing solutions with negative energy by starting with the classical relation:

$$(E + e\phi)^2 = m^2c^4 + c^2\pi^2; \tag{8.4}$$

the substitutions 8.3 and the introduction of a wave function then gives the Klein-Gordon equation. This too has a disadvantage in that the second derivative with respect to time appears with the consequence that the probability that the particle is somewhere in space is time dependent. In addition, the Klein-Gordon equation cannot account for the property of spin, so that, although it might be appropriate for other particles, it does not apply to the electron. It was this unsatisfactory situation that led Dirac to consider an alternative equation.

8.1 The Dirac equation for a free electron

In the Klein-Gordon equation the time and space coordinates appear symmetrically, but as second derivatives. Dirac argued that it is necessary for the time coordinate to appear as a first derivative and, in order to preserve the relativistic equivalence of space and time, it is necessary for the space coordinates to appear as first derivatives as well. This would also be in contrast to the Schrödinger equation in which the time derivative is first order, but the space derivatives are second order. To simplify the ensuing arguments we consider a free electron, that is one in the absence of electromagnetic potentials. In addition, the notation can be simplified by introducing the operator p_0 such that:

$$cp_0 = i\hbar(\partial/\partial t). \tag{8.5}$$

(The subscript 0 is used rather than 4, because p_0 is not the time component of the four-vector $p_\alpha = -i\hbar(\partial/\partial r_\alpha)$, although they are related by $p_4 = ip_0$). The equation that Dirac proposed has the form:

$$[p_0 - \alpha_x p_x - \alpha_y p_y - \alpha_z p_z - \beta mc]\psi = 0, \tag{8.6}$$

or in vector notation:

$$[p_0 - \boldsymbol{\alpha}.\mathbf{p} - \beta mc]\psi = 0, \qquad (8.6)$$

where the nature of $\boldsymbol{\alpha}$ and β needs to be considered more closely. In Equation 8.6 time and space coordinates do appear symmetrically, and one might expect this equation to be invariant to Lorentz transformations, a necessary requirement of a relativistic equation; that the equation is indeed Lorentz invariant will be confirmed later.

Since Equation 8.6 is to be an equation linear in space and time derivatives, $\boldsymbol{\alpha}$ and β must be independent of $\mathbf{p}$ and p_0. Also, we are considering the situation where there are no electromagnetic fields so that all points in space-time are equivalent and the operator $[p_0 - \boldsymbol{\alpha}.\mathbf{p} - \beta mc]$ cannot involve x, y, z or t, since this would introduce space-time inhomogeneity and anisotropy, although this operator can (and does) contain first derivatives with respect to these coordinates; consequently $\boldsymbol{\alpha}$ and β are also independent of x, y, z and t. To obtain further information about $\boldsymbol{\alpha}$ and β we return to the Klein-Gordon equation, which for a free particle can be written in our current notation as:

$$[p_0^2 - p_x^2 - p_y^2 - p_z^2 - m^2c^2]\psi = 0. \tag{8.7}$$

Since this is directly connected with the classical expression for energy, the solutions of the Dirac equation 8.6 are also expected to be solutions of the Klein-Gordon equation 8.7. This is not the same as requiring that these two equations have the same solutions; there will be solutions of the Klein-Gordon equation, for which the probability of the existence of the particle will be time dependent, and these cannot be solutions of the Dirac equation.

If we multiply the Dirac equation 8.6 on the left by the operator:

$$[p_0 + \alpha_x p_x + \alpha_y p_y + \alpha_z p_z + \beta mc], \tag{8.8}$$

we obtain a second-order equation in p_0 and the components of $\mathbf{p}$, which can be compared with the Klein-Gordon equation 8.7. In doing this the fact that p_0 and the components of $\mathbf{p}$ each commute with β and the components of $\boldsymbol{\alpha}$ may be used, since we have seen that $\boldsymbol{\alpha}$ and β must be independent of x, y, z and t. However, we must not assume that β, α_x, α_y and α_z commute with each other, since their properties are yet to be determined. The result of multiplying the Dirac equation by the operator 8.8 is thus:

$$\begin{aligned}[p_0 + \alpha_x p_x + \alpha_y p_y + \alpha_z p_z + \beta mc]\,[p_0 - \alpha_x p_x - \alpha_y p_y - \alpha_z p_z - \beta mc]\psi \\ = [p_0^2 - \alpha_x^2 p_x^2 - \alpha_y^2 p_y^2 - \alpha_z^2 p_z^2 - (\alpha_x\alpha_y + \alpha_y\alpha_x)p_x p_y \\ - (\alpha_y\alpha_z + \alpha_z\alpha_y)p_y p_z - (\alpha_z\alpha_x + \alpha_x\alpha_z)p_z p_x \\ - mc(\alpha_x\beta + \beta\alpha_x)p_x - mc(\alpha_y\beta + \beta\alpha_y)p_y \\ - mc(\alpha_z\beta + \beta\alpha_z)p_z - \beta^2 m^2c^2]\psi = 0.\end{aligned} \tag{8.9}$$

This is identical to the Klein-Gordon equation provided the following relations between β, α_x, α_y and α_z hold:

$$\begin{aligned}&\alpha_x^2 = \alpha_y^2 = \alpha_z^2 = \beta^2 = 1; \\ &(\alpha_x\alpha_y + \alpha_y\alpha_x) = (\alpha_y\alpha_z + \alpha_z\alpha_y) = (\alpha_z\alpha_x + \alpha_x\alpha_z) = 0; \\ &(\alpha_x\beta + \beta\alpha_x) = (\alpha_y\beta + \beta\alpha_y) = (\alpha_z\beta + \beta\alpha_z) = 0.\end{aligned} \tag{8.10}$$

If we let $\alpha_0 = \beta$ (α_0 is not the coefficient of p_0 and α_0 is not a component of a four-vector), then these relations may be summarized as one relation:

$$[\alpha_i, \alpha_j]_+ = \alpha_i\alpha_j + \alpha_j\alpha_i = 2\delta_{ij} \quad (i, j = 0, x, y, z), \tag{8.11}$$

where δ_{ij} is the Kronecker delta. This equation shows that the four α_i anti-commute with one another and the square of each is unity.

Thus, provided Equation 8.11 is satisfied, multiplication of the Dirac equation by the operator 8.8 gives the Klein-Gordon equation and solutions of the Dirac equation are also solutions of the Klein-Gordon equation. Although the Dirac equation does not suffer from many of the objections to the Klein-Gordon equation, there is still the possibility that solutions with negative energy exist. We will return to this point later, when we consider positrons.

8.2 The Dirac operators $\boldsymbol{\alpha}$ and β

It is clear that β (= α_0) and the components of $\boldsymbol{\alpha}$ cannot be numbers, since they must satisfy the anticommutation relations in Equation 8.11; they can be regarded as operators representing a new degree of freedom intrinsic to the electron. It will be recalled that the spin angular momentum operators, σ_x, σ_y and σ_z, of non-relativistic theory satisfy similar anticommutation relations (Equation 2.22):

$$[\sigma_i, \sigma_j]_+ = 2\delta_{ij}. \tag{8.12}$$

In addition, the properties of these operators can be represented by the three two-by-two Pauli spin matrices so that it is reasonable to explore the possibility that α_x, α_y, α_z and β can also have a matrix representation. However, we will see that we have to go to four-by-four matrices to achieve this.

The three two-by-two Pauli spin matrices:

$$\sigma_x = \begin{pmatrix} 0 & 1 \\ 1 & 0 \end{pmatrix}; \quad \sigma_y = \begin{pmatrix} 0 & -i \\ i & 0 \end{pmatrix}; \quad \sigma_z = \begin{pmatrix} 1 & 0 \\ 0 & -1 \end{pmatrix} \tag{8.13}$$

form a matrix representation of three quantities which anticommute with one another and whose squares are unity; the square of each of the matrices (Equations 8.13) is unity in the sense that it is equal to the two-by-two unit matrix. However, for a representation of the four α_i we need four matrices, and any attempt to find a fourth two-by-two matrix that anticommutes with σ_x, σ_y and σ_z is doomed to failure. This can be seen merely by noting that any two-by-two matrix can be expressed as a linear combination of σ_x, σ_y, σ_z and the two-by-two unit matrix. Consequently there is no two-by-two matrix that anti-commutes with all three components of $\boldsymbol{\sigma}$. (There is, of course, the trivial case of the null matrix, but its square is not unity.) From this fact it can also be seen there are an infinite number of sets of three anticommuting matrices, but that these are just orthogonal linear combinations of the components of $\boldsymbol{\sigma}$.

It might be thought that three-by-three matrices would be able to provide us with four matrices with the required properties, but this can be disproved by using some results of matrix algebra to show more generally that a matrix representation of the α_i can only be constructed from matrices of even order. Let us consider a matrix representation **A**, **B**, ... This can be changed into an equivalent representation **X**, **Y**, . . . , in which **X** is diagonal, by performing a similarity transformation on each of **A**, **B**, . . . :

$$\mathbf{X} = \mathbf{Q}^{-1}\mathbf{A}\mathbf{Q}; \qquad \mathbf{Y} = \mathbf{Q}^{-1}\mathbf{B}\mathbf{Q}; \qquad \ldots, \tag{8.14}$$

where **Q** is chosen to be the matrix that brings about diagonalization of **A**. Many of the properties of the matrices are unaltered by similarity transformations provided all matrices are subjected to the same transformation. If **A** and **B** anticommute, then so do **X** and **Y**:

$$\begin{aligned} \mathbf{XY} + \mathbf{YX} &= \mathbf{Q}^{-1}\mathbf{A}\mathbf{Q}\mathbf{Q}^{-1}\mathbf{B}\mathbf{Q} + \mathbf{Q}^{-1}\mathbf{B}\mathbf{Q}\mathbf{Q}^{-1}\mathbf{A}\mathbf{Q} \\ &= \mathbf{Q}^{-1}(\mathbf{AB} + \mathbf{BA})\mathbf{Q} = 0. \end{aligned} \tag{8.15}$$

Similarly, if $\mathbf{A}^2$ equals the unit matrix, then so does $\mathbf{X}^2$. This fact, combined with the fact that **X** has been chosen to be diagonal, means that the diagonal elements of **X** must be either + 1 or − 1.

The next step in our proof is to show that the trace (the sum of the diagonal elements) of the matrix **X** vanishes. To do this we use the fact that for two matrices **R** and **S** the trace of the product matrix **RS** is the same as the trace of the product matrix **SR**; this can easily be proved by expanding these products in terms of the elements of **R** and **S** (see Equation 2.108). Thus, the product matrix $\mathbf{XY}^2$ has the same trace as **YXY**, and also the same trace as **X**, since $\mathbf{Y}^2$ is required to be the unit matrix:

$$Tr\,(\mathbf{XY}^2) = Tr\,(\mathbf{YXY}) = Tr\,(\mathbf{X}). \tag{8.16}$$

However, since **X** and **Y** are to anticommute $\mathbf{XY}^2$ is equal to $-\mathbf{YXY}$, so that:

$$Tr(\mathbf{XY}^2) = -\,Tr(\mathbf{YXY}). \tag{8.17}$$

The only way in which Equations 8.16 and 8.17 can be reconciled is for the trace of $\mathbf{XY}^2$, and hence of **X**, to be zero. But it has already been shown that the diagonal elements of **X** can only be + 1 or − 1, so that the number of diagonal elements with value + 1 must be the same as the number with value − 1 and the order of the matrix must be even. Hence, only even-order matrices can satisfy our initial assumption of anticommuting matrices whose squares are the unit matrix. Consequently, we can move on to a consideration of four-by-four matrices.

As was indicated earlier, there is a set of four-by-four matrices that forms a representation of the α_i. To arrive at this set (or one related to it by a similarity transformation) in a rigorous fashion is somewhat involved, so that here a more intuitive approach is adopted. Since the α_i have properties similar to those of the components of $\boldsymbol{\sigma}$, it is not unreasonable to use them as our starting point, and we construct three four-by-four matrices by taking the direct product of each of σ_x, σ_y and σ_z (Equations 8.13) with the two-by-two unit matrix:

$$\sigma_x = \begin{pmatrix} 0 & 1 & 0 & 0 \\ 1 & 0 & 0 & 0 \\ 0 & 0 & 0 & 1 \\ 0 & 0 & 1 & 0 \end{pmatrix}; \quad \sigma_y = \begin{pmatrix} 0 & -i & 0 & 0 \\ i & 0 & 0 & 0 \\ 0 & 0 & 0 & -i \\ 0 & 0 & i & 0 \end{pmatrix};$$
$$\sigma_z = \begin{pmatrix} 1 & 0 & 0 & 0 \\ 0 & -1 & 0 & 0 \\ 0 & 0 & 1 & 0 \\ 0 & 0 & 0 & -1 \end{pmatrix}; \qquad (8.18)$$

it is common practice to use the same symbols for both the two-by-two and the corresponding four-by-four matrices, since the context should indicate which is intended. These four-by-four matrices satisfy the same relations (Equation 8.12) as the corresponding two-by-two matrices, that is they anticommute with one another and their squares are the unit (four-by-four) matrix. The possibility of constructing the α_i by introducing just two new matrices ρ and β is now examined. The proposed relations are:

$$\alpha_x = \rho\sigma_x; \qquad \alpha_y = \rho\sigma_y; \qquad \alpha_z = \rho\sigma_z; \qquad \alpha_0 = \beta. \qquad (8.19)$$

It can easily be shown that for the α_i (Equations 8.19) to satisfy the defining relations (Equation 8.11) ρ and β must have the properties that their squares are both equal to unity, that they anticommute with one another and that they both commute with the components of $\boldsymbol{\sigma}$:

$$\begin{aligned} \rho^2 &= \beta^2 = 1; \\ [\rho, \beta]_+ &= 0; \\ [\rho, \boldsymbol{\sigma}] &= [\beta, \boldsymbol{\sigma}] = 0. \end{aligned} \qquad (8.20)$$

For example:

$$\alpha_x\alpha_y + \alpha_y\alpha_x = \rho\sigma_x\rho\sigma_y + \rho\sigma_y\rho\sigma_x = \sigma_x\sigma_y + \sigma_y\sigma_x = 0, \quad (8.21)$$

or again:

$$\alpha_x\alpha_0 + \alpha_0\alpha_x = \rho\sigma_x\beta + \beta\rho\sigma_x = \sigma_x(\rho\beta + \beta\rho) = 0. \quad (8.22)$$

When a matrix representation of the α_i has been found, an equivalent representation can be produced merely by performing the same similarity transformation on each of the matrices in our original representation. However, the representation can be fixed by stipulating that our representation is such that β is diagonal. In the consideration of general matrices it was seen that, for the square of a diagonal matrix to equal unity, its diagonal elements can only be $+1$ or -1 and that, for it to anticommute with other matrices, its trace must be zero. Thus, β must have the same number of elements equal to $+1$ as equal to -1. The fact that β commutes with σ_x, σ_y and σ_z means that it can only be:

$$\beta = \begin{pmatrix} 1 & 0 & 0 & 0 \\ 0 & 1 & 0 & 0 \\ 0 & 0 & -1 & 0 \\ 0 & 0 & 0 & -1 \end{pmatrix} \quad (8.23)$$

or of course minus this; we choose it to be as shown in Equation 8.23.

It remains to find the matrix ρ, which is to anticommute with β and to commute with σ_x, σ_y and σ_z, all of which are now known explicitly. From its anticommutation property with β it can only have the form:

$$\rho = \begin{pmatrix} 0 & 0 & a & b \\ 0 & 0 & c & d \\ e & f & 0 & 0 \\ g & h & 0 & 0 \end{pmatrix}. \quad (8.24)$$

Commutation with σ_z demands that b, c, f and g are all zero, while commutation with σ_x (or σ_y) necessitates that $d = a$ and $h = e$. Finally, for ρ^2 to be the unit matrix $ae = 1$ and making the choice $a = e = 1$:

$$\rho = \begin{pmatrix} 0 & 0 & 1 & 0 \\ 0 & 0 & 0 & 1 \\ 1 & 0 & 0 & 0 \\ 0 & 1 & 0 & 0 \end{pmatrix}. \quad (8.25)$$

Substituting Equations 8.25, 8.23 and 8.18 into Equations 8.19 now gives us a set of four four-by-four matrices representing the α_i:

$$\alpha_x = \begin{pmatrix} 0 & 0 & 0 & 1 \\ 0 & 0 & 1 & 0 \\ 0 & 1 & 0 & 0 \\ 1 & 0 & 0 & 0 \end{pmatrix}; \quad \alpha_y = \begin{pmatrix} 0 & 0 & 0 & -i \\ 0 & 0 & i & 0 \\ 0 & -i & 0 & 0 \\ i & 0 & 0 & 0 \end{pmatrix};$$

$$\alpha_z = \begin{pmatrix} 0 & 0 & 1 & 0 \\ 0 & 0 & 0 & -1 \\ 1 & 0 & 0 & 0 \\ 0 & -1 & 0 & 0 \end{pmatrix}; \quad \alpha_0 = \beta = \begin{pmatrix} 1 & 0 & 0 & 0 \\ 0 & 1 & 0 & 0 \\ 0 & 0 & -1 & 0 \\ 0 & 0 & 0 & -1 \end{pmatrix} \tag{8.26}$$

and as would be expected they are all Hermitian.

Although we do not prove it here, this is, to within a similarity transformation, the only four-by-four matrix representation of the α_i. The particular representation chosen here is called the standard representation. However, it should be noted that as far as determining eigenvalues is concerned it does not matter which representation is used, since, although the operator in the Dirac equation:

$$[p_0 - \boldsymbol{\alpha}.\mathbf{p} - \beta mc]\psi = 0 \tag{8.27}$$

will have different forms in different representations, this is compensated by a change in the wave function. If a similarity transformation is applied to the matrices in the operator part of Equation 8.27:

$$\alpha_i \longrightarrow \alpha_i' = \mathbf{Q}^{-1}\alpha_i\mathbf{Q}, \tag{8.28}$$

where $\mathbf{Q}$ is independent of time and space coordinates, a new wave function ψ', is needed so that:

$$[p_0 - \boldsymbol{\alpha}'.\mathbf{p} - \beta' mc]\psi' = 0. \tag{8.29}$$

This is the same as Equation 8.27 multiplied on the left by $\mathbf{Q}^{-1}$, if ψ' is interpreted as:

$$\psi' = \mathbf{Q}^{-1}\psi. \tag{8.30}$$

Hence, use of a different representation merely transforms the wave function, but certain representations will be more convenient than others, and for chemical applications the standard representation (Equations 8.26) is most appropriate, as will be seen shortly.

In the previous paragraph it is implied that the wave function ψ can be multiplied by a four-by-four matrix $\mathbf{Q}^{-1}$. Also the total Dirac operator can be expressed as a four-by-four matrix, and this operates on ψ. This is only meaningful if ψ is represented by a four-component column vector or spinor:

$$\psi = \begin{pmatrix} \psi_1 \\ \psi_2 \\ \psi_3 \\ \psi_4 \end{pmatrix}, \tag{8.31}$$

and if p_0 is thought to be multiplied by a four-by-four unit matrix. Thus, when expanded fully, the Dirac equation consists of four coupled equations in the four components of ψ:

$$\left.\begin{aligned} (p_0 - mc)\psi_1 - p_z\psi_3 - (p_x - ip_y)\psi_4 &= 0 \\ (p_0 - mc)\psi_2 - (p_x + ip_y)\psi_3 + p_z\psi_4 &= 0 \\ -p_z\psi_1 - (p_x - ip_y)\psi_2 + (p_0 + mc)\psi_3 &= 0 \\ -(p_x + ip_y)\psi_1 + p_z\psi_2 + (p_0 + mc)\psi_4 &= 0 \end{aligned}\right\}. \tag{8.32}$$

That the wave function ψ has four components needs some explanation. In the spin theory due to Pauli the wave function of the electron is taken to have two components corresponding to the two different spin orientations; thus two-component wave functions would be expected in a theory that explained electron spin. That twice as many components appear, is due to the fact that solutions of the Dirac equation are also solutions of the Klein-Gordon equation, and negative energy solutions are possible. This can easily be seen from Equations 8.32, if we consider the time-independent case, where p_0 is replaced by E/c, and ψ is a function of space coordinates only. For a slowly moving electron the magnitude of the momentum is negligible compared with mc, so that Equations 8.32 have the approximate eigenvalues $+mc^2$ (twice) and $-mc^2$ (twice). It is the small terms involving the components of $\mathbf{p}$ that couple these so-called positive energy states and negative energy states making the theory a relativistic one and adding relatively small energy corrections to $\pm mc^2$. It is now seen that the choice of the standard representation for the Dirac operators provides the maximum separation of the positive and negative energy states. Although it is the positive energy states that are of interest to chemists, the negative energy states will be considered briefly later.

8.3 The introduction of an electromagnetic field

So far we have considered a free electron, that is in the absence of electromagnetic potentials. These can be introduced in the usual manner by allowing the replacements:

$$p_0 \longrightarrow p_0 + (e/c)\phi; \qquad \mathbf{p} \longrightarrow \boldsymbol{\pi} = \mathbf{p} + e\mathbf{A}, \tag{8.33}$$

where allowance has been made for the definition (Equation 8.5) of p_0 and its relation (Equation 8.3) to the energy. The resulting equation is:

$$[p_0 + (e/c)\phi - \boldsymbol{\alpha}.\boldsymbol{\pi} - \beta mc]\psi = 0. \tag{8.34}$$

From the considerations in Section 7.2 it is at once apparent that this equation is gauge invariant, since it contains only the operators $(\mathbf{p} + e\mathbf{A})$ and $(i\hbar(\partial/\partial t) + e\phi)$. A change in gauge merely multiplies the wave function by a phase factor, $\exp(-ief/\hbar)$, where f is the scalar function specifying the gauge transformation. In the special case of the time-independent problem Equation 8.34 reduces to:

$$[\beta mc^2 + c\boldsymbol{\alpha}.\boldsymbol{\pi} - e\phi]\psi = E\psi \tag{8.35}$$

and the operator on the left-hand side of this equation may be referred to as the Dirac Hamiltonian.

The Dirac equation has now been developed, but it has yet to be shown that it is appropriate to the electron. It has not been derived rigorously, nor can it be; it came about because of an insistence on a linear equation in p_0, p_x, p_y and p_z, and a consideration of the consequences. Its justification is the same as that for the Schrödinger equation – it works. Later in this chapter the Lorentz invariance of the Dirac equation will be demonstrated and the nature of the negative energy solutions will be considered. However, we will first show that the equation is appropriate to the electron, in that it leads to the concept of spin, since any further discussion would be pointless if this were not so.

8.4 Electron spin

Later in this chapter the relationship between the Dirac equation and the non-relativistic formulation of quantum mechanics will be considered in detail. Here, we will content ourselves with showing that, when the Dirac equation is reduced to a non-relativistic form, two new properties arise naturally. These properties are the intrinsic magnetic moment and the intrinsic angular momentum, that are associated with spin.

The time-independent Dirac equation 8.35 can be written as:

$$\mathcal{H}\psi = E\psi, \tag{8.36}$$

where the Dirac Hamiltonian is given by:

$$\mathcal{H} = \beta mc^2 + c\boldsymbol{\alpha}.\boldsymbol{\pi} - e\phi; \tag{8.37}$$

if a matrix representation is used, then ψ is a four-component vector and $\mathcal{H}$ is a four-by-four matrix. From Equation 8.37:

$$\begin{aligned}(\mathcal{H} + e\phi)^2/c^2 &= (\beta mc + \boldsymbol{\alpha}.\boldsymbol{\pi})^2 \\ &= m^2c^2 + (\boldsymbol{\alpha}.\boldsymbol{\pi})^2,\end{aligned} \tag{8.38}$$

where the properties (Equation 8.11) of the Dirac matrices have been used; in particular no cross terms arise on squaring $(\beta mc + \boldsymbol{\alpha}.\boldsymbol{\pi})$, since $\boldsymbol{\alpha}$ and β both commute with $\boldsymbol{\pi}$, but anticommute with each other. To expand Equation 8.38 further, we note that $\boldsymbol{\alpha} = \rho\boldsymbol{\sigma}$ (Equations 8.19), where ρ is the four-by-four matrix that has the properties (Equations 8.20) that it commutes with both $\boldsymbol{\sigma}$ and $\boldsymbol{\pi}$, and has the unit matrix as its square. Hence:

$$(\boldsymbol{\alpha}.\boldsymbol{\pi})^2 = (\boldsymbol{\sigma}.\boldsymbol{\pi})^2 = (\boldsymbol{\sigma}.\boldsymbol{\pi})(\boldsymbol{\sigma}.\boldsymbol{\pi}). \tag{8.39}$$

In Chapter 2 the relationship (Equation 2.43):

$$(\boldsymbol{\sigma}.\mathbf{B})(\boldsymbol{\sigma}.\mathbf{C}) = (\mathbf{B}.\mathbf{C}) + i\boldsymbol{\sigma}.(\mathbf{B}_\wedge\mathbf{C}), \tag{8.40}$$

was proved, where $\boldsymbol{\sigma}$ commutes with **B** and **C** and the components of $\boldsymbol{\sigma}$ are the two-by-two Pauli spin matrices. Although the $\boldsymbol{\sigma}$ used in Equation 8.39 are four-by-four matrices, they are constructed (Equations 8.18) from the Pauli matrices and satisfy similar commutation and anticommutation properties. Thus, since $\boldsymbol{\sigma}$ and $\boldsymbol{\pi}$ commute, Equation 8.40 can be used to expand Equation 8.39 provided it is remembered that four-by-four matrices are now involved:

$$(\boldsymbol{\sigma}.\boldsymbol{\pi})^2 = \pi^2 + i\boldsymbol{\sigma}.(\boldsymbol{\pi}_\wedge\boldsymbol{\pi}). \tag{8.41}$$

It is tempting to say that the second term in this equation is zero, since it involves the vector product of $\boldsymbol{\pi}$ with itself, but this is not true. In considering $(\boldsymbol{\pi}_\wedge\boldsymbol{\pi})$ the explicit form of $\boldsymbol{\pi}$ (Equations 8.33) must be used:

$$\begin{aligned}\boldsymbol{\pi}_\wedge\boldsymbol{\pi} &= (\mathbf{p} + e\mathbf{A})_\wedge(\mathbf{p} + e\mathbf{A}) \\ &= \mathbf{p}_\wedge\mathbf{p} + e(\mathbf{p}_\wedge\mathbf{A} + \mathbf{A}_\wedge\mathbf{p}) + e^2\mathbf{A}_\wedge\mathbf{A}.\end{aligned} \tag{8.42}$$

Although the first and last terms in this equation do vanish, the middle term does not, since **p** and **A** do not commute; this term may be simplified further, but in doing so it must be remembered that it is an operator and must always be thought of as operating on a wave function, even though the wave function is not included explicitly. Thus:

$$\boldsymbol{\pi}_\wedge \boldsymbol{\pi} = e[\mathbf{p}_\wedge \mathbf{A} + \mathbf{A}_\wedge \mathbf{p}]$$
$$= e[-\mathbf{A}_\wedge \mathbf{p} - i\hbar(\boldsymbol{\nabla}_\wedge \mathbf{A}) + \mathbf{A}_\wedge \mathbf{p}]$$
$$= -i\hbar e(\boldsymbol{\nabla}_\wedge \mathbf{A}) = -i\hbar e\mathbf{B}, \tag{8.43}$$

where **B** is the magnetic field associated with the potential **A** (Equation 5.20). In going from the first to the second line the vector relation 2.65 has been used to expand $(\boldsymbol{\nabla}_\wedge \mathbf{A}\psi)$:

$$(\boldsymbol{\nabla}_\wedge \mathbf{A}\psi) = \psi(\boldsymbol{\nabla}_\wedge \mathbf{A}) - (\mathbf{A}_\wedge \boldsymbol{\nabla})\psi. \tag{8.44}$$

Combining Equations 8.38 to 8.43 gives:

$$(\mathcal{H} + e\phi)^2/c^2 = m^2c^2 + \pi^2 + e\hbar\boldsymbol{\sigma}.\mathbf{B}. \tag{8.45}$$

Since we are interested in comparing the Dirac equation with non-relativistic theory, we wish to consider Equation 8.45 in the non-relativistic limit, that is when it is applied to a slowly moving electron. The Hamiltonian for a positive energy state can then be written as:

$$\mathcal{H} = mc^2 + \mathcal{H}', \tag{8.46}$$

where $\mathcal{H}'$ is small compared with mc^2. In expanding the left-hand side of Equation 8.45 it can then be assumed that $\mathcal{H}'^2/c^2$ is negligible as are other terms involving c^{-2}:

$$(\mathcal{H} + e\phi)^2/c^2 \simeq m^2c^2 + 2m\mathcal{H}' + 2me\phi. \tag{8.47}$$

Combining Equations 8.45 and 8.47 gives finally:

$$\mathcal{H}' \simeq -e\phi + \pi^2/2m + (e\hbar/2m)\boldsymbol{\sigma}.\mathbf{B}; \tag{8.48}$$

this confirms that $\mathcal{H}'$ is negligible compared with mc^2. This Hamiltonian is the same as the non-relativistic Hamiltonian for an electron except for the term $(e\hbar/2m)\boldsymbol{\sigma}.\mathbf{B}$, which may be interpreted as an additional potential energy due to an interaction with the external magnetic field. That is the particle possesses a magnetic moment:

$$-(e\hbar/2m)\boldsymbol{\sigma} = -2\mu_B\mathbf{s}, \tag{8.49}$$

where μ_B $(= e\hbar/2m)$ is the Bohr magneton and $\mathbf{s} = \boldsymbol{\sigma}/2$. This is just what is assumed in introducing spin into the non-relativistic Hamiltonian (see Chapter 1).

Of course, a matrix representation of the Hamiltonian (Equation 8.48) involves four-by-four matrices, but consideration of the specific forms of the components of $\boldsymbol{\sigma}$ (Equations 8.18) shows that, when it operates on a four-component wave function, the Hamiltonian factorizes into two identical pairs of coupled equations. Hence, Equation 8.48 with a four-by-four matrix representation of $\boldsymbol{\sigma}$ conveys no more information than the same Hamiltonian with a representation involving two-by-two Pauli matrices. This is in accord with the two-component formalism of non-relativistic spin theory.

To obtain the g value appropriate to a particle which satisfies the Dirac equation it is necessary to obtain the intrinsic angular momentum associated with the intrinsic magnetic moment, since the g value is related to the ratio of the magnitude of these quantities. The intrinsic magnetic moment gives rise to potential energy, but the intrinsic angular momentum does not and its presence must be demonstrated in a different fashion. To do this we consider a free electron, that is in the absence of electromagnetic fields. The Dirac Hamiltonian now has the simpler form:

$$\mathcal{H} = \beta mc^2 + c(\boldsymbol{\alpha}.\mathbf{p}). \tag{8.50}$$

In the non-relativistic quantum theory of a free particle the Hamiltonian, $\mathcal{H} = p^2/2m$, and the orbital angular momentum, $\hbar\mathbf{l} = \mathbf{r} \wedge \mathbf{p}$, commute with one another and $\hbar\mathbf{l}$ is a constant of motion. To prove this we consider the commutation of the component $\hbar l_i$ with p^2 and use the techniques of Chapter 2:

$$\begin{aligned} [\hbar l_i, p^2] &= [\epsilon_{ijk} r_j p_k, p_l p_l] \\ &= 2i\hbar\epsilon_{ijk}\delta_{lj} p_l p_k = 0. \end{aligned} \tag{8.51}$$

However, it will be seen that $\hbar\mathbf{l}$ does not commute with the Dirac Hamiltonian (Equation 8.50) since:

$$\begin{aligned} [\hbar l_i, \mathcal{H}] &= \hbar c[l_i(\boldsymbol{\alpha}.\mathbf{p}) - (\boldsymbol{\alpha}.\mathbf{p})l_i] \\ &= \hbar c\boldsymbol{\alpha}.[l_i, \mathbf{p}] \end{aligned} \tag{8.52}$$

and l_i does not commute with $\mathbf{p}$. In fact:

$$[l_i, p_j] = i\epsilon_{ijk} p_k, \tag{8.53}$$

so that:

$$\begin{aligned} [\hbar l_i, \mathcal{H}] &= i\hbar c\epsilon_{ijk}\alpha_j p_k \\ &= i\hbar c\rho(\boldsymbol{\sigma}\wedge\mathbf{p})_i, \end{aligned} \tag{8.54}$$

where $\boldsymbol{\alpha}$ has been replaced by $\rho\boldsymbol{\sigma}$ (Equations 8.19).

The orbital angular momentum $\hbar\mathbf{l}$ is no longer a constant of motion. This implies that $\hbar\mathbf{l}$ is not the total angular momentum associated with a particle satisfying the Dirac equation. The appearance of $\boldsymbol{\sigma}$ in Equation 8.54 suggests that a consideration of the commutator of $\boldsymbol{\sigma}$ with $\mathcal{H}$ might be pertinent. Remembering that β commutes with $\boldsymbol{\sigma}$ (Equations 8.20):

$$[\sigma_i, \mathcal{H}] = c[\sigma_i(\boldsymbol{\alpha}.\mathbf{p}) - (\boldsymbol{\alpha}.\mathbf{p})\sigma_i]$$

$$= c[\sigma_i, \boldsymbol{\alpha}].\mathbf{p} = c\rho[\sigma_i, \boldsymbol{\sigma}].\mathbf{p}, \tag{8.55}$$

where again Equations 8.19 and the properties of ρ (Equations 8.20) have been used. Now the commutation properties of the components of $\boldsymbol{\sigma}$ are (Equation 2.41):

$$[\sigma_i, \sigma_j] = 2i\epsilon_{ijk}\sigma_k; \tag{8.56}$$

this relationship holds whether a four-by-four or a two-by-two matrix representation is used for the components of $\boldsymbol{\sigma}$. Substitution in Equation 8.55 gives:

$$[\sigma_i, \mathcal{H}] = c\rho 2i\epsilon_{ijk}\sigma_k p_j$$

$$= -2ic\rho\,(\boldsymbol{\sigma}_\wedge \mathbf{p})_i, \tag{8.57}$$

showing that $\boldsymbol{\sigma}$ does not commute with the Dirac Hamiltonian either. However, comparison of Equations 8.54 and 8.57 shows that $(\hbar\mathbf{l} + \frac{1}{2}\hbar\boldsymbol{\sigma})$ does commute with $\mathcal{H}$:

$$[\hbar l_i + \tfrac{1}{2}\hbar\sigma_i, \mathcal{H}] = 0, \tag{8.58}$$

and is thus a new constant of motion. If we regard this as the total angular momentum, then in addition to the orbital angular momentum $\hbar\mathbf{l}$ there is an intrinsic angular momentum $\frac{1}{2}\hbar\boldsymbol{\sigma}$ and by identifying $\boldsymbol{\sigma}$ with 2s this intrinsic angular momentum of $\hbar\mathbf{s}$ is just what is assumed in non-relativistic theory (see Chapter 1).

The Dirac equation thus leads to an intrinsic magnetic moment of $-2\mu_B\mathbf{s}$ and an intrinsic angular momentum of $\hbar\mathbf{s}$. The ratio of these gives a g factor of 2. These results agree with the assumed electron properties that are introduced phenomonologically into the non-relativistic theory of the electron. It would appear that the Dirac equation is indeed an appropriate equation for the electron. However, we still have to verify the Lorentz invariance of the equation and examine the negative energy solutions more closely before its suitability is fully confirmed.

It is often said that electron spin is a consequence of special relativity, since it was first explained by the relativistic theory of Dirac. However, it is only fair to say that there is another point of view. The non-relativistic Hamiltonian (Equation 7.10) for a particle contains the operator $\pi^2/2m$. Replacement of this by $(\boldsymbol{\sigma}.\boldsymbol{\pi})^2/2m$, where the components of $\boldsymbol{\sigma}$ are the Pauli spin matrices, also leads to an intrinsic magnetic moment of the right magnitude, as may be seen from Equations 8.39 to 8.45. The same method may be used to develop a relativistic theory, the result being the Dirac equation; this approach is described in, for example, the book by Sakurai (see bibliography).

Finally, it should be pointed out that the present section has itself raised a problem in that **l** and **s**, which commute with the non-relativistic Hamiltonian, do not commute with the Dirac Hamiltonian; the reasons for this will be discussed in Section 8.10.

8.5 Lorentz invariance of the Dirac equation

Although the Dirac equation was constructed so that time and space coordinates appeared symmetrically, it is not at all obvious from Equation 8.34 that it is in fact Lorentz invariant. It need hardly be said that for the Dirac equation to be an acceptable relativistic equation it must be Lorentz invariant.

In most advanced texts the invariance with respect to a general Lorentz transformation is demonstrated. Here we will content ourselves with the specific transformation considered in Section 4.3. That is, we consider two inertial frames K and K', for which the axes x and x' coincide and the axes y and z are parallel to y' and z' respectively; the velocity of the K' frame is v relative to the K frame. A general Lorentz transformation can be reduced to this specific one merely be redefining the origins and the orientation of the axes within the two frames.

It will be recalled that the coordinates, x, y, z and t, in the K frame are related to x', y', z' and t', the coordinates in the K' frame, by Equations 4.18:

$$x = \gamma(x' + vt'); \quad y = y'; \quad z = z'; \quad t = \gamma[t' + (v/c^2)x'], \quad (8.59)$$

where $\gamma = (1 - v^2/c^2)^{-\frac{1}{2}}$. For the present purposes it is convenient to rewrite Equations 8.59 using Equations 4.9 and 4.11:

$$\begin{aligned} x &= x'\cos\theta - ict'\sin\theta; \qquad & y &= y'; \\ z &= z'; \qquad & t &= t'\cos\theta - i(x'/c)\sin\theta, \end{aligned} \quad (8.60)$$

where (Equation 4.14):

$$\tan\theta = iv/c; \quad (8.61)$$

θ is the parameter used in Chapter 4 and can be seen to be independent of the space and time coordinates. In Chapter 5 it was noted that the electromagnetic potentials, A_x, A_y, A_z and ϕ/c^2, transform in a similar fashion:

$$\begin{aligned} A_x &= A'_x\cos\theta - i(\phi'/c)\sin\theta; \qquad & A_y &= A'_y; \\ A_z &= A'_z; \qquad & \phi &= \phi'\cos\theta - icA'_x\sin\theta. \end{aligned} \quad (8.62)$$

The operators $\partial/\partial r_\alpha$ have also been shown to transform as the components of a four-vector (Equation 4.51), so that, recalling the definition of p_0 (Equation

8.5), the operators p_x, p_y, p_z and p_0 transform as:

$$\begin{aligned} p_x &= p'_x \cos\theta - ip'_0 \sin\theta; \qquad p_y = p'_y; \\ p_z &= p'_z; \qquad p_0 = p'_0 \cos\theta - ip'_x \sin\theta. \end{aligned} \tag{8.63}$$

By substituting Equations 8.62 and 8.63 into the Dirac Equation 8.34, we obtain, after some rearrangement:

$$\begin{aligned} [(\cos\theta + i\alpha_x \sin\theta)(p'_0 + (e/c)\phi') - \alpha_x(\cos\theta + i\alpha_x \sin\theta)\pi'_x \\ - \alpha_y \pi'_y - \alpha_z \pi'_z - \beta mc]\psi = 0, \end{aligned} \tag{8.64}$$

where ψ is still the wave function in the unprimed frame of reference. To show that the Dirac equation is Lorentz invariant it is necessary to demonstrate that it has the same form in both the primed and the unprimed frames. That is, we must show that Equation 8.64 is equivalent to:

$$[p'_0 + (e/c)\phi' - \boldsymbol{\alpha}.\boldsymbol{\pi}' - \beta mc]\psi' = 0, \tag{8.65}$$

where ψ' is the wave function appropriate to the primed frame of reference. This can be achieved in the following way. The first thing to note is that expressions such as sine and cosine can always be interpreted as power series in their arguments and thus there is no objection to their arguments involving matrices. Since sine and cosine are, respectively, odd and even functions of ~~theri~~ arguments and the square of α_x is the unit matrix we can write:

$$\cos\theta + i\alpha_x \sin\theta = \cos(\theta\alpha_x) + i\sin(\theta\alpha_x) = \exp(i\theta\alpha_x). \tag{8.66}$$

No problems arise about the order of θ and α_x in these expressions, since θ is a constant and commutes with α_x, and the interpretation of Equation 8.66 is unambiguous. Also, a term of the type $\exp(i\theta\alpha_x)$ commutes with α_x and the operators $(p'_0 + (e/c)\phi')$, π'_x, π'_y and π'_z, but it does not commute with α_y, α_z or β. By expanding the exponential as a power series and using the anticommutation properties (Equation 8.11) of α_x with α_y, α_z and β, it can be seen that:

$$\exp(i\theta\alpha_x)\alpha_y = \alpha_y \exp(-i\theta\alpha_x); \tag{8.67}$$

similar expressions hold for α_z and β. Thus, by multiplying Equation 8.64 on the left by $\exp(-i\theta\alpha_x/2)$ and altering the order of the terms using relations of the type in Equation 8.67, the transformed Dirac equation becomes:

$$[p'_0 + (e/c)\phi' - \boldsymbol{\alpha}.\boldsymbol{\pi}' - \beta mc] \exp(i\theta\alpha_x/2)\psi = 0, \tag{8.68}$$

which is identical to Equation 8.65 provided that ψ' is interpreted as $\exp(i\theta\alpha_x/2)\psi$. Hence, the Dirac equation in the K frame transforms into a similar equation in the K' frame, the wave function transforming as:

$$\psi \longrightarrow \psi' = \exp(i\theta\alpha_x/2)\psi. \tag{8.69}$$

Although the Dirac equation itself is invariant to the transformation, it is still necessary to show that the physical interpretation of its solutions is also independent of the frame of reference used. For example $\psi^+\psi$ is the probability density in the K frame, so that $\psi'^+\psi'$ should be the probability density in the K' frame. (The expression $\psi^+\psi$ is the scalar product of ψ, the four-component column vector (Equation 8.31), and its Hermitian conjugate ψ^+. The Hermitian conjugate of a matrix is obtained by taking its complex conjugate followed by transposition, so that ψ^+ is the four-component row vector (ψ_1^*, ψ_2^*, ψ_3^*, ψ_4^*); in general the Hermitian conjugate of a matrix with elements R_{ij} is the matrix with elements R_{ji}^*. For future use we note that Hermitian conjugation of a product matrix results in a change of order of the factors, that is $(AB)^+ = B^+A^+$, as may be confirmed by expanding the matrices in terms of their elements.) The two quantities, $\psi^+\psi$ and $\psi'^+\psi'$, are not expected to be the same in the two frames, since $\psi^+\psi$ is the probability of finding the electron at a point in space in the K frame and the Lorentz transformation mixes the space and time coordinates. Thus, the specific Lorentz transformation under discussion would change a volume element $\mathrm{d}x\mathrm{d}y\mathrm{d}z$ in the K frame to a mixture of $\mathrm{d}x'\mathrm{d}y'\mathrm{d}z'$ and $\mathrm{d}t'\mathrm{d}y'\mathrm{d}z'$ in the K' frame. That is, the probability density in the unprimed frame transforms into a mixture of the probability density and the x' component of the probability current density in the primed frame; the probability current density is just the probability of the electron crossing unit area in unit time. Thus, we have to show that $\psi^+\psi$ transforms like the time coordinate t, when subjected to a Lorentz transformation. Under the transformation:

$$\begin{aligned}\psi^+\psi \longrightarrow \psi'^+\psi' &= [\exp(i\theta\alpha_x/2)\psi]^+ \exp(i\theta\alpha_x/2)\psi \\ &= \psi^+ \exp(i\theta\alpha_x/2)\exp(i\theta\alpha_x/2)\psi, \end{aligned}\tag{8.70}$$

since α_x is real (Equations 8.26) and Equation 8.61 shows that θ is pure imaginary. Hence:

$$\psi^+\psi \longrightarrow \psi^+\exp(i\theta\alpha_x)\psi = \psi^+\psi\cos\theta + i\psi^+\alpha_x\psi\sin\theta, \tag{8.71}$$

and $\psi^+\psi$ does indeed transform like the coordinate t, while comparison with Equations 8.60 shows that it is $\psi^+c\alpha_x\psi$ that transforms like the space coordinate x; it will be seen in Section 8.10 that $c\boldsymbol{\alpha}$ is the electron's velocity operator, so that $\psi^+c\boldsymbol{\alpha}\psi$ is indeed the probability current density.

Finally, it must be demonstrated that the electron's charge is conserved, that is the change in the charge density in an infinitesimal volume element must be compensated by a flow of charge across the faces of the volume element. This conservation condition may be expressed as (Equation 2.55):

$$(\partial\rho/\partial t) + \boldsymbol{\nabla}.\mathbf{j} = 0. \tag{8.72}$$

In the present connection the charge density ρ and the current density $\mathbf{j}$ both differ by a factor $-e$ from the probability density $\psi^+\psi$ and the probability current density $\psi^+ c\boldsymbol{\alpha}\psi$, respectively, so that Equation 8.72 may be rewritten as:

$$\partial(\psi^+\psi)/\partial t + \boldsymbol{\nabla}.(\psi^+ c\boldsymbol{\alpha}\psi) = 0. \tag{8.73}$$

That the relationship is obeyed may be demonstrated by multiplying the Dirac equation 8.34 on the left by ψ^+:

$$\psi^+[(i\hbar/c)(\partial\psi/\partial t) + (e/c)\phi\psi] - \psi^+\boldsymbol{\alpha}.[-i\hbar\boldsymbol{\nabla}\psi + e\mathbf{A}\psi] - \beta mc\psi^+\psi = 0. \tag{8.74}$$

The Hermitian conjugate of this equation is:

$$[-(i\hbar/c)(\partial\psi^+/\partial t) + \psi^+(e/c)\phi]\psi - [i\hbar(\boldsymbol{\nabla}\psi^+) + \psi^+ e\mathbf{A}].\boldsymbol{\alpha}\psi - \beta mc\psi^+\psi = 0, \tag{8.75}$$

where use has been made of the fact that the components of $\boldsymbol{\alpha}$ are Hermitian so that $(\psi^+\boldsymbol{\alpha})^+ = \boldsymbol{\alpha}\psi$. By subtracting Equation 8.75 from Equation 8.74 and dividing by $(i\hbar/c)$:

$$\psi^+(\partial\psi/\partial t) + (\partial\psi^+/\partial t)\psi + \psi^+ c\boldsymbol{\alpha}.(\boldsymbol{\nabla}\psi) + (\boldsymbol{\nabla}\psi^+).c\boldsymbol{\alpha}\psi = 0, \tag{8.76}$$

which is merely an expanded form of Equation 8.73. Thus, the Dirac equation is consistent with the conservation of charge.

8.6 The negative energy solutions – positrons

We have already seen that the Dirac equation for a free electron has four solutions, which is twice as many as is needed to explain spin. This multiplicity of solutions arose because half the solutions have negative energies. In fact, this happens in any relativistic theory, even classical ones, since the energy always involves a square root (see for example Equation 8.1), which can have both positive and negative values. Classically the positive square root is taken, the presence of negative solutions causing no embarrassment. This is because classically the energy must vary continuously, and energies between $+mc^2$ and $-mc^2$ are not permitted, so that, if the energy is positive at one time it cannot become negative at a later time. However, quantum mechanically discontinuous transitions can occur so that the negative energy solutions cannot be ignored.

The solutions of the Dirac equation are represented by a wave function which is a four-component vector or spinor:

$$\psi = \begin{pmatrix} \psi_1 \\ \psi_2 \\ \psi_3 \\ \psi_4 \end{pmatrix} \tag{8.77}$$

Since the dominant term in the equation is βmc^2 and β is given by:

$$\beta = \begin{pmatrix} 1 & 0 & 0 & 0 \\ 0 & 1 & 0 & 0 \\ 0 & 0 & -1 & 0 \\ 0 & 0 & 0 & -1 \end{pmatrix} \tag{8.78}$$

in the matrix representation that we are using, ψ_1 and ψ_2 may be thought of as positive energy components and ψ_3 and ψ_4 as negative energy components. For solutions with positive energy the main components will be ψ_1 and ψ_2, but small amounts of ψ_3 and ψ_4 will be mixed in by the terms $\boldsymbol{\alpha}.\boldsymbol{\pi}$ which couple positive and negative energy components. In this section the relationship between the positive energy solutions and the negative energy solutions will be examined, and it will be shown that if the positive energy solutions correspond to a particle with electric charge $-e$, then the negative energy solutions may be thought of as being for a particle with the same mass, but with charge $+e$.

The Dirac equation for a particle with charge $-e$:

$$[p_0 + (e/c)\phi - \boldsymbol{\alpha}.(\mathbf{p} + e\mathbf{A}) - \beta mc]\psi = 0 \tag{8.79}$$

has as its complex conjugate:

$$[-p_0 + (e/c)\phi - \boldsymbol{\alpha}^*.(-\mathbf{p} + e\mathbf{A}) - \beta mc]\psi^* = 0; \tag{8.80}$$

although β, α_x and α_z are real, α_y is imaginary (Equations 8.26) so that the complex conjugate of $\boldsymbol{\alpha}$ must be written as $\boldsymbol{\alpha}^*$. A new real matrix γ (not to be confused with the number γ in Equations 8.59) is now introduced:

$$\gamma = i\beta\alpha_y. \tag{8.81}$$

Using the anticommutation relations (Equation 8.11) of β and the components of $\boldsymbol{\alpha}$, together with the fact that $\alpha_y^* = -\alpha_y$, the following properties of γ may be demonstrated:

$$\gamma\beta = -\beta\gamma; \qquad \gamma\boldsymbol{\alpha}^* = \boldsymbol{\alpha}\gamma. \tag{8.82}$$

Multiplication of Equation 8.80 on the left by $-\gamma$ followed by the use of these relations leads to:

$$[p_0 - (e/c)\phi - \boldsymbol{\alpha}.(\mathbf{p} - e\mathbf{A}) - \beta mc]\gamma\psi^* = 0. \tag{8.83}$$

Comparison of this equation with Equation 8.79 shows that, if ψ is a solution of the Dirac equation, then $\gamma\psi^*$ is a solution of the same equation, except that the sign of the charge has been reversed. That is, if ψ is a solution of the Dirac equation for a particle with charge $-e$ subject to electromagnetic potentials, then $\gamma\psi^*$ is a solution of the Dirac equation for a particle of charge $+e$ subject to the same electromagnetic potentials. For this reason γ is referred to as the charge conjugation operator.

It can easily be shown that if ψ is a positive energy solution then $\gamma\psi^*$ is a negative energy solution, since from Equations 8.26, 8.77 and 8.81:

$$\gamma\psi^* = i\beta\alpha_y\psi^* = \begin{pmatrix} 0 & 0 & 0 & 1 \\ 0 & 0 & -1 & 0 \\ 0 & -1 & 0 & 0 \\ 1 & 0 & 0 & 0 \end{pmatrix}\begin{pmatrix} \psi_1^* \\ \psi_2^* \\ \psi_3^* \\ \psi_4^* \end{pmatrix} = \begin{pmatrix} \psi_4^* \\ -\psi_3^* \\ -\psi_2^* \\ \psi_1^* \end{pmatrix}. \tag{8.84}$$

ψ being a positive energy solution, ψ_1 and ψ_2 are the main components, so that for $\gamma\psi^*$ it is the negative energy components that predominate and $\gamma\psi^*$ is a negative energy solution.

This demonstrates that the negative energy solutions of the Dirac equation may be thought of as corresponding to a new particle having the same mass as the electron, but opposite charge. These particles are given the name positrons and were first suggested by Dirac in 1930. Their discovery in 1933 by Anderson can be regarded as a triumph of the Dirac theory. Since then a number of other so-called antiparticles have been recognised.

However, there are still a number of problems associated with the concept of positrons and the situation cannot be as simple as it has been described above, since the theory so far implies that a positron has a negative kinetic energy. Dirac avoided this difficulty by supposing that nearly all the negative energy states are occupied with one electron in each state, in accordance with the Pauli exclusion principle. An unoccupied negative energy state then appears as having positive energy, since to remove it an electron with negative energy must be introduced. Dirac suggested that these unoccupied negative energy states or holes are, in fact, positrons. This picture suggests that, if an electron makes a transition from a positive energy state to a negative energy state, assuming that there is one unoccupied, then this would be interpreted as the simultaneous disappearance of both an electron and a positron with the release of an enormous amount of energy as electromagnetic radiation. The reverse process would involve the creation of an electron and a positron out of electromagnetic radiation. It can be seen that this theory is symmetrical between electrons and positrons, and that a theory using the positron as the basic particle could be developed. These ideas lead to the concept of antimatter universes.

Although this assumption of occupied negative energy states solves some problems, it introduces further difficulties, since it implies that there is an infinitely dense distribution of electrons everywhere, and that for a perfect vacuum all the negative energy states are filled while all the positive energy states are unoccupied. For Maxwell's equations to hold it is necessary to assume that this infinitely dense distribution of electrons does not contribute to electromagnetic fields, and that Maxwell's equations only apply to departures from the vacuum state.

It can thus be seen that this theory is by no means satisfactory, but all the difficulties can be avoided by using quantum electrodynamics. However, chemistry is concerned with positive energy states and this topic will be pursued no further.

8.7 The non-relativistic approximation of the Dirac equation

The Dirac equation is an appropriate relativistic quantum mechanical equation for the electron. However, it suffers from the disadvantage of being a four-component equation; that is, it involves four-by-four matrices and the wave function is a four-component vector. In this it differs from non-relativistic equations, whose wave functions have two components corresponding to the two possible orientations of the electron's spin. The Dirac equation has, in addition, two negative energy components, which are coupled to the positive energy components. Since chemistry is concerned only with the positive energy solutions it would be advantageous to decouple the positive and negative energy components to give a two-component equation for the positive energy solutions which would be similar to the non-relativistic equations and could be used in the same way; by decoupling the positive and negative energy components the Dirac equation would in effect be reduced to a non-relativistic form. Two methods of performing this decoupling will be examined. In the next section the method of small components is briefly considered, since this is the method normally used in the few places that the Dirac equation is discussed in the chemical literature, and a number of texts do not appreciate an important point. The Foldy-Wouthuysen approach is described in detail in Section 8.9.

In principle both the methods may be used to decouple the positive and negative energy solutions completely, but in practice this is only possible in special cases (see Section 8.10). In both methods the Hamiltonian is obtained as a power series, which converges for electrons moving with velocities much less than the speed of light. Thus there are a number of problems shared by the two methods, the choice of expansion parameter for the power series, the estimation of the size of the various terms in the series and the decision on how far to take the expansion. The remainder of this section will be devoted to these questions.

A number of quantities have been used as the power series expansion parameter, c^{-1}, (u/c) and α, where u is the speed of the electron and α is the fine structure constant (not to be confused with the Dirac operator $\boldsymbol{\alpha}$ or its components). The quantity c^{-1} is frequently used, but it is not really suitable, since it is not dimensionless and consequently its value depends on the system of units employed. Both (u/c) and α are dimensionless and we will shortly see that they are in a sense equivalent. However, we will make the more professional choice of the fine structure constant α, which is defined by:

$$\alpha = e^2\mu_0 c/2h = e^2/2h\epsilon_0 c = 7{\cdot}29735\ (\pm\ 0{\cdot}00001) \times 10^{-3}; \quad (8.85)$$

it is a fundamental constant of importance in atomic spectroscopy and, since it is dimensionless, its value is independent of the units used.

Now the largest term in the Dirac Hamiltonian (Equation 8.37) is mc^2, so that it is appropriate to estimate the magnitude of other terms relative to this. It will be seen that the order of magnitude of each of the other terms in the Hamiltonian, $e\phi$, $c\boldsymbol{\alpha}.\mathbf{p}$ and $ce\boldsymbol{\alpha}.\mathbf{A}$, can be expressed as $mc^2\alpha^n$. Since chemists are concerned with atoms and molecules, it is appropriate to consider their magnitude in an atomic or molecular environment and for this purpose we choose the simplest possible situation for an electron, the hydrogen atom.

In the first Bohr orbit the electron's speed is $u = e^2/2\epsilon_0 h = c\alpha$, and it is now seen that α and (u/c) are comparable expansion parameters; in addition the radius of the first Bohr orbit is $a_0 = 4\pi\epsilon_0\hbar^2/me^2$. Thus, classically:

$$c\boldsymbol{\alpha}.\mathbf{p} \sim cmu = mc^2\alpha, \quad (8.86)$$

while from Equation 5.61:

$$e\phi \sim e^2/4\pi\epsilon_0 a_0 = mc^2\alpha^2. \quad (8.87)$$

The magnitude of the electric field encountered by an electron in the first Bohr orbit is, from Equation 5.60, $e/4\pi\epsilon_0 a_0^2 \sim 5 \times 10^{11}\,\mathrm{Vm}^{-1}$, which is considerably larger than electric fields found in the laboratory; consequently, external electric fields do not affect this estimate that $e\phi$ is of order $mc^2\alpha^2$.

To find the order of magnitude of the term $ce\boldsymbol{\alpha}.\mathbf{A}$, we use the magnitude $eu/4\pi\epsilon_0 c^2 a_0$ (Equations 5.72) of the magnetic potential at the proton due to the motion of an electron in the first Bohr orbit. Hence:

$$ce\boldsymbol{\alpha}.\mathbf{A} \sim e^2u/4\pi\epsilon_0 ca_0 = mc^2\alpha^3. \quad (8.88)$$

The contribution to $ce\boldsymbol{\alpha}.\mathbf{A}$ from the electron spin magnetic moment is of the same magnitude, while nuclear contributions are smaller than this by a factor of the order of (m/M), where M is the mass of the proton. In addition, the internal magnetic field in the hydrogen atom is of the order of $eu/4\pi\epsilon_0 c^2 a_0^2 \sim$ 10T (10^5G), which is an order of magnitude larger than external magnetic fields normally realizable in the laboratory. It is concluded that $ce\boldsymbol{\alpha}.\mathbf{A}$ is of the order of $mc^2\alpha^3$.

The two parts of $c\boldsymbol{\alpha}.\boldsymbol{\pi} = c\boldsymbol{\alpha}.(\mathbf{p} + e\mathbf{A})$ are thus of different orders of magnitude and could be treated separately, but it is advantageous to keep them together so that any resulting Hamiltonian is gauge invariant (Section 7.2). In any application of the Hamiltonian negligible terms in **A** may be omitted after the gauge has been chosen and the magnitude of these terms has been estimated.

It will be seen that the magnitude of any term that arises in the expansion of the Dirac Hamiltonian in a non-relativistic form can be estimated from these results. It has been noted that in principle the positive and negative energy solutions can be completely decoupled, but in practice the manipulations involved become tedious in going beyond an approximation correct to order $mc^2\alpha^4$. However, this is not a serious disadvantage, since higher order corrections are in general not at present observable and, in addition, at this order radiative corrections become important and a quantum electrodynamic treatment becomes essential. The positive and negative energy states can be completely separated in some specific situations; the free electron is one of these special cases and is treated in detail in Section 8.10, since it gives an insight into the connection between the Dirac equation and its non-relativistic approximation.

8.8 The method of small components

In the method of small components the starting point is the time-dependent Dirac equation 8.34. Since it is an electron with positive energy that is of interest, we remove the rest mass energy for positive energy states ($+\,mc^2$) by introducing an appropriate phase factor into the wave function:

$$\psi \longrightarrow \psi' = \exp(-imc^2t/\hbar)\psi. \quad (8.89)$$

Substituting in Equation 8.34 the Dirac equation then becomes:

$$[p_0 + (e/c)\phi - (\boldsymbol{\alpha}.\boldsymbol{\pi}) - (\beta - 1)mc]\psi = 0. \quad (8.90)$$

The wave function ψ is still a four-component vector (Equation 8.31), but it can be rewritten as:

$$\psi = \begin{pmatrix} \psi_+ \\ \psi_- \end{pmatrix}, \tag{8.91}$$

where ψ_+ and ψ_- are two-component vectors:

$$\psi_+ = \begin{pmatrix} \psi_1 \\ \psi_2 \end{pmatrix}; \qquad \psi_- = \begin{pmatrix} \psi_3 \\ \psi_4 \end{pmatrix}; \tag{8.92}$$

ψ_+ has as its components the positive energy components, ψ_1 and ψ_2, while ψ_- has ψ_3 and ψ_4, the negative energy components. The solutions of interest are those in which the positive energy components ψ_+ predominate, and consequently the two components of ψ_+ are called the large components while those of ψ_- are the small components; it is the latter that are to be eliminated. Using the substitution (Equation 8.91) and the explicit form of the Dirac matrices (Equations 8.26), Equation 8.90 can be rewritten in terms of the two-by-two Pauli spin matrices (Equations 8.13):

$$\left.\begin{aligned} [p_0 + (e/c)\phi]\psi_+ - (\boldsymbol{\sigma}.\boldsymbol{\pi})\psi_- &= 0 \\ [p_0 + (e/c)\phi]\psi_- - (\boldsymbol{\sigma}.\boldsymbol{\pi})\psi_+ + 2mc\psi_- &= 0 \end{aligned}\right\}. \tag{8.93}$$

These equations are coupled and each involves two-by-two matrices and two-component wave functions.

The second of these coupled equations can be used to express the small components ψ_- in terms of the large components ψ_+:

$$\psi_- = [2mc + p_0 + (e/c)\phi]^{-1}(\boldsymbol{\sigma}.\boldsymbol{\pi})\psi_+. \tag{8.94}$$

The presence of $2mc$ in the denominator justifies the description of ψ_- as the small components. Substitution of Equation 8.94 into the first of the Equations 8.93 gives an equation involving only ψ_+:

$$[p_0 + (e/c)\phi - (1/2mc)(\boldsymbol{\sigma}.\boldsymbol{\pi})K(\boldsymbol{\sigma}.\boldsymbol{\pi})]\psi_+ = 0, \tag{8.95}$$

where the Hermitian operator K is given by:

$$K = [1 + (p_0 + (e/c)\phi)/2mc]^{-1}. \tag{8.96}$$

Identifying cp_0 with $i\hbar(\partial/\partial t)$ (Equation 8.5), Equation 8.95 can be written as:

$$\mathcal{H}\psi_+ = i\hbar(\partial\psi_+/\partial t) \tag{8.97}$$

with the Hamiltonian given by:

$$\mathcal{H} = -e\phi + (1/2m)(\boldsymbol{\sigma}.\boldsymbol{\pi})K(\boldsymbol{\sigma}.\boldsymbol{\pi}). \tag{8.98}$$

For the case of a slowly moving electron K can be approximated by unity and the Hamiltonian becomes:

$$\mathcal{H} = -e\phi + (1/2m)\pi^2 + (e\hbar/2m)(\boldsymbol{\sigma}.\mathbf{B}), \tag{8.99}$$

where we have employed manipulations similar to those used in obtaining Equation 8.45 from Equation 8.38. Again we note that the Dirac equation leads to a term in the Hamiltonian that demonstrates that the electron has an intrinsic magnetic moment. The non-relativistic Hamiltonian (Equation 8.99) has already been obtained in Section 8.4, but the present method can be used to obtain further relativistic correction terms by expanding K as a power series. However, the manipulations involved may be facilitated by two results, which will first be derived.

For any operator G, we may write:

$$[G, K] = K[K^{-1}, G]K = K[p_0 + (e/c)\phi, G]K/2mc, \tag{8.100}$$

where Equation 8.96 has been substituted for K^{-1}. This relationship may be used to move the operator K to the right in any operator expression, and this is done at the expense of introducing the additional but smaller term given in Equation 8.100; in practice the new terms are two orders of magnitude smaller in α and the convergence is quite dramatic, as will be seen.

When K has been moved to the right-hand end of an operator expression, some sort of substitution is needed for it. This may be found by noting that in this problem the Hamiltonian (Equation 8.98) only operates on ψ_+ so that an approximation for $K\psi_+$ is needed. From Equation 8.96:

$$p_0 + (e/c)\phi = 2mc(K^{-1} - 1), \tag{8.101}$$

and substitution of this into Equation 8.95 followed by premultiplication throughout by K gives, after rearrangement:

$$K\psi_+ = [1 - K(\boldsymbol{\sigma}.\boldsymbol{\pi})K(\boldsymbol{\sigma}.\boldsymbol{\pi})/4m^2c^2]\psi_+. \tag{8.102}$$

This may be expanded by using Equation 8.100 and Equation 8.102 itself, together with the identity:

$$[p_0 + (e/c)\phi, \boldsymbol{\sigma}.\boldsymbol{\pi}] = -(i\hbar e/c)(\boldsymbol{\sigma}.\mathbf{E}), \tag{8.103}$$

which is obtained with the help of Equation 5.21. To the order of accuracy that we require the result is:

$$K\psi_+ = [1 - (\boldsymbol{\sigma}.\boldsymbol{\pi})^2/4m^2c^2 + O(\alpha^4)]\psi_+. \tag{8.104}$$

Equations 8.100, 8.103 and 8.104 may now be used to expand the Hamiltonian (Equation 8.98), which becomes after some manipulation:

$$\mathcal{H} = -e\phi + (\boldsymbol{\sigma}.\boldsymbol{\pi})^2/2m - (\boldsymbol{\sigma}.\boldsymbol{\pi})^4/8m^3c^2$$
$$+ (i\hbar e/4m^2c^2)(\boldsymbol{\sigma}.\boldsymbol{\pi})(\boldsymbol{\sigma}.\mathbf{E}) + O(mc^2\alpha^6); \qquad (8.105)$$

the order of magnitude of the terms has been estimated using the information given in the previous section. This Hamiltonian cannot be correct, since the last term is not Hermitian. However, this defect may be remedied by noting that ψ_+, the wave function associated with Equation 8.105, is not normalized; for the Dirac equation it is ψ (Equation 8.91) that is normalized. To obtain a Hermitian Hamiltonian, renormalization is essential and it is this necessity that is not appreciated in some texts.

Renormalization may be accomplished by introducing a new normalized two-component wave function χ_+:

$$\chi_+ = N\psi_+. \qquad (8.106)$$

The operator N may be determined by requiring the probability density associated with χ_+ to be the same as that for the four-component wave function ψ (Equation 8.91):

$$\psi^+\psi = \psi_+^+\psi_+ + \psi_-^+\psi_- = \chi_+^+\chi_+. \qquad (8.107)$$

Substitution for ψ_- (Equation 8.94), χ_+ (Equation 8.106) and use of Equation 8.96 gives:

$$N^+N = 1 + (\boldsymbol{\sigma}.\boldsymbol{\pi})K^2(\boldsymbol{\sigma}.\boldsymbol{\pi})/4m^2c^2, \qquad (8.108)$$

so that N is the Hermitian operator:

$$N = [1 + (\boldsymbol{\sigma}.\boldsymbol{\pi})K^2(\boldsymbol{\sigma}.\boldsymbol{\pi})/4m^2c^2]^{\frac{1}{2}}$$
$$= 1 + (\boldsymbol{\sigma}.\boldsymbol{\pi})^2/8m^2c^2 + O(\alpha^4). \qquad (8.109)$$

The wave function ψ_+ satisfies the time-dependent Schrödinger equation 8.97, but as we have seen the Hamiltonian $\mathcal{H}$ is not Hermitian. The new normalized wave function χ_+ satisfies a similar equation:

$$\mathcal{H}'\chi_+ = i\hbar(\partial\chi_+/\partial t), \qquad (8.110)$$

but the transformed Hamiltonian $\mathcal{H}'$ is expected to be Hermitian. The Hamiltonian $\mathcal{H}'$ can be related to $\mathcal{H}$ by substituting the inverse of Equation 8.106, $\psi_+ = N^{-1}\chi_+$, into Equation 8.97:

$$\mathcal{H}N^{-1}\chi_+ = i\hbar(\partial(N^{-1}\chi_+)/\partial t)$$
$$= i\hbar N^{-1}(\partial\chi_+/\partial t) + i\hbar[\partial/\partial t, N^{-1}]\chi_+. \qquad (8.111)$$

Premultiplication by the operator N followed by rearrangement now gives Equation 8.110 with $\mathcal{H}'$ given by:

$$\mathcal{H}' = N\mathcal{H}N^{-1} - i\hbar N[\partial/\partial t, N^{-1}]. \tag{8.112}$$

To obtain the final Hamiltonian for the positive energy states it only remains to substitute for N (Equation 8.109) and $\mathcal{H}$ (Equation 8.105) in Equation 8.112 and to perform some manipulations. Since N commutes with $(\boldsymbol{\sigma}.\boldsymbol{\pi})$ and the result is only required to be accurate to order $mc^2\alpha^4$, we may write:

$$\begin{aligned}\mathcal{H}' &= \mathcal{H} + N[-e\phi, N^{-1}] - i\hbar N[\partial/\partial t, N^{-1}] + O(mc^2\alpha^6)\\ &= \mathcal{H} - cN[p_0 + (e/c)\phi, N^{-1}] + O(mc^2\alpha^6)\\ &= \mathcal{H} + N[p_0 + (e/c)\phi, (\boldsymbol{\sigma}.\boldsymbol{\pi})^2]/8m^2c + O(mc^2\alpha^6).\end{aligned} \tag{8.113}$$

Now the commutator involved can be evaluated by noting that:

$$\begin{aligned}[p_0 + (e/c)\phi, (\boldsymbol{\sigma}.\boldsymbol{\pi})^2] &= [p_0 + (e/c)\phi, (\boldsymbol{\sigma}.\boldsymbol{\pi})](\boldsymbol{\sigma}.\boldsymbol{\pi})\\ &\quad + (\boldsymbol{\sigma}.\boldsymbol{\pi})[p_0 + (e/c)\phi, (\boldsymbol{\sigma}.\boldsymbol{\pi})]\end{aligned} \tag{8.114}$$

and using Equation 8.103, so that:

$$\mathcal{H}' = \mathcal{H} - (i\hbar e/8m^2c^2)[(\boldsymbol{\sigma}.\mathbf{E})(\boldsymbol{\sigma}.\boldsymbol{\pi}) + (\boldsymbol{\sigma}.\boldsymbol{\pi})(\boldsymbol{\sigma}.\mathbf{E})] + O(mc^2\alpha^6). \tag{8.115}$$

The Hamiltonian appropriate to the normalized two-component wave function χ_+ is thus:

$$\begin{aligned}\mathcal{H}' &= -e\phi + (\boldsymbol{\sigma}.\boldsymbol{\pi})^2/2m - (\boldsymbol{\sigma}.\boldsymbol{\pi})^4/8m^3c^2\\ &\quad - (i\hbar e/8m^2c^2)[(\boldsymbol{\sigma}.\mathbf{E}), (\boldsymbol{\sigma}.\boldsymbol{\pi})] + O(mc^2\alpha^6).\end{aligned} \tag{8.116}$$

As anticipated this Hamiltonian is Hermitian. It could now be expanded further to give a Hamiltonian that may be compared directly with the usual non-relativistic Hamiltonian, but this is postponed until the Foldy-Wouthuysen method has been discussed in the next section; in addition the physical interpretation of the individual terms is not given till then.

8.9 The Foldy-Wouthuysen transformation

In this section the positive and negative energy states are decoupled using a method due to Foldy and Wouthuysen. The non-relativistic Hamiltonian is obtained by a series of unitary transformations, by which the coupling terms are progressively eliminated. However, before the details of the transformation are considered it is necessary to look more closely at the properties of the coupling terms.

The Dirac equation may be written as $\mathcal{H}\psi = i\hbar(\partial\psi/\partial t)$, where ψ is a four-component vector and $\mathcal{H}$ is the Dirac Hamiltonian:

$$\begin{aligned}\mathcal{H} &= \beta mc^2 - e\phi + c\boldsymbol{\alpha}.\boldsymbol{\pi} \\ &= \beta mc^2 + \mathcal{E} + \mathcal{O}. \end{aligned} \tag{8.117}$$

The symbol $\mathcal{E}$ $(= -e\phi)$ represents an even operator, that is one that has no matrix elements between positive and negative energy components, while $\mathcal{O}$ $(= c\boldsymbol{\alpha}.\boldsymbol{\pi})$ is an odd operator having only matrix elements between positive and negative energy components. The term βmc^2 is an even operator, but it is expedient to keep it separate from the other even operator $(-e\phi)$, since they are of different orders of magnitude. This is also true of the two parts of $c\boldsymbol{\alpha}.\boldsymbol{\pi}$, but keeping them together ensures the gauge invariance of any resulting Hamiltonian.

Whether an operator is even or odd is determined by the matrix involved in its representation. Operators involving β or the unit matrix, which are both diagonal, are even ($\mathcal{E}$), while those involving α_x, α_y or α_z (Equations 8.26) are odd ($\mathcal{O}$). Other matrices encountered in this chapter can also be classified as even or odd, σ_x, σ_y and σ_z being even and ρ being odd. The conditions for an operator to be even or odd can be expressed more formally; an even operator must commute with β:

$$[\mathcal{E}, \beta] = 0, \tag{8.118}$$

while an odd operator must anticommute with β:

$$[\mathcal{O}, \beta]_+ = 0. \tag{8.119}$$

From these conditions it is easy to show that the product of two even operators or two odd operators must be even, but that the product of an even operator with an odd operator is odd. These facts will be used later. Finally, it may be shown that any operator G can always be broken down into the sum of an even and an odd operator:

$$G = \tfrac{1}{2}(G + \beta G\beta) + \tfrac{1}{2}(G - \beta G\beta); \tag{8.120}$$

conditions 8.118 and 8.119 show that, since $\beta^2 = 1$, the first part of Equation 8.120 is even, while the second part is odd.

In the Foldy-Wouthuysen transformation the odd operator $\mathcal{O}$ in the Dirac Hamiltonian (Equation 8.117) is removed by a unitary transformation. Other odd operators will be introduced by the transformation, but provided it is chosen correctly, they will be of higher order in α. This process can be repeated to eliminate the new odd operators until the Hamiltonian contains no odd operators to the desired order in α; the following discussion is aimed at a Hamiltonian correct to order $mc^2\alpha^4$.

A unitary transformation may be applied to the wave function ψ to give a new wave function ψ':

$$\psi' = \exp(iS)\psi, \tag{8.121}$$

where S is a Hermitian operator and by $\exp(iS)$ is meant the power series expansion:

$$\exp(iS) = 1 + iS + (i^2/2!)S^2 + \ldots; \tag{8.122}$$

the inverse of Equation 8.121 is just $\psi = \exp(-iS)\psi'$. The Hamiltonian must also be transformed and reference to the preceding section and Equations 8.106 and 8.112 in particular shows that the new Hamiltonian is:

$$\mathcal{H}_1 = \exp(iS)\mathcal{H}\exp(-iS) - i\hbar\exp(iS)[\partial/\partial t, \exp(-iS)]. \tag{8.123}$$

This may be expanded as a series by substituting Equation 8.122:

$$\begin{aligned}\mathcal{H}_1 &= (1 + iS + (i^2/2!)S^2 + \ldots)\mathcal{H}(1 - iS + (i^2/2!)S^2 + \ldots)\\ &\quad - i\hbar(1 + iS + (i^2/2!)S^2 + \ldots)[\partial/\partial t, (1 - iS + (i^2/2!)S^2 + \ldots)]\\ &= \mathcal{H} + i(S\mathcal{H} - \mathcal{H}S) + (i^2/2!)(S^2\mathcal{H} - 2S\mathcal{H}S + \mathcal{H}S^2) + \ldots\\ &\quad - \hbar\{[\partial/\partial t, S] + (i/2)(S[\partial/\partial t, S] - [\partial/\partial t, S]S)\\ &\quad + (i^2/2!)(S^2[\partial/\partial t, S] - 2S[\partial/\partial t, S]S + [\partial/\partial t, S]S^2)/3 + \ldots\}\\ &= \mathcal{H} - \hbar[\partial/\partial t, S] + i[S, \mathcal{H} - \tfrac{1}{2}\hbar[\partial/\partial t, S]]\\ &\quad + (i^2/2!)[S, [S, \mathcal{H} - (\hbar/3)[\partial/\partial t, S]]] + \ldots;\end{aligned} \tag{8.124}$$

a developing pattern of nested commutators is easily detected if further terms in this expansion are needed. In the above manipulations it must always be remembered that S is an operator yet to be specified and must not be assumed to commute with anything.

If S has the specific form:

$$S = -i\beta\mathcal{O}/2mc^2, \tag{8.125}$$

the first commutator in Equation 8.124 contributes a term:

$$\begin{aligned}i[S, \beta mc^2] &= \tfrac{1}{2}[\beta\mathcal{O}, \beta]\\ &= \tfrac{1}{2}(\beta\mathcal{O}\beta - \beta^2\mathcal{O})\\ &= \tfrac{1}{2}(-\beta^2\mathcal{O} - \beta^2\mathcal{O})\\ &= -\mathcal{O},\end{aligned} \tag{8.126}$$

where the facts that $\mathcal{O}$ anticommutes with β (Equation 8.119) and $\beta^2 = 1$ have been used. Thus, with this choice of S part of the first commutator cancels the odd operator appearing in $\mathcal{H}$, the first term of Equation 8.124. Complete substitution for S and $\mathcal{H}$ in Equation 8.124 now gives the transformed Hamiltonian; in doing this the expressions can be simplified so that β only appears linearly and only terms up to order $mc^2\alpha^4$ are retained, the order being determined on the basis $\mathcal{E} \sim mc^2\alpha^2$ and $\mathcal{O} \sim mc^2\alpha$. The result may be written as:

$$\mathcal{H}_1 = \beta mc^2 + \mathcal{E}_1 + \mathcal{O}_1, \qquad (8.127)$$

where

$$\mathcal{E}_1 = \mathcal{E} + \beta\mathcal{O}^2/2mc^2 - [\mathcal{O}, [i\hbar\partial/\partial t - \mathcal{E}, \mathcal{O}]]/8m^2c^4$$
$$- \beta\mathcal{O}^4/8m^3c^6 + O(mc^2\alpha^6) \qquad (8.128)$$

and

$$\mathcal{O}_1 = \beta[i\hbar\partial/\partial t - \mathcal{E}, \mathcal{O}]/2mc^2 - \mathcal{O}^3/3m^2c^4 + O(mc^2\alpha^5). \qquad (8.129)$$

It can be seen that the explicit operators in $\mathcal{O}_1$, the odd part of the transformed Hamiltonian, are both of order $mc^2\alpha^3$. Thus, the odd operator $\mathcal{O}$ in the Dirac equation, which is of order $mc^2\alpha$, has been removed by the transformation at the expense of introducing odd operators of order $mc^2\alpha^3$ and higher.

This process may be repeated to remove the new odd operators. That is by using:

$$S_1 = -i\beta\mathcal{O}_1/2mc^2 \qquad (8.130)$$

another transformation may be performed on Equation 8.127 to remove the odd operator $\mathcal{O}_1$. The resulting Hamiltonian is:

$$\mathcal{H}_2 = \beta mc^2 + \mathcal{E}_2 + \mathcal{O}_2, \qquad (8.131)$$

where

$$\mathcal{E}_2 = \mathcal{E} + \beta\mathcal{O}^2/2mc^2 - [\mathcal{O}, [i\hbar\partial/\partial t - \mathcal{E}, \mathcal{O}]]/8m^2c^4$$
$$- \beta\mathcal{O}^4/8m^3c^6 + O(mc^2\alpha^6) \qquad (8.132)$$

and

$$\mathcal{O}_2 = O(mc^2\alpha^5). \qquad (8.133)$$

Only two applications of a unitary transformation have been necessary to reduce the Hamiltonian to one in which the largest odd terms are of order $mc^2\alpha^5$, and these may be removed by a further unitary transformation to give the Hamiltonian $\mathcal{H}_3$, which is correct to order $mc^2\alpha^5$. This reduced Hamiltonian $\mathcal{H}_3$ operates on the transformed wave function ψ'''. In the subsequent discussion, the primes and subscripts will be dropped, since it will always be obvious when the Hamiltonian is a non-relativistic approximation to the Dirac Hamiltonian. However, it should be noted that the wave function appropriate to a non-relativistic approximation is not the same as that corresponding to the Dirac Hamiltonian; they differ by a unitary transformation. Thus, the transformed Dirac equation is:

$$\mathcal{H}\psi = i\hbar(\partial\psi/\partial t), \tag{8.134}$$

where

$$\mathcal{H} = \beta mc^2 + \mathcal{E} + \beta\mathcal{O}^2/2mc^2 - [\mathcal{O}, [i\hbar\partial/\partial t - \mathcal{E}, \mathcal{O}]]/8m^2c^4$$
$$- \beta\mathcal{O}^4/8m^3c^6 + \mathrm{O}(mc^2\alpha^6). \tag{8.135}$$

It is now necessary to substitute $\mathcal{E} = -e\phi$ and $\mathcal{O} = c(\boldsymbol{\alpha}.\boldsymbol{\pi})$ into Equation 8.135. The first term for which the substitution is not trivial contains $\mathcal{O}^2$, but we saw earlier (Equations 8.39 to 8.45) that:

$$(\boldsymbol{\alpha}.\boldsymbol{\pi})^2 = (\boldsymbol{\sigma}.\boldsymbol{\pi})^2 = \pi^2 + e\hbar\boldsymbol{\sigma}.\mathbf{B}. \tag{8.136}$$

Thus

$$\beta\mathcal{O}^2/2mc^2 = \beta\pi^2/2m + (e\hbar/2m)\beta(\boldsymbol{\sigma}.\mathbf{B}). \tag{8.137}$$

From this result the term involving $\mathcal{O}^4$ is easily seen to be:

$$-\beta\mathcal{O}^4/8m^3c^6 = -\beta\pi^4/8m^3c^2 - e\hbar\beta(\boldsymbol{\sigma}.\mathbf{B})\pi^2/4m^3c^2 + \mathrm{O}(mc^2\alpha^8); \tag{8.138}$$

although the second term in this equation is formally of order $mc^2\alpha^6$, its consequences have been observed experimentally and we choose to retain it. For the remaining term in Equation 8.135, we need:

$$[i\hbar\partial/\partial t - \mathcal{E}, \mathcal{O}] = [i\hbar\partial/\partial t + e\phi, c\boldsymbol{\alpha}.\boldsymbol{\pi}]$$
$$= i\hbar ec\boldsymbol{\alpha}.(\nabla\phi + \partial\mathbf{A}/\partial t)$$
$$= -i\hbar ec\boldsymbol{\alpha}.\mathbf{E}, \tag{8.139}$$

where the electric field $\mathbf{E}$ is given by Equation 5.21; this result is similar to that quoted in the last section (Equation 8.103). Finally:

$$[\mathcal{O}, [i\hbar\partial/\partial t - \mathcal{E}, \mathcal{O}]] = -i\hbar ec^2[(\boldsymbol{\alpha}.\boldsymbol{\pi}), (\boldsymbol{\alpha}.\mathbf{E})] = -i\hbar ec^2[(\boldsymbol{\sigma}.\boldsymbol{\pi}), (\boldsymbol{\sigma}.\mathbf{E})]$$
$$= -i\hbar ec^2[\boldsymbol{\pi}.\mathbf{E} - \mathbf{E}.\boldsymbol{\pi} + i\boldsymbol{\sigma}.(\boldsymbol{\pi}_\wedge\mathbf{E} - \mathbf{E}_\wedge\boldsymbol{\pi})]$$
$$= -\hbar^2ec^2(\nabla.\mathbf{E}) + \hbar ec^2\boldsymbol{\sigma}.(\boldsymbol{\pi}_\wedge\mathbf{E} - \mathbf{E}_\wedge\boldsymbol{\pi}), \tag{8.140}$$

where the commutator has been expanded in a similar way to $(\boldsymbol{\alpha}.\boldsymbol{\pi})^2$ using Equations 8.39 and 8.40. Substitution of Equations 8.137, 8.138 and 8.140 into Equation 8.135 now gives:

$$\begin{aligned}\mathcal{H} = {} & \beta mc^2 - e\phi + \beta\pi^2/2m + (e\hbar/2m)\beta(\boldsymbol{\sigma}.\mathbf{B}) \\ & + \hbar^2 e(\nabla.\mathbf{E})/8m^2c^2 - \hbar e\boldsymbol{\sigma}.(\boldsymbol{\pi}\wedge\mathbf{E} - \mathbf{E}\wedge\boldsymbol{\pi})/8m^2c^2 \\ & - \beta\pi^4/8m^3c^2 - e\hbar\beta(\boldsymbol{\sigma}.\mathbf{B})\pi^2/4m^3c^2 + O(mc^2\alpha^6). \qquad (8.141)\end{aligned}$$

This Hamiltonian still involves four-by-four matrices and operates on a four-component wave function, but to order $mc^2\alpha^5$ there are no operators connecting positive and negative energy components. Thus, if Equation 8.141 and the explicit form of the matrix β (Equation 8.23) are substituted in Equation 8.134, we obtain two pairs of coupled equations, one for the positive energy states and one for the negative energy states. The Hamiltonian for the positive energy states may be obtained from Equation 8.141 simply by replacing β by $+1$, interpreting the components of $\boldsymbol{\sigma}$ as two-by-two matrices and remembering that it now operates on a two-component wave function. If, in addition, $\boldsymbol{\sigma}$ is replaced by 2s, the notation of non-relativistic quantum mechanics, the final Hamiltonian is:

$$\left.\begin{aligned}\mathcal{H} = {} & mc^2 - e\phi + \pi^2/2m && \text{(a)} \\ & + (e\hbar/m)(\mathbf{s}.\mathbf{B}) && \text{(b)} \\ & - (e\hbar/4m^2c^2)\mathbf{s}.(\boldsymbol{\pi}\wedge\mathbf{E} - \mathbf{E}\wedge\boldsymbol{\pi}) && \text{(c)} \\ & - (1/8m^3c^2)\pi^4 - (e\hbar/2m^3c^2)(\mathbf{s}.\mathbf{B})\pi^2 && \text{(d)} \\ & + (e\hbar^2/8m^2c^2)(\nabla.\mathbf{E}) && \text{(e)} \\ & + O(mc^2\alpha^6). &&\end{aligned}\right\} \qquad (8.142)$$

The appropriate Hamiltonian for the negative energy states may be obtained from Equation 8.141 by replacing β by -1.

The Hamiltonian 8.142 for the positive energy states could have been derived using the method of small components. Indeed, the Hamiltonian obtained by substitution of Equations 8.136 and 8.140 into Equation 8.116 of the previous section only differs from Equation 8.142 by the rest energy mc^2, but it will be recalled that in the method of small components only the positive energy states were considered and the energy zero was adjusted. In fact, the method of small components can also be used to obtain the non-relativistic Hamiltonian for the negative energy states and there is no reason to suppose that this method does not give the same results as the Foldy-Wouthuysen transformation to any order of accuracy; this has been demonstrated explicitly for the special case of the free electron, for which the positive and negative energy states can be completely decoupled.

The different terms in the non-relativistic approximation (Equation 8.142) of the Dirac Hamiltonian for the positive energy states may now be identified. The terms in (a) are, respectively, the rest energy of the electron, the energy due to the interaction of the charge of the electron with an electric field and the kinetic energy of the electron. The energy of the interaction of the spin magnetic moment of the electron with a magnetic field is represented by (b); this term, which has been encountered already in Section 8.4, is often referred to as the spin Zeeman term.

The next term (c) is responsible for the interaction of the spin magnetic moment of the electron with an electric field. This arises because, as we saw in Section 5.4, an electron which is moving in an inertial frame in which there is an electric field but no magnetic field, will nevertheless experience a magnetic field in its own rest frame, and that this magnetic field is perpendicular to the electric field and the direction of motion of the electron. When the electric field originates at a nucleus, it is this term that is responsible for spin-orbit coupling. The classical expression for this term was found in Section 5.10, but it will be recalled that a simple calculation gives a result that is a factor of two too large for agreement with experiment, and that the Thomas precession had to be taken into account; the Dirac equation accounts automatically for the Thomas correction factor of $\frac{1}{2}$.

The term (d) gives relativistic corrections to the kinetic energy in (a) and the Zeeman term in (b); the first of these appears in classical relativistic expressions for the energy. Finally, (e) is the so-called Darwin term which is peculiar to relativistic quantum mechanics and has no classical analogue; its origin will become clear when the case of the free electron is considered in detail in the next section.

8.10 The free electron

Although of no chemical interest, a consideration of the free electron provides an insight into the nature of electron spin and the connection between the Dirac equation and its non-relativistic approximation. The Foldy-Wouthuysen transformation is used, since we wish to work with both positive and negative energy states, although the method of small components could be employed. However, before doing this we will solve the Dirac equation itself:

$$\mathcal{H}\psi = i\hbar(\partial\psi/\partial t), \tag{8.143}$$

where

$$\mathcal{H} = \beta mc^2 + c(\boldsymbol{\alpha}.\mathbf{p}). \tag{8.144}$$

Since $\mathcal{H}$ and $\mathbf{p}$ commute, the Hamiltonian and the momentum operator have common eigenfunctions, which may be written:

$$\psi = u(\mathbf{k}) \exp [i(\mathbf{k}.\mathbf{r} - Et/\hbar)], \tag{8.145}$$

where $\hbar\mathbf{k}$ and E are the eigenvalues of $\mathbf{p}$ and $\mathcal{H}$ $(= i\hbar\partial/\partial t)$, respectively, and $u(\mathbf{k})$ is a four-component vector, which is independent of the position $\mathbf{r}$. Substitution in Equation 8.143 gives:

$$[\beta mc^2 + \hbar c(\boldsymbol{\alpha}.\mathbf{k}) - E]u(\mathbf{k}) = 0, \tag{8.146}$$

which is a set of four coupled equations for the components of $u(\mathbf{k})$. Since E and the components of $\mathbf{k}$ are numbers, Equation 8.146 represents an ordinary eigenvalue problem; by expanding the determinant of the four-by-four matrix the eigenvalues are found to be:

$$E_{\pm} = \pm (m^2c^4 + \hbar^2c^2k^2)^{\frac{1}{2}}, \tag{8.147}$$

where each solution occurs twice. This result may be obtained more easily by multiplying Equation 8.146 on the left by the matrix $(\beta mc^2 + \hbar c(\boldsymbol{\alpha}.\mathbf{k}) + E)$, the resulting matrix being diagonal with all four elements equal to $(m^2c^4 + \hbar^2c^2k^2 - E^2)$. The coefficients $u(\mathbf{k})$ may be found by substituting $E_{\pm}$ back into Equation 8.146 and as expected the positive energy components predominate in the positive energy solutions with a small amount of the negative energy components mixed in, while the reverse is true for the negative energy solutions.

Thus, even if both the energy and momentum eigenvalues are specified, a double degeneracy remains and some other operator is necessary to distinguish between two degenerate solutions. It will be recalled from Section 8.4 that, in contrast to non-relativistic quantum mechanics, the orbital angular momentum $(\mathbf{r}_\wedge \mathbf{p})$ does not commute with the Dirac Hamiltonian, but that the operator $[(\mathbf{r}_\wedge \mathbf{p}) + \frac{1}{2}\hbar\boldsymbol{\sigma}]$ does. This implies that the electron has an intrinsic angular momentum of $\frac{1}{2}$ in units of $\hbar$. Although $\boldsymbol{\sigma}$ does not commute with the Hamiltonian, its projection on the momentum, $(\boldsymbol{\sigma}.\mathbf{p}/p)$, does and can be used to distinguish between two solutions with the same energy. The square of this operator is just unity by Equation 8.40 and hence its eigenvalues are ± 1. The two positive energy solutions of the Dirac equation can then be chosen so that they are eigenfunctions of $(\boldsymbol{\sigma}.\mathbf{p}/p)$, one with eigenvalue $+1$ and the other with eigenvalue -1; the negative energy solutions can be chosen in a similar fashion. To obtain the components of $u(\mathbf{k})$ corresponding to these solutions requires some algebraic manipulation, the inclusion of which here would serve little purpose.

Some peculiarities of the solution of the Dirac equation have already been noted, in particular the existence of negative energy states and the intrinsic angular momentum of the electron. However, a still stranger result is obtained, if the velocity operator is determined. From Equation 1.24 this is given by:

$$\dot{\mathbf{r}} = (\mathrm{d}\mathbf{r}/\mathrm{d}t) = (i/\hbar)[\mathcal{H}, \mathbf{r}] = c\boldsymbol{\alpha}; \tag{8.148}$$

this fact was referred to in Section 8.5. Thus the operator appropriate to a component of the velocity $\dot{\mathbf{r}}$ is proportional to the corresponding component of $\boldsymbol{\alpha}$. Now the square of a component of $\boldsymbol{\alpha}$ is equal to unity, so that the eigenvalues of a component of $\boldsymbol{\alpha}$ can only be ± 1. This suggests that any determination of the component of velocity of an electron must give $\pm c$, in contrast to the non-relativistic result that the velocity operator is $\mathbf{p}/m$. This result further suggests that the instantaneous speed of the electron can only be $\sqrt{3}c$, but the components of $\boldsymbol{\alpha}$ do not commute with one another, so that two components of the velocity cannot be precisely measured simultaneously. In addition, it might be thought that the result that an instantaneous component of velocity can only be $\pm c$ is in contradiction to reality, but it should be remembered that any experimental determination must involve an average over an appreciable time interval. Thus, if the electron's velocity is oscillating very rapidly about a mean corresponding to the observed value, an instantaneous component of velocity can still be $\pm c$.

This suggestion can be substantiated by considering how the velocity operator varies with time:

$$\dot{\boldsymbol{\alpha}} = (\mathrm{d}\boldsymbol{\alpha}/\mathrm{d}t) = (i/\hbar)[\mathcal{H},\boldsymbol{\alpha}] = (2i/\hbar)(c\mathbf{p}-\boldsymbol{\alpha}\mathcal{H}), \qquad (8.149)$$

where the anticommutation properties (Equation 8.11) have been used. Now for a free particle $\mathcal{H}$ and $\mathbf{p}$ are independent of time, so that Equation 8.149 may be treated as a differential equation in $\boldsymbol{\alpha}$ and solved provided the order of $\mathcal{H}$ and $\boldsymbol{\alpha}$ are not changed. Solution of this equation gives:

$$\boldsymbol{\alpha}(t) = c\mathbf{p}\mathcal{H}^{-1} + c\mathbf{p}\mathcal{H}^{-1}b \exp(-2i\mathcal{H}t/\hbar), \qquad (8.150)$$

as can be verified by back substitution; b is a dimensionless integration constant. When it is remembered that the velocity operator is $c\boldsymbol{\alpha}$, use of Equations 4.38 and 4.39 shows that the first term in Equation 8.150 has the same form as a classical relativistic expression for velocity. However, the second term is oscillatory with a frequency of $(2\mathcal{H}/h)$; the magnitude of the energy must be at least mc^2 so this frequency is at least $(2mc^2/h)$, which is of the order of $2 \times 10^{20}\,\mathrm{s}^{-1}$! Thus, the electron has a highly oscillatory motion about a mean position. An experimental determination of the velocity taking much longer than $(h/2mc^2) \sim 5 \times 10^{-21}$ s gives a measure of the motion of the mean position, but an instantaneous velocity component can still be $\pm c$, because of the presence of the oscillatory term.

Equation 8.150 may be integrated once more to give for the position operator:

$$\mathbf{r}(t) = \mathbf{a} + c^2\mathcal{H}^{-1}\mathbf{p}t + \tfrac{1}{2}i\hbar c^2\mathbf{p}b \exp(-2i\mathcal{H}t/\hbar)\mathcal{H}^{-2}, \qquad (8.151)$$

where **a** is an integration constant and b is the same constant as before. As well as the constant and a term with the classical relativistic form, there is again an oscillatory term with a frequency of at least $(2mc^2/h)$. The amplitude of this oscillation may be estimated by expressing it as:

$$\tfrac{1}{2}i\hbar(c\boldsymbol{\alpha}(t) - c^2\mathcal{H}^{-1}\mathbf{p})\mathcal{H}^{-1}, \tag{8.152}$$

where Equation 8.150 has been used to eliminate b and the exponential term. In Equation 8.152 the term in parenthesis has a magnitude of about c, since $c\boldsymbol{\alpha}$ is the instantaneous velocity operator whose components have eigenvalues $\pm c$, whereas $(c^2\mathcal{H}^{-1}\mathbf{p})$ corresponds to the velocity that is determined experimentally. Thus, the amplitude (Equation 8.152) of the oscillatory motion is of the order of $(\hbar/mc) = \alpha a_0 \sim 4 \times 10^{-13}$ m, where α is the fine structure constant (Equation 8.85).

We are thus led to the conclusion that the motion of a free electron consists of two parts. The first part is an average motion that can be observed experimentally. The other part is a highly oscillatory motion that has a frequency of order $(2mc^2/h)$ and an amplitude of order $(\hbar/mc)$. This rapidly varying motion was given the name Zitterbewegung (literally, trembling motion) by Schrödinger.

To see how the Zitterbewegung manifests itself in the non-relativistic approximation of the Dirac equation, the Foldy-Wouthuysen transformation for a free electron is now considered. It has already been stated that the positive and negative energy states can be completely decoupled in the special case of a free electron. This is due to the even operator $(\mathcal{E} = -e\phi)$ being zero when electromagnetic potentials are absent. In addition, the absence of an external magnetic potential means that the odd operator $(\mathcal{O} = c\boldsymbol{\alpha}.\boldsymbol{\pi})$ has no specific time dependence so that the operator S in the unitary transformation:

$$\psi' = \exp(iS)\psi \tag{8.153}$$

may also be chosen to be specifically time independent; this is still true if there is a constant magnetic field present and complete separation can also be achieved in this case. The transformed Hamiltonian (Equation 8.123) now takes the simpler form:

$$\mathcal{H}' = \exp(iS)\mathcal{H}\exp(-iS). \tag{8.154}$$

Reference to the previous section now shows that removal of an odd operator of order $mc^2\alpha^n$ by a unitary transformation leads to a transformed Hamiltonian in which the largest odd operator is of order $mc^2\alpha^{n+2}$. The important thing is that there is only one odd operator of this order, in contrast to the case of the electron in an electromagnetic field where the number of odd operators increases rapidly with each transformation. Specifically:

$$\mathcal{O} = c(\boldsymbol{\alpha}.\mathbf{p});$$

$$\mathcal{O}_1 = -\mathcal{O}^3/3m^2c^4 + O(mc^2\alpha^5) = -(\boldsymbol{\alpha}.\mathbf{p})p^2/3m^2c + O(mc^2\alpha^5);$$

$$\mathcal{O}_2 = +\mathcal{O}^5/5m^4c^8 + O(mc^2\alpha^7) = +(\boldsymbol{\alpha}.\mathbf{p})p^4/5m^4c^3 + O(mc^2\alpha^7), \qquad (8.155)$$

where use has been made of the result:

$$(\boldsymbol{\alpha}.\mathbf{p})^2 = p^2, \qquad (8.156)$$

which follows from Equations 8.39 and 8.40.

The largest odd operator after n applications of the unitary transformation is thus:

$$\mathcal{O}_n = (-1)^n(\boldsymbol{\alpha}.\mathbf{p})p^{2n}/[(2n+1)m^{2n}c^{2n-1}] \qquad (8.157)$$

and the Hermitian operator S_n for the $(n+1)$th unitary transformation may be written as:

$$S_n = -\tfrac{1}{2}i\beta(\boldsymbol{\alpha}.\mathbf{p})p^{-1}(-1)^n(2n+1)^{-1}(p/mc)^{2n+1} \qquad (8.158)$$

from Equation 8.125. The presence of the operator p^{-1} should not cause concern, since it will only appear in expressions where it is multiplied by positive odd powers of p; it is introduced so that abbreviated notation may be used for power series in the following manipulations. It is now apparent that the unitary transformations may be "summed" and applied all at once, the total unitary operator being:

$$\prod_{n=0}^{\infty} [\exp(iS_n)] = \exp\left[\sum_{n=0}^{\infty} iS_n\right]. \qquad (8.159)$$

This equality is only true because the operators S_n commute with one another; if they did not, ambiguities would arise about the order of the operators when the exponential term was expanded as a power series, and this is one of the reasons why this technique cannot be used for the case of an electron in a general electromagnetic field. Thus, the complete removal of odd operators may be accomplished with one unitary transformation defined by the operator:

$$S' = \sum_{n=0}^{\infty} S_n = -\tfrac{1}{2}i\beta(\boldsymbol{\alpha}.\mathbf{p})p^{-1}\sum_{n=0}^{\infty}(-1)^n(2n+1)^{-1}(p/mc)^{2n+1}$$

$$= -\tfrac{1}{2}i\beta(\boldsymbol{\alpha}.\mathbf{p})p^{-1}f, \qquad (8.160)$$

where

$$f = \tan^{-1}(p/mc) = \sum_{n=0}^{\infty}(-1)^n(2n+1)^{-1}(p/mc)^{2n+1} \qquad (8.161)$$

is a function of p only and commutes with $\boldsymbol{\alpha}$ and β.

Now S' can easily be shown to anticommute with the Hamiltonian (Equation 8.144), so that the transformed Hamiltonian is:

$$\mathcal{H}' = \exp(iS')\mathcal{H}\exp(-iS') = \exp(i2S')\mathcal{H} = (\cos(2S') + i\sin(2S'))\mathcal{H}, \tag{8.162}$$

as can be confirmed by expanding the exponentials as power series. In addition $(2S')^2$ is equal to f^2, since use of the anticommutation properties of β and the components of $\boldsymbol{\alpha}$ (Equation 8.11), together with Equations 8.160 and 8.156 shows that:

$$\begin{aligned}(2S')^2 &= -p^{-2}f^2[\beta(\boldsymbol{\alpha}.\mathbf{p})]^2\\ &= +p^{-2}f^2(\boldsymbol{\alpha}.\mathbf{p})^2\\ &= f^2.\end{aligned} \tag{8.163}$$

Since cosine is an even function of its argument:

$$\cos(2S') = \cos f, \tag{8.164}$$

while for the odd function sine:

$$\sin(2S') = -i\beta(\boldsymbol{\alpha}.\mathbf{p})p^{-1}\sin f. \tag{8.165}$$

The transformed Hamiltonian (Equation 8.162) can now be written as:

$$\begin{aligned}\mathcal{H}' &= (\cos f + \beta(\boldsymbol{\alpha}.\mathbf{p})p^{-1}\sin f)(\beta mc^2 + c(\boldsymbol{\alpha}.\mathbf{p}))\\ &= \beta(mc^2\cos f + cp\sin f) + mc^2(\boldsymbol{\alpha}.\mathbf{p})p^{-1}((p/mc)\cos f - \sin f).\end{aligned} \tag{8.166}$$

It only remains to express $\sin f$ and $\cos f$ in terms of (p/mc). From Equation 8.161, $\tan f = (p/mc)$ so that, using trigonometric relations:

$$\cos f = (mc/p)\sin f = [1 + (p/mc)^2]^{-\frac{1}{2}}, \tag{8.167}$$

and the second term in Equation 8.166, the one involving odd operators, vanishes. The remaining term involves only even operators so that the completely decoupled Hamiltonian is:

$$\begin{aligned}\mathcal{H}' &= \beta mc^2[1 + (p/mc)^2]\cos f\\ &= \beta mc^2[1 + (p/mc)^2]^{\frac{1}{2}},\end{aligned} \tag{8.168}$$

from Equation 8.167. The Hamiltonian for the positive energy states is obtained by setting β equal to $+1$, while substitution of β equals -1 gives the Hamiltonian for the negative energy states. Consequently Equation 8.168 confirms our previous result (Equation 8.147).

All this manipulation to obtain a result we already had may seem rather unnecessary. However, the important thing to note is that when the positive energy states are decoupled by a unitary transformation (Equation 8.153), all operators must be subjected to the same transformation, not just the Hamiltonian. By determining how operators transform, we will be able to see how operators before and after the transformation are related to one another; the description of the electron before the transformation may be referred to as the Dirac representation, and that after the transformation as the Foldy-Wouthuysen representation. For example, the momentum operator $\mathbf{p}$ is the same in both representations, since it commutes with S', but the position operator $\mathbf{r}$ in the Dirac representation has for its operator in the Foldy-Wouthuysen representation $\mathbf{r}'$, which is given by:

$$\begin{aligned}\mathbf{r}' &= \exp(iS')\mathbf{r}\exp(-iS') \\ &= \mathbf{r} + i[S', \mathbf{r}] + (i^2/2!)[S', [S', \mathbf{r}]] + \ldots, \end{aligned} \tag{8.169}$$

where S' is given by Equation 8.160. This expression may be expanded, but, since it involves considerable though straightforward manipulation, only an outline will be given here. The commutator of S' and $\mathbf{r}$ is first established to be:

$$[S', \mathbf{r}] = -\tfrac{1}{2}\hbar mc^3(E_+p)^{-2}\beta(\boldsymbol{\alpha}.\mathbf{p})\mathbf{p} + \tfrac{1}{2}\hbar p^{-3}f\beta[(\boldsymbol{\alpha}.\mathbf{p})\mathbf{p} - p^2\boldsymbol{\alpha}], \tag{8.170}$$

where

$$E_+ = +[m^2c^4 + c^2p^2]^{\frac{1}{2}}. \tag{8.171}$$

Substitution of Equation 8.170 into Equation 8.169 and evaluation of the commutators leads to an expression, which may be split up into a number of summable power series, giving:

$$\begin{aligned}\mathbf{r}' = \mathbf{r} &- \tfrac{1}{2}i\hbar p^{-1}\beta\boldsymbol{\alpha}\sin f - \tfrac{1}{2}i\hbar p^{-2}[E_+^{-2}mc^3 - p^{-1}\sin f]\beta(\boldsymbol{\alpha}.\mathbf{p})\mathbf{p} \\ &+ \tfrac{1}{2}\hbar(\cos f - 1)(\boldsymbol{\sigma}\wedge\mathbf{p})/p^2. \end{aligned} \tag{8.172}$$

Finally, substitution for the functions of f (Equation 8.167) gives:

$$\begin{aligned}\mathbf{r}' = \mathbf{r} &- \tfrac{1}{2}i\hbar cE_+^{-1}\beta\boldsymbol{\alpha} + \tfrac{1}{2}i\hbar c^3E_+^{-2}(E_+ + mc^2)^{-1}\beta(\boldsymbol{\alpha}.\mathbf{p})\mathbf{p} \\ &- \tfrac{1}{2}\hbar c^2E_+^{-1}(E_+ + mc^2)^{-1}(\boldsymbol{\sigma}\wedge\mathbf{p}). \end{aligned} \tag{8.173}$$

Once again the appearance of reciprocal operators causes no problems, since they are merely a shorthand notation for the corresponding power series. Equation 8.173 shows that the position operator $\mathbf{r}$ in the Dirac representation has, as its operator in the Foldy-Wouthuysen representation, $\mathbf{r}$ plus three other terms, which depend on the momentum operator, thus indicating that the transformation is a non-local one; that is a point in the Dirac representation is not transformed into a point in the Foldy-Wouthuysen representation, but into an extended region of space.

In the Foldy-Wouthuysen representation **r** must represent some new physical observable which is called the mean position for reasons to become apparent shortly. In the Dirac representation the mean position has as its operator **R**, which is given by:

$$\mathbf{r} = \exp(iS')\mathbf{R}\exp(-iS') \tag{8.174}$$

or

$$\mathbf{R} = \exp(-iS')\mathbf{r}\exp(iS'). \tag{8.175}$$

By reference to Equations 8.169 and 8.173 this can be expanded immediately:

$$\mathbf{R} = \mathbf{r} + \tfrac{1}{2}i\hbar cE_+^{-1}\beta\boldsymbol{\alpha} - \tfrac{1}{2}i\hbar c^3E_+^{-2}(E_+ + mc^2)^{-1}\beta(\boldsymbol{\alpha}.\mathbf{p})\mathbf{p}$$
$$-\tfrac{1}{2}\hbar c^2E_+^{-1}(E_+ + mc^2)^{-1}(\boldsymbol{\sigma}\wedge\mathbf{p}). \tag{8.176}$$

The determination of the time derivative of **R** in the Dirac representation enables this new position operator to be interpreted. From Equation 1.24:

$$(\mathrm{d}\mathbf{R}/\mathrm{d}t) = (i/\hbar)[\mathcal{H},\mathbf{R}] = (i/\hbar)[\beta mc^2 + c(\boldsymbol{\alpha}.\mathbf{p}),\mathbf{R}], \tag{8.177}$$

where **R** is given in Equation 8.176. This can be shown, after some manipulation, to be:

$$(\mathrm{d}\mathbf{R}/\mathrm{d}t) = \mathcal{H}E_+^{-2}c^2\mathbf{p} = \pm c^2E_+^{-1}\mathbf{p}, \tag{8.178}$$

where use is made of the fact that $\mathcal{H}E_+^{-1}$ has the eigenvalue $+1$ or -1, respectively, for positive and negative energy states. Reference to Equation 4.38 now shows that it is $\dot{\mathbf{R}}$, the mean velocity, which has the classical relativistic relation to the momentum. Finally, it can be noted that the components of $\dot{\mathbf{R}}$ commute unlike the components of $c\boldsymbol{\alpha}$, the operator corresponding to the time derivative of the position in the Dirac representation.

Thus, there are two important position operators, the Dirac position and the mean position. It is the latter which is to be identified with the position operator in non-relativistic theory and is separately defined for positive energy states. The Dirac position has no classical analogue, since it requires both positive and negative energy states for its definition. All this agrees with the results of the earlier part of this section. Equation 8.176 shows that the position operator **r** in the Dirac representation consists of two parts, a mean position **R**, which is related to the classical concept of position and the motion of which is experimentally observable, and another part, $\mathbf{r}-\mathbf{R}$, which corresponds to the Zitterbewegung encountered earlier.

The peculiarities involving the angular momentum can also be explained. It will be recalled from Section 8.3 that the orbital and spin angular momenta do not commute separately with the Dirac Hamiltonian; that is, they are not physical observables, although they are in the non-relativistic Schrödinger treatment. However, the operators $(\mathbf{R}\wedge\mathbf{p})$ and $\mathbf{\Sigma}$ which are given by:

$$(\mathbf{R}\wedge\mathbf{p}) = \exp(-iS')(\mathbf{r}\wedge\mathbf{p})\exp(iS') \tag{8.179}$$

and

$$\begin{aligned}\mathbf{\Sigma} &= \exp(-iS')\boldsymbol{\sigma}\exp(iS')\\ &= \boldsymbol{\sigma} - icE_+^{-1}\beta(\boldsymbol{\alpha}\wedge\mathbf{p}) - c^2E_+^{-1}(E_+ + mc^2)^{-1}\mathbf{p}\wedge(\boldsymbol{\sigma}\wedge\mathbf{p})\end{aligned} \tag{8.180}$$

and whose analogues in the Foldy-Wouthuysen representation are $(\mathbf{r}\wedge\mathbf{p})$ and $\boldsymbol{\sigma}$, do commute with the Dirac Hamiltonian and may be thought of as the mean orbital angular momentum and the mean spin angular momentum (in units of $\frac{1}{2}\hbar$). It is, of course, these quantities that are thought of as the orbital and spin angular momenta in the non-relativistic theory. Finally, we recall that the total angular momentum, $[(\mathbf{r}\wedge\mathbf{p}) + \frac{1}{2}\hbar\boldsymbol{\sigma}]$, does commute with the Dirac Hamiltonian and, as might be expected, combination of Equations 8.173 and:

$$\begin{aligned}\boldsymbol{\sigma}' &= \exp(iS')\boldsymbol{\sigma}\exp(-iS')\\ &= \boldsymbol{\sigma} + icE_+^{-1}\beta(\boldsymbol{\alpha}\wedge\mathbf{p}) - c^2E_+^{-1}(E_+ + mc^2)\mathbf{p}\wedge(\boldsymbol{\sigma}\wedge\mathbf{p}),\end{aligned} \tag{8.181}$$

readily shows that it has the same operator form in both the Dirac and Foldy-Wouthuysen representations.

In the previous section the Foldy-Wouthuysen transformation was applied to an electron in the presence of external electromagnetic potentials. The resulting non-relativistic approximation of the Dirac Hamiltonian is correct to order $mc^2\alpha^5$, and contains a number of terms whose presence can now be explained in the light of the foregoing discussion for the free electron. In the Dirac representation the electron interacts with the electromagnetic field at its position. However, in the Foldy-Wouthuysen representation the electron position becomes spread out over a small region around the mean position, and the electron oscillates very rapidly about this mean position. As we have seen, the interaction of the electron with the electromagnetic field in the Foldy-Wouthuysen representation is expressed in terms of $\mathbf{r}$, the mean position, and consequently correction terms will appear in the non-relativistic approximation of the Dirac Hamiltonian to allow for the superimposed Zitterbewegung.

The circulation of the electron about its mean position gives rise to both the intrinsic angular momentum and the intrinsic magnetic moment associated with the spin of the electron. Indeed a classical model for the Zitterbewegung may be constructed that gives the correct spin angular momentum for the electron. At the beginning of this section it was seen that the Zitterbewegung is a highly oscillatory motion that has a frequency of order $2mc^2/h$ and an amplitude of order $\hbar/mc$. If we consider an electron moving in a circle of radius $\hbar/2mc$ with a frequency of $2mc^2/h$, its velocity must be just c in agreement with the earlier assertions that the instantaneous value of a component of the electron's velocity can only be $\pm c$. The angular momentum associated with this motion is just the product of the linear momentum mc and the radius of the circle $\hbar/2mc$, so that the intrinsic or spin angular momentum of the electron is $\hbar/2$ in agreement with experiment. Unfortunately this model cannot be correct, since, although the electric current associated with this motion gives rise to an intrinsic magnetic moment, the predicted g factor is unity as in most classical models.

Except for the Darwin term (Equation 8.142e) all the terms in the non-relativistic approximation to the Dirac Hamiltonian (Equation 8.142) can now be understood, since they are related to electron spin or have classical relativistic counterparts. However, the Darwin term may be understood in terms of the Zitterbewegung, since it is just a correction to the electrostatic interaction between the electron and the electric potential to allow for the smearing out of the electron's charge by the Zitterbewegung.

The position operator $\mathbf{r}$ in the Dirac representation has as its operator in the Foldy-Wouthuysen representation the mean position $\mathbf{r}$ plus the oscillatory term, which will be given the symbol $\boldsymbol{\delta}$. Corresponding to the electrostatic interaction term $-e\phi(\mathbf{r})$ in the Dirac representation is the term $-e\phi(\mathbf{r}+\boldsymbol{\delta})$ in the Foldy-Wouthuysen representation, and this must be averaged over the Zitterbewegung. Now $\phi(\mathbf{r}+\boldsymbol{\delta})$ may be expanded as a Taylor series in $\boldsymbol{\delta}$ about $\mathbf{r}$:

$$\phi(\mathbf{r}+\boldsymbol{\delta}) = [1 + (\boldsymbol{\delta}.\boldsymbol{\nabla}) + \tfrac{1}{2}(\boldsymbol{\delta}.\boldsymbol{\nabla})^2 + \ldots]\phi(\mathbf{r}), \qquad (8.182)$$

and the time average of this is:

$$\langle\phi(\mathbf{r}+\boldsymbol{\delta})\rangle = \phi(\mathbf{r}) + \tfrac{1}{2}\langle(\boldsymbol{\delta}.\boldsymbol{\nabla})^2\rangle\phi(\mathbf{r}) + \ldots, \qquad (8.183)$$

since $\langle\boldsymbol{\delta}\rangle$ vanishes. The expression $\langle(\boldsymbol{\delta}.\boldsymbol{\nabla})^2\rangle$ may be evaluated by expanding it in terms of vector components and noting that $\langle\delta_i\delta_j\rangle = \delta_{ij}\langle\delta^2\rangle/3$, so that:

$$\begin{aligned}\langle(\boldsymbol{\delta}.\boldsymbol{\nabla})^2\rangle &= \langle\delta_i\nabla_i\delta_j\nabla_j\rangle = \langle\delta_i\delta_j\rangle\nabla_i\nabla_j \\ &= \langle\delta^2\rangle\delta_{ij}\nabla_i\nabla_j/3 = \langle\delta^2\rangle\nabla^2/3. \qquad (8.184)\end{aligned}$$

Thus, the electrostatic interaction energy in the Dirac representation is given in the Foldy-Wouthuysen representation by:

$$-e\phi(\mathbf{r}+\boldsymbol{\delta}) = -e\phi(\mathbf{r}) - e\langle\delta^2\rangle\nabla^2\phi(\mathbf{r})/6 + \ldots. \qquad (8.185)$$

Now in the absence of magnetic potentials $\mathbf{E} = -\boldsymbol{\nabla}\phi$ (Equation 5.21), so that the energy correction due to the Zitterbewegung is just $+e\langle\delta^2\rangle\nabla.\mathbf{E}/6$; this agrees with the Darwin term (Equation 8.142e) provided that:

$$\langle\delta^2\rangle = 3(\hbar/2mc)^2, \tag{8.186}$$

which is of the order of magnitude one would expect. The agreement is more striking, if a slowly moving electron ($\mathbf{p} = 0$, $E_+ = mc^2$) is considered. Equation 8.173 then gives:

$$\boldsymbol{\delta} = -(i\hbar/2mc)\beta\boldsymbol{\alpha}, \tag{8.187}$$

which on squaring becomes:

$$\begin{aligned}\delta^2 &= -(\hbar/2mc)^2(\beta\boldsymbol{\alpha}).(\beta\boldsymbol{\alpha}) \\ &= (\hbar/2mc)^2\beta^2\alpha^2 = 3(\hbar/2mc)^2,\end{aligned} \tag{8.188}$$

since $\beta^2 = 1$ and $\alpha^2 = 3$ from Equations 8.10.

Bibliography

The Dirac equation

Dirac, P. A. M. (1958), *Principles of Quantum Mechanics*, Clarendon Press, Oxford: Chapter 11.
Foldy: pages 4-39.
Rose: in particular Chapters 3 and 4.
Sakurai: Chapter 3 includes the alternative approach to the Dirac equation mentioned at the end of Section 8.4.

The method of small components

Löwdin, P-O. (1964), 'Studies in perturbation theory. Part VIII. Separation of Dirac equation and study of the spin-orbit coupling and Fermi contact term', *J. Mol. Spectry.*, **14**, 131.
McWeeny, R., and Sutcliffe, B.T. (1969), *Methods of Molecular Quantum Mechanics*, Academic Press, New York: Appendix 4.

The Foldy-Wouthuysen transformation

Foldy, L. L., and Wouthuysen, S. A. (1950), 'On the Dirac theory of spin $\frac{1}{2}$ particles and its non-relativistic limit', *Phys. Rev.*, **78**, 29.
Foldy: pages 15-20.
Rose: pages 87-92 and 123-130.

The relativistic correction (Equation 8.142d) to the spin Zeeman term

Abragam, A., and Van Vleck, J. H. (1953), 'Theory of the microwave Zeeman effect in atomic oxygen', *Phys. Rev.* **92**, 1448.
Atkins, P. W., and Jamieson, A. M. (1967), 'Perturbation corrections to the *g* tensor', *Molec. Phys.*, **14**, 425.

Complete separation of the Dirac equation

Case, K. M. (1954), 'Some generalizations of the Foldy-Wouthuysen transformation', *Phys. Rev.*, **95**, 1323: complete separation for an electron subject only to a constant magnetic field.

Foldy, L. L., and Wouthuysen, S. A. (1950), 'On the Dirac theory of spin $\frac{1}{2}$ particles and its non-relativistic limit', *Phys. Rev.*, **78**, 29.

Löwdin, P-O. (1964), 'Studies in perturbation theory. Part VIII. Separation of Dirac equation and study of the spin-orbit coupling and Fermi contact term', *J. Mol. Spectry.*, **14**, 131.

Moss, R. E. (1971), 'The reduction of the Dirac equation to a non-relativistic form', *Am. J. Phys.*, **39**, 1169: a demonstration that for a free electron the Foldy-Wouthuysen transformation and the method of small components are formally equivalent.

Rose: pages 87-88.

Zitterbewegung and the mean position operator

Foldy, L. L., and Wouthuysen, S. A. (1950), 'On the Dirac theory of spin $\frac{1}{2}$ particles and its non-relativistic limit', *Phys. Rev.*, **78**, 29.

Foldy: pages 20-23.

Rose: pages 90-91 and 129-130.

CHAPTER NINE

The Wave Equation for Many Electrons

Although we now have a suitable wave equation for one electron in the presence of external electromagnetic fields, we are not yet in a position to write down the Hamiltonian for a general molecule. In this chapter we move towards this goal by first considering the problem of just two electrons before generalizing the theory to the case of many electrons.

In fact, a correct relativistic wave equation for two electrons cannot be written down in closed form, but it is possible to obtain an approximate Hamiltonian provided the kinetic energy of the electrons is very much less than their rest energy, mc^2; this is just the situation that pertains in molecules.

There are a number of possible approaches to this problem of two electrons. One way is to start with the non-relativistic approximation of the Dirac Hamiltonian for each electron. The interaction between the two electrons then arises, since each contributes to the electromagnetic potentials experienced by the other. Although this method has a number of associated difficulties, it does give a physical picture of the origin of the individual terms in the final Hamiltonian. A more sophisticated approach is to seek an approximate wave equation for two electrons, which is the counterpart of the Dirac equation for one electron. Such an equation is the Breit equation, but to be useful it has to be reduced to a non-relativistic form.

Both these approaches give the same non-relativistic approximation to the Hamiltonian for two electrons, although the policies adopted to circumvent problems in the first method are chosen to ensure this agreement. Of course, the most satisfactory approach is to use quantum electrodynamics, but we content ourselves with noting that the Hamiltonian derived here is the same as that obtained by Itoh. Even so, it is necessary to include certain radiative corrections in a phenomenological manner.

9.1 The electromagnetic potentials due to a moving electron

If the problem of two electrons is approached from the non-relativistic approximation (Equation 8.142) of the Dirac equation, the electromagnetic potentials at one electron due to the other are needed. It is the purpose of this section to obtain them in a suitable form. Before considering their explicit forms it is necessary to decide how accurately they need to be known in the context of a molecular environment. This may be done by considering the individual terms in Equation 8.142, using the results of Section 8.7 and recalling that the present aim is to obtain a Hamiltonian, which is correct to order $mc^2\alpha^4$.

The electrostatic potential ϕ appears as $-e\phi$ in 8.142a, so that ϕ is required to be correct to order $mc^2\alpha^4/e$. The terms 8.142c and 8.142e also involve ϕ through the electric field **E**, but as they are already small, the scalar potential only needs to be correct to order $mc^2\alpha^2/e$ for these terms to be correct to order $mc^2\alpha^4$.

The largest contribution made by the vector potential is through the terms 8.142b and $\pi^2/2m$ in 8.142a, and in both cases it must be known to an accuracy of $mc\alpha^3/e$. In Section 8.7 we saw that in any case the order of magnitude of the vector potentials in a molecule is only $mc\alpha^3/e$, so that only the leading term of any power series expansion of **A** will be needed. There are other smaller terms in Equation 8.142, but these are too small for the vector potential to make a significant contribution to them.

Now, if the Coulomb gauge is chosen, the potentials due to a charge $-e$ moving with velocity **u** are given by Equations 5.73 as:

$$\begin{aligned} \mathbf{A} &= -(e/8\pi\epsilon_0 c^2)[\mathbf{u}/r + (\mathbf{u.r})\mathbf{r}/r^3] + O(mc\alpha^5/e); \\ \phi &= -e/4\pi\epsilon_0 r + O(mc^2\alpha^6/e), \end{aligned} \tag{9.1}$$

where **r** is the position of the electron relative to the observer; these potentials are corrected for retardation effects. The expressions 9.1 are not yet in a suitable form, since they are to be used in a quantum mechanical theory and should be expressed in terms of momenta rather than velocities. This may be remedied by replacing **u** by $\mathbf{p}/m$, since, although they are not strictly equal (Equation 4.38), the correction terms do not contribute significantly to the potentials:

$$\begin{aligned} \mathbf{A} &= -(e/8\pi\epsilon_0 mc^2)[\mathbf{p}/r + (\mathbf{p.r})\mathbf{r}/r^3] + O(mc\alpha^5/e); \\ \phi &= -e/4\pi\epsilon_0 r + O(mc^2\alpha^6/e). \end{aligned} \tag{9.2}$$

These potentials cannot be used straightforwardly in a quantum mechanical theory for two electrons, since **p** and **r**, which must be interpreted as operators, do not commute. Thus, the order in which the various factors appear in the expression for the vector potential is important. The choice may be guided by insisting that the resulting Hamiltonian for two electrons be Hermitian and symmetric in the two electrons. By using the order given above these conditions will be satisfied, but the real justification is that the final Hamiltonian is the same as that obtained by more sophisticated methods.

The charge of the electron is responsible for the potentials (Equations 9.2), but the spin magnetic moment of the electron also provides a contribution to the vector potential and this is found by considering a point magnetic moment, although it is not clear that this is permissible. As our model we take a slowly moving particle of charge q and mass m moving in a circle about the origin; the associated magnetic moment is (see Section 7.1):

$$\boldsymbol{\mu} = (q/2m)(\mathbf{a} \wedge \mathbf{p}) \simeq \tfrac{1}{2}q(\mathbf{a} \wedge \mathbf{u}), \tag{9.3}$$

where $\mathbf{a}$, $\mathbf{p}$ and $\mathbf{u}$ are the position, canonical momentum and velocity of the particle. The average vector potential at position $\mathbf{r}$, where r is very much larger than a, may be expressed in terms of $\boldsymbol{\mu}$. The vector potential due to a point magnetic moment is then obtained in the limit that $\mathbf{a}$ vanishes.

For simplicity Equations 5.72 rather than Equations 5.73 are used for the vector potential in this model, since the gauge may always be adjusted later if desired. Now for a slowly moving particle:

$$\mathbf{A} = q\,\mathbf{u}/4\pi\epsilon_0 c^2|\mathbf{r}-\mathbf{a}|, \tag{9.4}$$

and this may be expanded as a power series in $\mathbf{a}$ about $\mathbf{r}$ by noting that for any function $f(\mathbf{r}-\mathbf{a})$:

$$f(\mathbf{r}-\mathbf{a}) = f(\mathbf{r}) - \mathbf{a}.\nabla f(\mathbf{r}) + \ldots, \tag{9.5}$$

so that using Equation 2.78:

$$\begin{aligned} 1/|\mathbf{r}-\mathbf{a}| &= 1/r - \mathbf{a}.\nabla(1/r) + \ldots \\ &= 1/r + (\mathbf{a}.\mathbf{r})/r^3 + \ldots. \end{aligned} \tag{9.6}$$

Substitution in Equation 9.4 followed by rearrangement now gives:

$$\begin{aligned} \mathbf{A} &= (q/4\pi\epsilon_0 c^2)[\mathbf{u}/r + \mathbf{u}(\mathbf{a}.\mathbf{r})/r^3 + \ldots] \\ &= (q/4\pi\epsilon_0 c^2)[(\mathrm{d}/\mathrm{d}t)(\mathbf{a}/r + \mathbf{a}(\mathbf{a}.\mathbf{r})/2r^3) \\ &\quad + \tfrac{1}{2}(\mathbf{u}(\mathbf{a}.\mathbf{r}) - \mathbf{a}(\mathbf{u}.\mathbf{r}))/r^3 + \ldots], \end{aligned} \tag{9.7}$$

where use has been made of the facts that $\mathbf{u} = \mathrm{d}\mathbf{a}/\mathrm{d}t$ and $\mathbf{r}$ is a constant vector.

Equation 9.7 gives the instantaneous potential at position $\mathbf{r}$. When the average is taken the first term vanishes, since it is the time derivative of a periodic function; this may easily be checked explicitly for the model employed here. The remainder of Equation 9.7 may be written in terms of $\boldsymbol{\mu}$ by noting that from Equation 2.38:

$$(\mathbf{a}\wedge\mathbf{u})\wedge\mathbf{r} = \mathbf{u}(\mathbf{a}.\mathbf{r}) - \mathbf{a}(\mathbf{u}.\mathbf{r}), \tag{9.8}$$

so that the average vector potential is:

$$\bar{\mathbf{A}} = (1/4\pi\epsilon_0 c^2)(\bar{\boldsymbol{\mu}}\wedge\mathbf{r})/r^3 + \ldots. \tag{9.9}$$

When $\mathbf{a}$ is allowed to tend to zero, the vector potential at a position $\mathbf{r}$ relative to a constant magnetic moment $\boldsymbol{\mu}$ is obtained:

$$\mathbf{A} = (1/4\pi\epsilon_0 c^2)(\boldsymbol{\mu}\wedge\mathbf{r})/r^3. \tag{9.10}$$

In the case of the electron the spin magnetic moment is (Equation 8.49) $\boldsymbol{\mu} = -(e\hbar/m)\mathbf{s}$, so that the associated vector potential is:

$$\mathbf{A} = -(e\hbar/4\pi\epsilon_0 mc^2)(\mathbf{s}\wedge\mathbf{r})/r^3. \tag{9.11}$$

This potential satisfies the Coulomb gauge condition (Equation 5.35), since it is divergenceless. In a molecular environment it is of order $mc\alpha^3/e$, confirming the assertion made in Section 8.7.

The vector potential due to a spin magnetic moment is taken to have the form of Equation 9.11. However, one might expect that in a frame in which the electron is moving the observed vector potential will not quite be as given in Equation 9.11 and, in addition, a small scalar potential will be experienced. The correction to the vector potential is of order $mc\alpha^5/e$ and may safely be ignored as may retardation effects, but the associated electric potential, which is given approximately by $\mathbf{u}.\mathbf{A}$, is of order $mc^2\alpha^4/e$ and cannot be neglected. This contribution to the total scalar potential is:

$$\phi = -(e\hbar/4\pi\epsilon_0 m^2c^2)\mathbf{s}.(\mathbf{r}\wedge\mathbf{p})/r^3. \tag{9.12}$$

It will be seen later that, if this contribution is included, agreement is not obtained with the Hamiltonian derived from the Breit equation later in this chapter. It is necessary to argue that a classical model for the spin magnetic moment is not quite appropriate and that this refinement should be ignored. This is a further difficulty associated with the present approach to the problem of two electrons.

The total potentials, $\mathbf{A}_i$ and ϕ_i, at an electron i in the presence of a second electron j may now be written down to the desired accuracy (subject to the remarks of the previous paragraph). In doing this the vector:

$$\mathbf{r}_{ij} = \mathbf{r}_i - \mathbf{r}_j, \tag{9.13}$$

is introduced; it is noted that $\mathbf{r}$ in Equation 9.11 is just $\mathbf{r}_{ij}$, but that $\mathbf{r}$ in Equations 9.2 is $-\mathbf{r}_{ij}$. In addition, the subscripts are added to the charges and masses of the electrons, so that the origin of the terms in the final Hamiltonian may be seen more easily. If $\mathbf{A}$ and ϕ are now the external potentials:

$$\begin{aligned} \mathbf{A}_i &= \mathbf{A} - (e_j/8\pi\epsilon_0 m_j c^2)[\mathbf{p}_j/r_{ij} + (\mathbf{p}_j.\mathbf{r}_{ij})\mathbf{r}_{ij}/r_{ij}^3] \\ &\quad - (e_j\hbar/4\pi\epsilon_0 m_j c^2)(\mathbf{s}_j \wedge \mathbf{r}_{ij}/r_{ij}^3) + \mathrm{O}(mc^4/e) \end{aligned} \tag{9.14}$$

and

$$\phi_i = \phi - e_j/4\pi\epsilon_0 r_{ij} + \mathrm{O}(mc^2\alpha^5/e); \tag{9.15}$$

these potentials satisfy the Coulomb gauge. Although the highest neglected terms in the expression for the vector potential are formally of order $mc\alpha^5/e$, we have only taken Equation 9.14 to be correct to the required accuracy of $mc\alpha^3/e$, because of the uncertainties involved in this derivation of the Hamiltonian. Similarly the scalar potential (Equation 9.15) has only been assumed to be correct to order $mc^2\alpha^4/e$.

9.2 The Hamiltonian for two electrons

The non-relativistic approximation to the Dirac equation for an electron i is $\mathcal{H}_i\psi = i\hbar\partial\psi/\partial t$, where (Equation 8.142):

$$\left.\begin{aligned} \mathcal{H}_i &= m_ic^2 - e_i\phi_i + \pi_i^2/2m_i && \text{(a)} \\ &\quad + (e_i\hbar/m_i)(\mathbf{s}_i.\mathbf{B}_i) && \text{(b)} \\ &\quad - (e_i\hbar/4m_i^2c^2)\mathbf{s}_i.(\boldsymbol{\pi}_i \wedge \mathbf{E}_i - \mathbf{E}_i \wedge \boldsymbol{\pi}_i) && \text{(c)} \\ &\quad - (1/8m_i^3c^2)\pi_i^4 - (e_i\hbar/2m_i^3c^2)(\mathbf{s}_i.\mathbf{B}_i)\pi_i^2 && \text{(d)} \\ &\quad + (e_i\hbar^2/8m_i^2c^2)(\boldsymbol{\nabla}.\mathbf{E}_i) && \text{(e)} \\ &\quad + \mathrm{O}(mc^2\alpha^6) \end{aligned}\right\} \tag{9.16}$$

and the wave function ψ has two components.

For two electrons the Hamiltonian can be written as:

$$\mathcal{H} = \mathcal{H}_1 + \mathcal{H}_2 = \sum_{i=1,2} \mathcal{H}_i, \tag{9.17}$$

where each electron contributes to the potentials experienced by the other, that is the potentials experienced by electron i are given by Equations 9.14 and 9.15. The wave function associated with this Hamiltonian has four components, since each electron has two possible spin orientations.

All that is necessary is to substitute for the potentials, and simplify the resulting terms, so that in the final Hamiltonian any potentials that appear are just the external potentials. However, at this stage a further difficulty appears, since for any effect that one electron has on the other there will be a reciprocal effect, and a decision has to be made as to which interactions should be counted twice and which should not. That some interactions should only be counted once, may be seen by referring to Section 5.9. There the Lagrangian (Equation 5.76) for one charged particle contains terms arising from the interaction of its charge with the potentials due to the charge of another particle. However, in the Lagrangian (Equation 5.77) for both particles together these same terms appear once, not twice, since the interaction is symmetric between the two particles. If the Lagrangians for the individual particles had been added together a correction factor of $\frac{1}{2}$ would have been needed in these interaction terms. Similar remarks apply to the corresponding Hamiltonians. In deciding which interaction terms should only be counted once (or which should have correction factors of $\frac{1}{2}$), we may be guided by their symmetry with respect to interchange of the two particles, but the ultimate decision will be made so that agreement with other methods is achieved.

For simplicity it is supposed that the external potentials, $\mathbf{A}$ and ϕ, vanish, since they can be reintroduced later by the usual replacement:

$$\mathbf{p}_i \longrightarrow \boldsymbol{\pi}_i = \mathbf{p}_i + e_i\mathbf{A}; \qquad \mathcal{H} \longrightarrow \mathcal{H} + \sum_i e_i\phi. \tag{9.18}$$

The influence of the scalar potentials due to the electrons is discussed first.

The term $-e_i\phi_i$ in 9.16a makes a contribution:

$$e_1e_2/4\pi\epsilon_0 r_{12} \tag{9.19}$$

for each of the electrons. This is the usual Coulomb interaction term, and being symmetric between the two electrons should only be counted once, as is confirmed by the non-quantum-mechanical theory in Section 5.9.

This scalar potential also makes a significant contribution through the electric field $\mathbf{E}_i$ in the terms 9.16c and 9.16e. Of course the electric field also depends on the vector potential (Equation 5.21), but this contribution is negligible, so that:

$$\mathbf{E}_i = -\nabla\phi_i - \partial\mathbf{A}_i/\partial t = \nabla(e_j/4\pi\epsilon_0 r_{ij}). \tag{9.20}$$

Now, since ∇ operates on the position of electron i only, Equation 2.78 shows that:

$$\mathbf{E}_i = -(e_j/4\pi\epsilon_0)\mathbf{r}_{ij}/r_{ij}^3. \tag{9.21}$$

In the term 9.16c the vector potential also appears by virtue of the presence of $\boldsymbol{\pi}_i$, but again the contribution is negligible.

Hence the contribution of the term 9.16c to $\mathcal{H}_1$ due to the presence of electron 2 is:

$$(e_1e_2\hbar/16\pi\epsilon_0m_1^2c^2)\mathbf{s}_1.[\mathbf{p}_1\wedge(\mathbf{r}_{12}/r_{12}^3)-(\mathbf{r}_{12}/r_{12}^3)\wedge\mathbf{p}_1]+O(mc^2\alpha^5). \quad (9.22)$$

This may be simplified by noting that $\mathbf{p}_1\wedge(\mathbf{r}_{12}/r_{12}^3)$ is just equal to $-(\mathbf{r}_{12}/r_{12}^3)\wedge\mathbf{p}_1$, as may easily be shown using the techniques of Chapter 2 and in particular Equation 2.96. There is a similar contribution to $\mathcal{H}_2$ due to the presence of electron 1 and, since each of these terms is not symmetric between the two electrons, they must both be counted. The total contribution of the term 9.16c to the Hamiltonian for two electrons is thus:

$$-(e_1e_2\hbar/8\pi\epsilon_0c^2r_{12}^3)[\mathbf{s}_1.(\mathbf{r}_{12}\wedge\mathbf{p}_1)/m_1^2-\mathbf{s}_2.(\mathbf{r}_{12}\wedge\mathbf{p}_2)/m_2^2]+O(mc^2\alpha^5), \quad (9.23)$$

since $\mathbf{r}_{21}=-\mathbf{r}_{12}$ (Equation 9.13). This expression represents the spin-orbit interaction, which is the interaction between the spin and orbital motion of the same electron. It arises from the electric field produced by the other electron, since by virtue of its motion the electron experiences a magnetic field which interacts with its spin magnetic moment.

The term 9.16e may be dealt with more easily, since from Equations 9.21 and 2.95:

$$\nabla.\mathbf{E}_i = -(e_j/4\pi\epsilon_0)\nabla.(\mathbf{r}_{ij}/r_{ij}^3) = -(e_j/\epsilon_0)\delta(\mathbf{r}_{ij}). \quad (9.24)$$

Again it is clear that each of the two possible contributions to the Hamiltonian is not symmetric between the two electrons and that both must be counted, so that the term 9.16e is responsible for:

$$-(e_1e_2\hbar^2/8\epsilon_0c^2)\delta(\mathbf{r}_{12})(1/m_1^2+1/m_2^2)+O(mc^2\alpha^5). \quad (9.25)$$

This term is similar to the Darwin term (8.142e), which is a correction to the interaction of the charge of an electron with an electric field due to the Zitterbewegung; in this case the other electron is responsible for the electric field.

All the significant terms that arise from the scalar potentials of the electrons have now been considered. It has already been noted that the vector potentials of the electrons only contribute significantly to the final Hamiltonian through the term 9.16b and the term $\pi_i^2/2m_i$ in 9.16a. For potentials that satisfy the Coulomb gauge condition this latter term can be expanded (Equation 7.14):

$$\pi_i^2/2m_i = p_i^2/2m_i+e_i\mathbf{A}_i.\mathbf{p}_i/m_i+e_i^2A_i^2/2m_i; \quad (9.26)$$

the last term is negligible and only the middle term need be considered further. The orbital parts of the vector potentials due to electrons 2 and 1 give rise to similar terms when substituted in $\mathcal{H}_1$ and $\mathcal{H}_2$, respectively. Hence, as is indicated by the non-quantal theory of Section 5.9, this contribution to the Hamiltonian should only be counted once:

$$-(e_1e_2/8\pi\epsilon_0 m_1 m_2 c^2)[\mathbf{p}_1.(r_{12}^{-1}\mathbf{p}_2) + (\mathbf{p}_1.\mathbf{r}_{12})r_{12}^{-3}(\mathbf{r}_{12}.\mathbf{p}_2)] + O(mc^2\alpha^5). \qquad (9.27)$$

In the present approach there is no way of determining the correct order of operators in this expression, although it can be required to be Hermitian and symmetric to interchange of the two electrons. Both these conditions are satisfied by the term 9.27 and it will be seen later that the order of the operators as written here is in fact correct. The interpretation of the term 9.27 is that it represents the interaction of one electron with the magnetic field produced by the orbital motion of the other and vice versa; it is referred to as the orbit-orbit interaction and includes a correction for retardation.

The vector potential due to the spin of electron j provides, through the second term in Equation 9.26, a contribution:

$$-(e_ie_j\hbar/4\pi\epsilon_0 m_i m_j c^2)(\mathbf{s}_j \wedge \mathbf{r}_{ij}/r_{ij}^3).\mathbf{p}_i + O(mc^2\alpha^5). \qquad (9.28)$$

Since it is the spin of j but the momentum of i that appears in this term, it is not symmetric between the two electrons, so that the total contribution to the Hamiltonian is:

$$+(e_1e_2\hbar/4\pi\epsilon_0 m_1 m_2 c^2 r_{12}^3)[\mathbf{s}_1.(\mathbf{r}_{12} \wedge \mathbf{p}_2) - \mathbf{s}_2.(\mathbf{r}_{12} \wedge \mathbf{p}_1)] + O(mc^2\alpha^5). \qquad (9.29)$$

This represents a spin-other-orbit interaction, since one electron interacts with the magnetic field produced by the spin of the other.

Finally the term 9.16b must be considered and for this the magnetic field at electron i due to electron j is needed; from Equation 5.20 it is given by $\mathbf{B}_i = \nabla \wedge \mathbf{A}_i$. Direct use of the orbital part of Equation 9.14 together with liberal use of the techniques of Chapter 2 shows that the orbital contribution to this magnetic field is:

$$(e_j/4\pi\epsilon_0 m_j c^2 r_{ij}^3)(\mathbf{r}_{ij} \wedge \mathbf{p}_j). \qquad (9.30)$$

This result may be obtained more easily by using Equations 5.72 instead of Equations 5.73, since these are related by a gauge transformation and only differ by the gradient of a scalar, the contribution of which to the field vanishes because the curl of a gradient is zero (Equation 2.70).

The spin contribution to $\mathbf{B}_i$ is:

$$-(e_j\hbar/4\pi\epsilon_0 m_j c^2)\nabla \wedge (\mathbf{s}_j \wedge \mathbf{r}_{ij})/r_{ij}^3, \qquad (9.31)$$

where ∇ operates on the position of electron i. Use of Equation 2.68 shows that:

$$\nabla \wedge (\mathbf{s}_j \wedge \mathbf{r}_{ij})/r_{ij}^3 = \mathbf{s}_j\nabla.(\mathbf{r}_{ij}/r_{ij}^3) - (\mathbf{s}_j.\nabla)\mathbf{r}_{ij}/r_{ij}^3 , \qquad (9.32)$$

since $\mathbf{s}_j$ and $\boldsymbol{\nabla}$ commute. The first term is just $4\pi\delta(\mathbf{r}_{ij})\mathbf{s}_j$ from Equation 2.95, while the second term may be evaluated using Equation 2.96:

$$(\mathbf{s}_j.\boldsymbol{\nabla})\mathbf{r}_{ij}/r_{ij}^3 = (4\pi/3)\delta(\mathbf{r}_{ij})\mathbf{s}_j + \mathbf{s}_j/r_{ij}^3 - 3(\mathbf{s}_j.\mathbf{r}_{ij})\mathbf{r}_{ij}/r_{ij}^5, \qquad (9.33)$$

where it should be remembered that in the present context the subscripts i and j refer to electrons not vector components. The contribution of the spin magnetic moment of electron j to the magnetic field at electron i is thus:

$$-(e_j\hbar/4\pi\epsilon_0 m_j c^2)[(8\pi/3)\delta(\mathbf{r}_{ij})\mathbf{s}_j - \mathbf{s}_j/r_{ij}^3 + 3(\mathbf{s}_j.\mathbf{r}_{ij})\mathbf{r}_{ij}/r_{ij}^5]. \qquad (9.34)$$

When the orbital part (9.30) of $\mathbf{B}_i$ is substituted into 9.16b, two different terms arise, one for each electron, so that the contribution to the Hamiltonian is:

$$+(e_1e_2\hbar/4\pi\epsilon_0 m_1 m_2 c^2 r_{12}^3)[\mathbf{s}_1.(\mathbf{r}_{12}\wedge\mathbf{p}_2) - \mathbf{s}_2.(\mathbf{r}_{12}\wedge\mathbf{p}_1)] + O(mc^2\alpha^5). \qquad (9.35)$$

This term is the same as 9.29 and provides another contribution to the spin-other-orbit interaction. In this case the spin of each electron interacts with the magnetic field produced by the orbital motion of the other, whereas the term 9.29 is interpreted as being due to the motion of each electron in the presence of the magnetic field due to the spin of the other. It may be argued that these two interactions are really one and the same and that only one of the terms, 9.29 and 9.35, should be counted. This is in fact necessary if agreement is to be obtained with the Hamiltonian obtained from the Breit equation.

Finally, the spin part (9.34) of $\mathbf{B}_i$ must be substituted into 9.16b and for each electron an identical term arises, which is consequently only included once in the final Hamiltonian:

$$+(e_1e_2\hbar^2/4\pi\epsilon_0 m_1 m_2 c^2)[(\mathbf{s}_1.\mathbf{s}_2)/r_{12}^3 - 3(\mathbf{s}_1.\mathbf{r}_{12})(\mathbf{r}_{12}.\mathbf{s}_2)/r_{12}^5$$
$$-(8\pi/3)\delta(\mathbf{r}_{12})(\mathbf{s}_1.\mathbf{s}_2)] + O(mc^2\alpha^5). \qquad (9.36)$$

This arises from the interaction of the spin magnetic moment of each electron with the magnetic field of the spin magnetic moment of the other. It thus represents the spin-spin interaction, the first two terms in 9.36 being analogous to the classical dipole-dipole interaction, while the term involving the Dirac delta function represents a Fermi-contact-type interaction for which the two electrons must coincide.

All the significant interaction terms in the two-electron Hamiltonian have now been derived. They represent the Coulomb (9.19), the orbit-orbit (9.27), the spin-orbit (9.23), the spin-other-orbit (9.29 and 9.35) and the spin-spin (9.36) interactions; in addition there is a term (9.25) that is analogous to the Darwin term in the one-electron Hamiltonian. If external electromagnetic fields are reintroduced by the replacements 9.18, the total Hamiltonian for two electrons is given by:

$$
\begin{aligned}
\mathcal{H} = \sum_{i=1,2} & [m_i c^2 - e_i\phi_i + \pi_i^2/2m_i + (e_i\hbar/m_i)(\mathbf{s}_i.\mathbf{B}_i) \\
& - (e_i\hbar/4m_i^2c^2)\mathbf{s}_i.(\boldsymbol{\pi}_i \wedge \mathbf{E}_i - \mathbf{E}_i \wedge \boldsymbol{\pi}_i) - (1/8m_i^3c^2)\pi_i^4 \\
& - (e_i\hbar/2m_i^3c^2)(\mathbf{s}_i.\mathbf{B}_i)\pi_i^2 + (e_i\hbar^2/8m_i^2c^2)(\boldsymbol{\nabla}.\mathbf{E}_i)] \\
& + (e_1e_2/4\pi\epsilon_0 r_{12}) \\
& - (e_1e_2/8\pi\epsilon_0 m_1 m_2 c^2)[\boldsymbol{\pi}_1.(r_{12}^{-1}\boldsymbol{\pi}_2) + (\boldsymbol{\pi}_1.\mathbf{r}_{12})r_{12}^{-3}(\mathbf{r}_{12}.\boldsymbol{\pi}_2)] \\
& - (e_1e_2\hbar/8\pi\epsilon_0 c^2 r_{12}^3)[\mathbf{s}_1.(\mathbf{r}_{12} \wedge \boldsymbol{\pi}_1)/m_1^2 - \mathbf{s}_2.(\mathbf{r}_{12} \wedge \boldsymbol{\pi}_2)/m_2^2] \\
& + (e_1e_2\hbar/4\pi\epsilon_0 m_1 m_2 c^2 r_{12}^3)[\mathbf{s}_1.(\mathbf{r}_{12} \wedge \boldsymbol{\pi}_2) - \mathbf{s}_2.(\mathbf{r}_{12} \wedge \boldsymbol{\pi}_1)] \\
& + (e_1e_2\hbar^2/4\pi\epsilon_0 m_1 m_2 c^2)[(\mathbf{s}_1.\mathbf{s}_2)/r_{12}^3 - 3(\mathbf{s}_1.\mathbf{r}_{12})(\mathbf{r}_{12}.\mathbf{s}_2)/r_{12}^5 \\
& \qquad - (8\pi/3)\delta(\mathbf{r}_{12})(\mathbf{s}_1.\mathbf{s}_2)] \\
& - (e_1e_2\hbar^2/8\epsilon_0 c^2)\delta(\mathbf{r}_{12})(1/m_1^2 + 1/m_2^2) + O(mc^2\alpha^5). \qquad (9.37)
\end{aligned}
$$

The potentials ϕ_i and $\mathbf{A}_i$ and the fields $\mathbf{E}_i$ and $\mathbf{B}_i$ at electron i are now due entirely to external sources and not to the other electron.

In Equation 9.37 only one of the two contributions (9.29 and 9.35) to the spin-other-orbit interaction has been included. It is recalled that another difficulty associated with this method involves the order of the operators in the orbit-orbit interaction. In addition, the contribution (Equation 9.12) of the spin magnetic moment of one electron to the scalar potential at another has been omitted; if it had been retained there would have been an additional term in Equation 9.37 equal to twice the spin-orbit term (9.23). For these reasons this approach is not entirely satisfactory, but it does have the advantage that it exhibits the origin of each term in the Hamiltonian and allows a physical interpretation to be given to each.

9.3 The Breit equation

In the previous section a non-relativistic approximation to the Hamiltonian for two electrons was sought by starting with the non-relativistic approximation to the Dirac Hamiltonian. We now enquire whether it is possible to write down a relativistic Hamiltonian for two electrons.

Such a Hamiltonian would be expected to have the form:

$$\mathcal{H} = \mathcal{H}_1 + \mathcal{H}_2 + \mathcal{H}_{12}, \tag{9.38}$$

where $\mathcal{H}_i$ is the Dirac Hamiltonian for electron i:

$$\mathcal{H}_i = \beta_i m_i c^2 + c\boldsymbol{\alpha}_i.\boldsymbol{\pi}_i - e_i\phi_i \tag{9.39}$$

and $\mathcal{H}_{12}$ is an appropriate interaction term. The corresponding wave function has sixteen components, since the wave function for a single electron has four components. This means that the operators in the Hamiltonian (Equation 9.38) are represented by matrices of order sixteen and, for example, the Dirac operators β_i and $\boldsymbol{\alpha}_i$ in Equation 9.39 are now to be thought of as an appropriate direct product of β and $\boldsymbol{\alpha}$ with the unit matrix (see the next section). Consequently a relativistic wave equation for two electrons is rather complicated but one might expect it to be reducible to a non-relativistic form appropriate to two electrons that are both in positive energy states; the associated wave function would then have four components, since there are two spin orientations for each electron.

The form of the interaction term $\mathcal{H}_{12}$ in the Hamiltonian (Equation 9.38) must now be considered. As a leading term one would expect the Coulomb interaction, but a Hamiltonian:

$$\mathcal{H} = \mathcal{H}_1 + \mathcal{H}_2 + (e_1 e_2/4\pi\epsilon_0 r_{12}) \tag{9.40}$$

cannot be correct, since the wave equation with this Hamiltonian is not even approximately Lorentz invariant. The correction terms can only be obtained rigorously by using quantum electrodynamics. In principle they may be obtained to any order of accuracy, but in practice they can only be obtained approximately. It is beyond the scope of this book to go into the details, but essentially second-order perturbation theory is used to describe the emission of a virtual photon by one electron and its absorption by the other and vice versa; to do this the eigenvalues and eigenfunctions of the approximate Hamiltonian (Equation 9.40) are used. The result is an additional interaction term:

$$-(e_1 e_2/8\pi\epsilon_0 r_{12})[(\boldsymbol{\alpha}_1.\boldsymbol{\alpha}_2) + (\boldsymbol{\alpha}_1.\mathbf{r}_{12})(\boldsymbol{\alpha}_2.\mathbf{r}_{12})/r_{12}^2], \tag{9.41}$$

which is known as the Breit operator. Because perturbation theory is used and because of the approximations involved in its derivation, this operator must only be used for slowly moving electrons and then only in first order. The matrix representation of the operator (Equation 9.41) also involves matrices of order 16; in this case they are related to direct products of the matrix representations of the components of the Dirac operators $\boldsymbol{\alpha}_1$ and $\boldsymbol{\alpha}_2$.

Although we are not in a position to derive the Breit operator, it can be seen that it is approximately correct by referring to the classical (non-quantum-mechanical) Hamiltonian for two electrons, that was derived in Section 5.9. Equation 5.81 shows that the classical interaction Hamiltonian consists of the Coulomb term together with:

$$-(e_1e_2/8\pi\epsilon_0 m_1 m_2 r_{12}c^2)[(\mathbf{p}_1.\mathbf{p}_2) + (\mathbf{p}_1.\mathbf{r}_{12})(\mathbf{p}_2.\mathbf{r}_{12})/r_{12}^2] \quad (9.42)$$

or in terms of the electrons' velocities, $\mathbf{u}_1$ and $\mathbf{u}_2$:

$$-(e_1e_2/8\pi\epsilon_0 r_{12}c^2)[(\mathbf{u}_1.\mathbf{u}_2) + (\mathbf{u}_1.\mathbf{r}_{12})(\mathbf{u}_2.\mathbf{r}_{12})/r_{12}^2]. \quad (9.43)$$

This is a classical expression and in going over to a relativistic quantum mechanical theory it is appropriate to replace the velocity $\mathbf{u}_i$ of electron i by its Dirac velocity operator $c\boldsymbol{\alpha}_i$ (Equation 8.148). With this replacement, 9.43 becomes identical to the operator 9.41.

The Breit equation is thus $\mathcal{H}\psi = i\hbar(\partial\psi/\partial t)$ with the Breit Hamiltonian given by:

$$\begin{aligned}\mathcal{H} &= \mathcal{H}_1 + \mathcal{H}_2 + (e_1e_2/4\pi\epsilon_0 r_{12}) \\ &\quad - (e_1e_2/8\pi\epsilon_0 r_{12})[(\boldsymbol{\alpha}_1.\boldsymbol{\alpha}_2) + (\boldsymbol{\alpha}_1.\mathbf{r}_{12})(\boldsymbol{\alpha}_2.\mathbf{r}_{12})/r_{12}^2] + O(mc^2\alpha^5), \quad (9.44)\end{aligned}$$

where $\mathcal{H}_i$ is the Dirac Hamiltonian for electron i (Equation 9.39). It is clear that like the Dirac equation the Breit equation is gauge invariant, since it can be written in terms of the operators $\boldsymbol{\pi}_i$ and $(i\hbar\partial/\partial t + \sum_i e_i\phi_i)$ (see Section 7.2).

Although both the Coulomb term and the Breit operator are formally of order $mc^2\alpha^2$ in a molecular environment, the Breit Hamiltonian (Equation 9.44) is in fact correct to order $mc^2\alpha^4$. It must be emphasized that it is an approximate Hamiltonian, since it is not Lorentz invariant and the effects of quantum electrodynamics have been introduced using perturbation theory. Again we note that because of this the Breit operator should only be used to first order in any application. In addition, the Breit equation should only be used for electrons moving with velocities much less than the speed of light.

9.4 Reduction of the Breit equation to non-relativistic form

It is appropriate to reduce the Breit equation to a non-relativistic form, since it is only valid for electrons with low velocities. In addition, a four-component equation is more convenient than one with sixteen components and the resulting Hamiltonian can be treated with well-established techniques. Both the method of small components (Section 8.8) and a Foldy-Wouthuysen-type reduction may be used, but the latter has attracted more attention and, being more straightforward, it is this approach that is presented here.

As might be imagined the labour involved is considerably greater than that needed for the reduction of the Dirac equation and only an outline of the method is given. For simplicity the time-independent problem is discussed, but there is no reason to suppose that the time-dependent Hamiltonian does not have the same form.

In the Foldy-Wouthuysen reduction (Section 8.9) of the Dirac equation the operators in the Dirac Hamiltonian are classified as even ($\mathcal{E}$) or odd ($\mathcal{O}$) depending on whether they have no matrix elements or only matrix elements between the positive and negative energy components of the wave function. More formally even operators commute and odd operators anticommute with the Dirac operator β. The odd operators are removed progressively by the application of a sequence of unitary transformations.

A more complicated classification of operators must be used for two electrons, since the wave function has sixteen components, which cannot be labelled simply as positive or negative energy components. However, the wave function can be written as:

$$\psi = \begin{pmatrix} \psi_{++} \\ \psi_{+-} \\ \psi_{-+} \\ \psi_{--} \end{pmatrix}, \tag{9.45}$$

where, for example, ψ_{++} has four components for which both electrons have positive energy, while for ψ_{+-} electron 1 has positive energy but electron 2 has negative energy. We are interested in decoupling ψ_{++} from ψ_{+-}, ψ_{-+} and ψ_{--} to obtain a four-component Hamiltonian for two electrons, which both have positive energy; this Hamiltonian is required to be correct to order $mc^2\alpha^4$.

It will be seen that the Breit Hamiltonian (Equation 9.44) can now be written:

$$\mathcal{H} = \beta_1 m_1 c^2 + \beta_2 m_2 c^2 + (\mathcal{E}\mathcal{E}) + (\mathcal{E}\mathcal{O}) + (\mathcal{O}\mathcal{E}) + (\mathcal{O}\mathcal{O}). \tag{9.46}$$

by analogy with Equation 8.117. This is just an extension of the Foldy-Wouthuysen notation, so that, for example, an even-even operator ($\mathcal{E}\mathcal{E}$) such as β_1 or β_2 is even with respect to both electrons 1 and 2. On the other hand the components of $\boldsymbol{\alpha}_2$ are even-odd operators ($\mathcal{E}\mathcal{O}$) and are even and odd with respect to electrons 1 and 2, respectively. As a further illustration, ψ_{++} is only connected with ψ_{-+} by an odd-even operator ($\mathcal{O}\mathcal{E}$).

These operators may be defined more formally. For an operator to be even with respect to electron 1 it must commute with β_1, but if it anticommutes with β_1 it is odd with respect to electron 1. Thus:

$$\begin{aligned}
[(\mathcal{E}\mathcal{E}),\beta_1] &= [(\mathcal{E}\mathcal{O}),\beta_1] = [(\mathcal{E}\mathcal{E}),\beta_2] = [(\mathcal{O}\mathcal{E}),\beta_2] = 0;\\
[(\mathcal{O}\mathcal{E}),\beta_1]_+ &= [(\mathcal{O}\mathcal{O}),\beta_1]_+ = [(\mathcal{E}\mathcal{O}),\beta_2]_+ = [(\mathcal{O}\mathcal{O}),\beta_2]_+ = 0.
\end{aligned} \tag{9.47}$$

It is now easy to show that the product of say an odd-even operator with an odd-odd operator gives an even-odd operator:

$$(\mathcal{O}\mathcal{E})(\mathcal{O}\mathcal{O}) = (\mathcal{E}\mathcal{O}) \tag{9.48}$$

and so on. With these definitions the various Dirac operators can now be classified:

$$\begin{aligned}
&(\mathcal{E}\mathcal{E}): && \beta_1, \beta_2, \text{ the } 16 \times 16 \text{ unit matrix};\\
&(\mathcal{E}\mathcal{O}): && \alpha_{2x}, \alpha_{2y}, \alpha_{2z};\\
&(\mathcal{O}\mathcal{E}): && \alpha_{1x}, \alpha_{1y}, \alpha_{1z};\\
&(\mathcal{O}\mathcal{O}): && \text{any product of a component of } \boldsymbol{\alpha}_1 \text{ with a component of } \boldsymbol{\alpha}_2.
\end{aligned} \tag{9.49}$$

All these assertions may be checked explicitly. With the choice of wave function (Equation 9.45) the matrix representation of the Dirac operators for electron 1 are:

$$\beta_1 = \begin{pmatrix} 1 & 0 & 0 & 0 \\ 0 & 1 & 0 & 0 \\ 0 & 0 & -1 & 0 \\ 0 & 0 & 0 & -1 \end{pmatrix}; \quad \boldsymbol{\alpha}_1 = \begin{pmatrix} 0 & 0 & \boldsymbol{\sigma}_1 & 0 \\ 0 & 0 & 0 & \boldsymbol{\sigma}_1 \\ \boldsymbol{\sigma}_1 & 0 & 0 & 0 \\ 0 & \boldsymbol{\sigma}_1 & 0 & 0 \end{pmatrix}, \tag{9.50}$$

where each element is itself a four-by-four matrix and in particular, if $\mathbf{e}_x$, $\mathbf{e}_y$ and $\mathbf{e}_z$ are unit orthogonal vectors:

$$\boldsymbol{\sigma}_1 = \begin{pmatrix} \mathbf{e}_z & 0 & (\mathbf{e}_x - i\mathbf{e}_y) & 0 \\ 0 & \mathbf{e}_z & 0 & (\mathbf{e}_x - i\mathbf{e}_y) \\ (\mathbf{e}_x + i\mathbf{e}_y) & 0 & -\mathbf{e}_z & 0 \\ 0 & (\mathbf{e}_x + i\mathbf{e}_y) & 0 & -\mathbf{e}_z \end{pmatrix}. \tag{9.51}$$

The corresponding matrices for electron 2 are:

$$\beta_2 = \begin{pmatrix} 1 & 0 & 0 & 0 \\ 0 & -1 & 0 & 0 \\ 0 & 0 & 1 & 0 \\ 0 & 0 & 0 & -1 \end{pmatrix}; \quad \boldsymbol{\alpha}_2 = \begin{pmatrix} 0 & \boldsymbol{\sigma}_2 & 0 & 0 \\ \boldsymbol{\sigma}_2 & 0 & 0 & 0 \\ 0 & 0 & 0 & \boldsymbol{\sigma}_2 \\ 0 & 0 & \boldsymbol{\sigma}_2 & 0 \end{pmatrix}, \tag{9.52}$$

where

$$\boldsymbol{\sigma}_2 = \begin{pmatrix} \mathbf{e}_z & (\mathbf{e}_x - i\mathbf{e}_y) & 0 & 0 \\ (\mathbf{e}_x + i\mathbf{e}_y) & -\mathbf{e}_z & 0 & 0 \\ 0 & 0 & \mathbf{e}_z & (\mathbf{e}_x - i\mathbf{e}_y) \\ 0 & 0 & (\mathbf{e}_x + i\mathbf{e}_y) & -\mathbf{e}_z \end{pmatrix} \qquad (9.53)$$

These matrix representations satisfy all the relationships they are expected to. The operators β_i and the components of $\boldsymbol{\alpha}_i$ all anticommute with one another and the square of each is the unit matrix of order sixteen. In addition, β_1 and the components of $\boldsymbol{\alpha}_1$ all commute with β_2 and the components of $\boldsymbol{\alpha}_2$, as they must, since they operate on different electrons.

We are now in a position to identify the operators in Equation 9.46:

$$\begin{aligned} (\mathcal{E}\mathcal{E}) &= -e_1\phi_1 - e_2\phi_2 + e_1e_2/4\pi\epsilon_0 r_{12} \sim mc^2\alpha^2; \\ (\mathcal{E}\mathcal{O}) &= c\boldsymbol{\alpha}_2.\boldsymbol{\pi}_2 \sim mc^2\alpha; \\ (\mathcal{O}\mathcal{E}) &= c\boldsymbol{\alpha}_1.\boldsymbol{\pi}_1 \sim mc^2\alpha; \\ (\mathcal{O}\mathcal{O}) &= -(e_1e_2/8\pi\epsilon_0 r_{12})[(\boldsymbol{\alpha}_1.\boldsymbol{\alpha}_2) + (\boldsymbol{\alpha}_1.\mathbf{r}_{12})(\boldsymbol{\alpha}_2.\mathbf{r}_{12})/r_{12}^2] \sim mc^2\alpha^2. \end{aligned} \qquad (9.54)$$

From this identification it is clear that:

$$[(\mathcal{E}\mathcal{O}), (\mathcal{O}\mathcal{E})] = 0; \qquad [(\mathcal{E}\mathcal{E}), (\mathcal{O}\mathcal{O})] = 0, \qquad (9.55)$$

commutation relations that are useful in the subsequent development.

With these preliminaries settled, we may return to the original problem of reducing to non-relativistic form the Hamiltonian (Equation 9.46), which operates on the wave function (Equation 9.45). To achieve this, unwanted operators are removed by the Foldy-Wouthuysen method of subjecting the wave function to a unitary transformation:

$$\psi' = \exp(iS)\psi. \qquad (9.56)$$

At the same time the Hamiltonian is modified and for the time-independent problem the transformed Hamiltonian is, from Equations 8.123 and 8.124:

$$\begin{aligned} \mathcal{H}_1 &= \exp(iS)\mathcal{H}\exp(-iS) \\ &= \mathcal{H} + i[S, \mathcal{H}] + (i^2/2!)[S, [S, \mathcal{H}]] + \dots . \end{aligned} \qquad (9.57)$$

This transformation introduces new undesirable operators into the Hamiltonian, but, if the Hermitian operator S is correctly chosen, these are formally smaller than the operators removed. This procedure is repeated until the Hamiltonian is correct to the desired accuracy, in this case $mc^2\alpha^4$.

It might be thought that it is necessary to remove all even-odd, odd-even and odd-odd operators from the Hamiltonian so that ψ_{++}, ψ_{+-}, ψ_{-+} and ψ_{--} are completely decoupled from one another. To do this the operator S must be:

$$S = -i\beta_1(\mathcal{O}\mathcal{E})/2m_1c^2 - i\beta_2(\mathcal{E}\mathcal{O})/2m_2c^2$$
$$-i(\beta_1 m_1 - \beta_2 m_2)(\mathcal{O}\mathcal{O})/2(m_1^2 - m_2^2); \qquad (9.58)$$

the first two terms arise by analogy to the one electron case (Equation 8.125) while the effectiveness of the third term in removing $(\mathcal{O}\mathcal{O})$ can easily be checked. This operator has the disadvantage that for two electrons $m_1 = m_2$ and the third term in Equation 9.58 has a zero denominator. This problem may be circumvented by pretending that the electrons have different masses until after β_1 and β_2 have been set equal to the unit four-by-four matrix to obtain a Hamiltonian appropriate to both electrons having positive energy. Although the resulting Hamiltonian does not contain infinities and in fact it is the same as the one to be derived shortly, it is not very satisfying to ignore this difficulty; in addition, the problem is not avoided if, for example, β_1 is set equal to $+1$ and β_2 to -1 in an attempt to obtain the Hamiltonian for two electrons, one with positive energy and the other with negative energy.

However, to obtain a Hamiltonian appropriate to two electrons, which both have positive energy, it is not necessary to seek a reduced Hamiltonian containing only even-even operators. This completely decouples ψ_{++}, ψ_{+-}, ψ_{-+} and ψ_{--} from each other, but to achieve our objective it is only necessary to decouple ψ_{++} from the other three; it does not matter if ψ_{+-}, ψ_{-+} and ψ_{--} are not uncoupled from each other. In fact Equation 9.58 is just one of a family of operators that achieve this more limited aim, it just happens to go further than is necessary and in so doing introduces the problem associated with the zero denominator in Equation 9.58.

Pictorially, use of Equation 9.58 leads to a transformed Hamiltonian in which the operators have only matrix representations with the structure:

$$\begin{pmatrix} X & 0 & 0 & 0 \\ 0 & X & 0 & 0 \\ 0 & 0 & X & 0 \\ 0 & 0 & 0 & X \end{pmatrix}, \qquad (9.59)$$

where the X's indicate any matrix of order four having non-zero elements. It is only necessary to obtain a Hamiltonian whose matrix representation has the structure:

$$\begin{pmatrix} X & 0 & 0 & 0 \\ 0 & X & X & X \\ 0 & X & X & X \\ 0 & X & X & X \end{pmatrix}. \tag{9.60}$$

As it happens the transformation that achieves this least permissible separation is quite complicated and the one given here is the simplest that does not run into problems with zero denominators. This simplest transformation gives a Hamiltonian with the structure:

$$\begin{pmatrix} X & 0 & 0 & 0 \\ 0 & X & X & 0 \\ 0 & X & X & 0 \\ 0 & 0 & 0 & X \end{pmatrix}, \tag{9.61}$$

as will be seen shortly.

For the transformation used here:

$$S = -i\beta_1(\mathcal{O}\mathcal{E})/2m_1c^2 - i\beta_2(\mathcal{E}\mathcal{O})/2m_2c^2$$
$$-i(\beta_1+\beta_2)(\mathcal{O}\mathcal{O})/4(m_1+m_2)c^2. \tag{9.62}$$

The first two terms lead to the removal of the $(\mathcal{O}\mathcal{E})$ and $(\mathcal{E}\mathcal{O})$ operators from the Hamiltonian (Equation 9.46), although this is at the expense of introducing smaller operators of the same type. However, the last term, in addition to introducing smaller undesirable operators, leaves in the transformed Hamiltonian a term:

$$(1-\beta_1\beta_2)(\mathcal{O}\mathcal{O})/2. \tag{9.63}$$

Although this is an odd-odd operator it has the structure:

$$\begin{pmatrix} 0 & 0 & 0 & 0 \\ 0 & 0 & X & 0 \\ 0 & X & 0 & 0 \\ 0 & 0 & 0 & 0 \end{pmatrix}. \tag{9.64}$$

In particular in extracting a Hamiltonian appropriate to two electrons, which both have positive energy, β_1 and β_2 are both set equal to the unit matrix of order four and the operator 9.63 vanishes.

Although Equation 9.62 is the simplest acceptable choice for S, its substitution in Equation 9.57 and the subsequent expansion is quite laborious even when the commutation relations 9.55 are used; consequently only the result is given here. Two applications of the transformation are necessary to obtain a Hamiltonian, which is of the desired type and correct to order $mc^2\alpha^4$. The order of magnitude of the terms is estimated on the basis (Equations 9.54) that the $(\mathcal{E}\mathcal{E})$ and $(\mathcal{O}\mathcal{O})$ operators in Equation 9.46 are of order $mc^2\alpha^2$, while the $(\mathcal{E}\mathcal{O})$ and $(\mathcal{O}\mathcal{E})$ operators are both of order $mc^2\alpha$. The result is:

$$\begin{aligned}\mathcal{H}_2 = {} & \beta_1 mc^2 + \beta_2 mc^2 + (\mathcal{E}\mathcal{E}) + \beta_1(\mathcal{O}\mathcal{E})^2/2m_1c^2 + \beta_2(\mathcal{E}\mathcal{O})^2/2m_2c^2 \\ & - \beta_1(\mathcal{O}\mathcal{E})^4/8m_1^3c^6 - \beta_2(\mathcal{E}\mathcal{O})^4/8m_2^3c^6 - [(\mathcal{O}\mathcal{E}), [(\mathcal{O}\mathcal{E}), (\mathcal{E}\mathcal{E})]\,]/8m_1^2c^4 \\ & - [(\mathcal{E}\mathcal{O}), [(\mathcal{E}\mathcal{O}), (\mathcal{E}\mathcal{E})]\,]/8m_2^2c^4 + \beta_1\beta_2[(\mathcal{E}\mathcal{O}), [(\mathcal{O}\mathcal{E}), (\mathcal{O}\mathcal{O})]_+]_+/4m_1m_2c^4 \\ & + (\beta_1 + \beta_2)(\mathcal{O}\mathcal{O})^2/4(m_1 + m_2)c^2 + (1 - \beta_1\beta_2)(\mathcal{O}\mathcal{O})_1/2 + \mathrm{O}(mc^2\alpha^5).\end{aligned} \tag{9.65}$$

All the operators in this transformed Hamiltonian are of the even-even type except for the last, which is odd-odd and involves $(\mathcal{O}\mathcal{O})_1$, the sum of $(\mathcal{O}\mathcal{O})$ and other odd-odd operators introduced by the first unitary transformation.

The penultimate term in Equation 9.65 contains $(\mathcal{O}\mathcal{O})^2$, but $(\mathcal{O}\mathcal{O})$ is just the Breit operator and it has already been emphasized that it must only be used to first order; in fact retention of this term leads to discrepancies between theory and experiment for the atomic spectrum of helium. When this term is omitted and β_1 and β_2 are set equal to the four-by-four unit matrix, we obtain the four-component Hamiltonian appropriate to two electrons, which both have positive energy:

$$\begin{aligned}\mathcal{H} = {} & m_1c^2 + m_2c^2 + (\mathcal{E}\mathcal{E}) + (\mathcal{O}\mathcal{E})^2/2m_1c^2 + (\mathcal{E}\mathcal{O})^2/2m_2c^2 \\ & - (\mathcal{O}\mathcal{E})^4/8m_1^3c^6 - (\mathcal{E}\mathcal{O})^4/8m_2^3c^6 - [(\mathcal{O}\mathcal{E}), [(\mathcal{O}\mathcal{E}), (\mathcal{E}\mathcal{E})]\,]/8m_1^2c^4 \\ & - [(\mathcal{E}\mathcal{O}), [(\mathcal{E}\mathcal{O}), (\mathcal{E}\mathcal{E})]\,]/8m_2^2c^4 + [(\mathcal{E}\mathcal{O}), [(\mathcal{O}\mathcal{E}), (\mathcal{O}\mathcal{O})]_+]_+/4m_1m_2c^4 \\ & + \mathrm{O}(mc^2\alpha^5),\end{aligned} \tag{9.66}$$

where the subscript has been dropped from $\mathcal{H}$.

Of course the explicit forms of the operators $(\mathcal{E}\mathcal{E})$, $(\mathcal{E}\mathcal{O})$, $(\mathcal{O}\mathcal{E})$ and $(\mathcal{O}\mathcal{O})$ given in Equations 9.54 must now be substituted into Equation 9.66. Some of the work has already been done, since reference to Equations 8.137 and 8.138 shows that:

$$(\mathcal{O}\mathcal{E})^2/2m_1c^2 = \boldsymbol{\pi}_1^2/2m_1 + (e_1\hbar/2m_1)(\boldsymbol{\sigma}_1.\mathbf{B}_1) \tag{9.67}$$

and

$$-(\mathcal{O}\mathcal{E})^4/8m_1^3c^6 = -(1/8m_1^3c^2)\pi_1^4 - (e_1\hbar/4m_1^3c^2)(\boldsymbol{\sigma}_1.\mathbf{B}_1)\pi_1^2 + O(mc^2\alpha^6), \tag{9.68}$$

with similar expressions for the terms involving $(\mathcal{E}\mathcal{O})^2$ and $(\mathcal{E}\mathcal{O})^4$. However, the remaining terms are more complicated, but liberal use of the techniques of Chapter 2 and relations such as Equation 8.40 shows quite straightforwardly that:

$$\begin{aligned}
-[(\mathcal{O}\mathcal{E}),[(\mathcal{O}\mathcal{E}),(\mathcal{E}\mathcal{E})]]/8m_1^2c^4 &= (e_1\hbar^2/8m_1^2c^2)(\nabla.\mathbf{E}_1)\\
&\quad -(e_1\hbar/8m_1^2c^2)\boldsymbol{\sigma}_1.[\boldsymbol{\pi}_1\wedge\mathbf{E}_1-\mathbf{E}_1\wedge\boldsymbol{\pi}_1]\\
&\quad -(e_1e_2\hbar/16\pi\epsilon_0c^2r_{12}^3)\boldsymbol{\sigma}_1.(\mathbf{r}_{12}\wedge\boldsymbol{\pi}_1)/m_1^2\\
&\quad -(e_1e_2\hbar^2/8\epsilon_0c^2)\delta(\mathbf{r}_{12})/m_1^2,
\end{aligned} \tag{9.69}$$

with a similar expression for the term involving $[(\mathcal{E}\mathcal{O}),[(\mathcal{E}\mathcal{O}),(\mathcal{E}\mathcal{E})]]$ except that the subscripts 1 and 2 are interchanged. With even more labour it may also be shown that:

$$\begin{aligned}
&+[(\mathcal{E}\mathcal{O}),[(\mathcal{O}\mathcal{E}),(\mathcal{O}\mathcal{O})]_+]_+/4m_1m_2c^4\\
&= (e_1e_2/8\pi\epsilon_0m_1m_2c^2)[\boldsymbol{\pi}_1.(r_{12}^{-1}\boldsymbol{\pi}_2)+(\boldsymbol{\pi}_1.\mathbf{r}_{12})r_{12}^{-3}(\mathbf{r}_{12}.\boldsymbol{\pi}_2)]\\
&\quad +(e_1e_2\hbar/8\pi\epsilon_0m_1m_2c^2r_{12}^3)[\boldsymbol{\sigma}_1.(\mathbf{r}_{12}\wedge\boldsymbol{\pi}_2)-\boldsymbol{\sigma}_2.(\mathbf{r}_{12}\wedge\boldsymbol{\pi}_1)]\\
&\quad +(e_1e_2\hbar^2/16\pi\epsilon_0m_1m_2c^2)[(\boldsymbol{\sigma}_1.\boldsymbol{\sigma}_2)/r_{12}^3-3(\boldsymbol{\sigma}_1.\mathbf{r}_{12})(\mathbf{r}_{12}.\boldsymbol{\sigma}_2)/r_{12}^5\\
&\qquad -(8\pi/3)\delta(\mathbf{r}_{12})(\boldsymbol{\sigma}_1.\boldsymbol{\sigma}_2)].
\end{aligned} \tag{9.70}$$

The final Hamiltonian may now be assembled by substituting Equations 9.54 and 9.67 to 9.70 into Equation 9.66:

$$\begin{aligned}
\mathcal{H} = &\sum_{i=1,2}[m_ic^2-e_i\phi_i+\pi_i^2/2m_i+(e_i\hbar/2m_i)(\boldsymbol{\sigma}_i.\mathbf{B}_i)\\
&-(e_i\hbar/8m_i^2c^2)\boldsymbol{\sigma}_i.[\boldsymbol{\pi}_i\wedge\mathbf{E}_i-\mathbf{E}_i\wedge\boldsymbol{\pi}_i]-(1/8m_i^3c^2)\pi_i^4\\
&-(e_i\hbar/4m_i^3c^2)(\boldsymbol{\sigma}_i.\mathbf{B}_i)\pi_i^2+(e_i\hbar^2/8m_i^2c^2)(\nabla.\mathbf{E}_i)]\\
&+(e_1e_2/4\pi\epsilon_0r_{12})\\
&-(e_1e_2/8\pi\epsilon_0m_1m_2c^2)[\boldsymbol{\pi}_1.(r_{12}^{-1}\boldsymbol{\pi}_2)+(\boldsymbol{\pi}_1.\mathbf{r}_{12})r_{12}^{-3}(\mathbf{r}_{12}.\boldsymbol{\pi}_2)]\\
&-(e_1e_2\hbar/16\pi\epsilon_0c^2r_{12}^3)[\boldsymbol{\sigma}_1.(\mathbf{r}_{12}\wedge\boldsymbol{\pi}_1)/m_1^2-\boldsymbol{\sigma}_2.(\mathbf{r}_{12}\wedge\boldsymbol{\pi}_2)/m_2^2]\\
&+(e_1e_2\hbar/8\pi\epsilon_0m_1m_2c^2r_{12}^3)[\boldsymbol{\sigma}_1.(\mathbf{r}_{12}\wedge\boldsymbol{\pi}_2)-\boldsymbol{\sigma}_2.(\mathbf{r}_{12}\wedge\boldsymbol{\pi}_1)]
\end{aligned}$$

$$+ (e_1e_2\hbar^2/16\pi\epsilon_0 m_1 m_2 c^2)[(\boldsymbol{\sigma}_1.\boldsymbol{\sigma}_2)/r_{12}^3 - 3(\boldsymbol{\sigma}_1.\mathbf{r}_{12})(\mathbf{r}_{12}.\boldsymbol{\sigma}_2)/r_{12}^5 - (8\pi/3)\delta(\mathbf{r}_{12})(\boldsymbol{\sigma}_1.\boldsymbol{\sigma}_2)]$$

$$- (e_1e_2\hbar^2/8\epsilon_0 c^2)\delta(\mathbf{r}_{12})(1/m_1^2 + 1/m_2^2) + O(mc^2\alpha^5). \quad (9.71)$$

This Hamiltonian is identical to that obtained in Section 9.2, since it is recalled that $\boldsymbol{\sigma}_i$ has been replaced by $2\mathbf{s}_i$ in Equation 9.37. However, in deriving the Hamiltonian (Equation 9.37) certain problems were resolved by insisting that there should be eventual agreement with Equation 9.71. The advantage of the method of Section 9.2 is that the physical significance of the individual terms is quite clear from the way in which they are derived. In the present method the origin of some of the terms is not obvious and the amount of labour involved is substantial, but it is free of the objections to the method of Section 9.2. Finally we again note that the same Hamiltonian has been obtained using quantum electrodynamics.

9.5 Radiative corrections

The Dirac equation for the electron predicts a spin magnetic moment of $-(e\hbar/2m)\boldsymbol{\sigma}$ and a spin angular momentum of $\frac{1}{2}\hbar\boldsymbol{\sigma}$ (Section 8.4). The g factor of the electron is the ratio of the magnitude of these two quantities measured in units of $(e/2m)$. Consequently the Dirac equation leads to a g factor for the electron of exactly 2.

The transition to the notation of non-relativistic quantum theory can be made by replacing $\boldsymbol{\sigma}$ by $2\mathbf{s}$. This has been done in Equations 8.142 and 9.37, which are the non-relativistic approximations of the Dirac and Breit Hamiltonians, respectively. Now in these approximate Hamiltonians each of the terms involving the spin of an electron arises by virtue of its spin magnetic moment; the spin angular momentum does not appear as such in the Hamiltonian. Thus, rather than replace $\boldsymbol{\sigma}$ by $2\mathbf{s}$, we could have replaced the spin magnetic moment of the electron, $-(e\hbar/2m)\boldsymbol{\sigma}$, by $-g(e\hbar/2m)\mathbf{s}$, where g is exactly 2. Since $(e\hbar/2m)$ is just the Bohr magneton μ_B, this replacement may be written:

$$-(e\hbar/2m)\boldsymbol{\sigma} \longrightarrow -g\mu_B\mathbf{s}. \quad (9.72)$$

When Dirac first put forward his theory, the experimental value of the g factor was also 2. However, since then experiment has shown that it differs slightly but significantly from 2. It is now one of the most accurately known constants and one of the more recent experimental determinations gives it as:

$$g = 2{\cdot}002\,319\,315\,(\pm 7 \times 10^{-9}). \quad (9.73)$$

The discrepancy between Dirac theory and experiment is attributed to quantum electrodynamic effects, one of the latest theoretical estimates being 2·002319312 ($\pm 3 \times 10^{-9}$) in good agreement with the experimental value (Equation 9.73); the error is due to uncertainty in the experimental value of the fine structure constant (see later). Since it is not possible to go into the details here, the quantum electrodynamic correction to the electron's g factor is included phenomenologically. Thus the replacement 9.72 is used, but with the experimental value of Equation 9.73 for g instead of the Dirac value of 2.

Although the quantum electrodynamic theory of the correction to the g factor cannot be given here, an attempt can be made to indicate its origin. It has been stated in Section 9.3 that the Breit interaction between two electrons is attributable to the emission of a virtual photon by one electron and its subsequent absorption by the other. In the same way an isolated electron can interact with its own virtual radiation field by emitting a virtual photon, which it later reabsorbs. In addition, the electron can polarize the surrounding electromagnetic field creating a virtual electron-positron pair followed by annihilation of the pair with emission of a virtual photon which is absorbed by the electron. (These concepts are not as obscure as they appear, since this description is analogous to the interaction of a ground state configuration with excited configurations in the electronic theory of molecules, the complications arising because the radiation field is also involved).

A qualitative picture of how this produces a contribution to the electron's spin magnetic moment has been given and is as follows. The spins of the electron and positron in the virtual pair are generally antiparallel and tend to be oriented so that one is parallel and one is antiparallel to the spin direction of the original electron. However, the case where the spin of the virtual electron is parallel to that of the real electron is preferred; that the two situations are not equivalent may be seen by noting that annihilation generally takes place between an electron and a positron with opposite spins and in the preferred case the virtual positron can annihilate either the real or the virtual electron, while in the other case only the virtual electron can be involved in such an annihilation.

The two situations have different magnetic moments, since the electron has a negative magnetic moment, but the positron has a positive magnetic moment as a result of its positive charge (Section 8.6). The preferred situation with the spins of the real and virtual electrons parallel has a negative magnetic moment with three times the magnitude of an electron alone; the magnetic moment of the other situation is of the same magnitude as an electron but is positive. If the two situations were equally probable there would be no correction to the electron's magnetic moment, but since one is preferred there is a small negative contribution that accounts for the deviation of the g factor from 2.

When these quantum electrodynamic arguments are expressed quantitatively, the correct term is found to be α/π, where α is the familiar fine structure constant (Equation 8.85). Higher-order corrections have also been calculated, although a large number of interaction terms have to be considered. One of the more recent theoretical estimates for the g factor of the electron is:

$$g = 2 + (\alpha/\pi) - 0{\cdot}65696(\alpha/\pi)^2 + 1{\cdot}49(\alpha/\pi)^3 + O((\alpha/\pi)^4). \quad (9.74)$$

The coefficient of $(\alpha/\pi)^2$ is well established, but attention has only recently been directed to the $(\alpha/\pi)^3$ term. Equation 9.74 gives a value for the g factor that is in good agreement with the experimental value of Equation 9.73.

9.6 The many-electron Hamiltonian

A Hamiltonian for two electrons falls short of our goal of a molecular Hamiltonian. Here the extension to a many-electron Hamiltonian is considered; the inclusion of nuclei is discussed in the next chapter.

The two-electron Hamiltonian (Equation 9.71) may readily be extended to the case of many electrons provided it may be assumed that there are no interactions that involve three or more particles. This assumption cannot be correct, since one electron can influence a second by perturbing the motion of a third. However, the correction terms necessary may be shown to be of the order of $mc^2\alpha^6$. In addition, the work of Itoh, who used quantum electrodynamics, confirms that to order $mc^2\alpha^4$ there are no three-body terms.

In writing the many-electron Hamiltonian the subscripts i and j are used to distinguish electrons. If the opportunity is taken to introduce the radiative corrections discussed in the previous section by making the replacement 9.72, the resultant Hamiltonian is:

$$
\begin{aligned}
\mathcal{H} = \sum_i [\![& mc^2 - e\phi_i + \pi_i^2/2m && \text{(a)} \\
& + g\mu_B(\mathbf{s}_i.\mathbf{B}_i) && \text{(b)} \\
& - (g\mu_B/4mc^2)\mathbf{s}_i.[\boldsymbol{\pi}_i \wedge \mathbf{E}_i - \mathbf{E}_i \wedge \boldsymbol{\pi}_i] && \text{(c)} \\
& + (e\hbar^2/8m^2c^2)(\boldsymbol{\nabla}.\mathbf{E}_i) && \text{(d)} \\
& - (1/8m^3c^2)\pi_i^4 - (g\mu_B/2m^2c^2)(\mathbf{s}_i.\mathbf{B}_i)\pi_i^2 && \text{(e)} \\
& + \sum_{j\neq i} \{e^2/8\pi\epsilon_0 r_{ij} && \text{(f)}
\end{aligned}
$$

$$
\left.\begin{aligned}
&- (e^2/16\pi\epsilon_0 m^2c^2)[\boldsymbol{\pi}_i.(r_{ij}^{-1}\boldsymbol{\pi}_j) + (\boldsymbol{\pi}_i.\mathbf{r}_{ij})r_{ij}^{-3}(\mathbf{r}_{ij}.\boldsymbol{\pi}_j)] && \text{(g)}\\
&- (eg\mu_B/8\pi\epsilon_0 mc^2 r_{ij}^3)\mathbf{s}_i.(\mathbf{r}_{ij}\wedge\boldsymbol{\pi}_i) && \text{(h)}\\
&+ (eg\mu_B/4\pi\epsilon_0 mc^2 r_{ij}^3)\mathbf{s}_i.(\mathbf{r}_{ij}\wedge\boldsymbol{\pi}_j) && \text{(i)}\\
&+ (g^2\mu_B^2/8\pi\epsilon_0 c^2)[(\mathbf{s}_i.\mathbf{s}_j)/r_{ij}^3 - 3(\mathbf{s}_i.\mathbf{r}_{ij})(\mathbf{r}_{ij}.\mathbf{s}_j)/r_{ij}^5\\
&\qquad\qquad - (8\pi/3)\delta(\mathbf{r}_{ij})(\mathbf{s}_i.\mathbf{s}_j)] && \text{(j)}\\
&- (e^2\hbar^2/8\epsilon_0 m^2c^2)\delta(\mathbf{r}_{ij})\}]\!] && \text{(k)}\\
&+ \mathrm{O}(mc^2\alpha^5).
\end{aligned}\right\} \quad (9.75)
$$

Since this Hamiltonian involves a summation over $j \neq i$, it has been necessary to introduce a factor of $\frac{1}{2}$ into (f), (g) and (j), so that these interactions are not counted twice.

In Equation 9.73 the subscripts have been dropped from the constants m, e and g, since the origin of each term may be ascertained by referring back to Equations 9.37 or 9.71. However, for completeness the identification of each term is repeated here:

(a) an electron's rest energy, its energy of interaction with the external electric potential and its kinetic energy, respectively;

(b) the Zeeman term representing the interaction of an electron's spin magnetic moment with the external magnetic field;

(c) the interaction of an electron's spin magnetic moment with the magnetic field it experiences by virtue of its motion relative to the external electric field;

(d) the Darwin term representing the correction to the second term in (a) due to an electron's Zitterbewegung;

(e) relativistic corrections to the kinetic energy in (a) and the Zeeman term (b);

(f) the Coulomb interaction between pairs of electrons;

(g) the retarded orbit-orbit interaction of one electron with the electromagnetic field due to the relative motion of another;

(h) the spin-orbit term representing the interaction of the spin magnetic moment of an electron with its own orbital motion due to the electric field of another electron;

(i) the spin-other-orbit term representing the interaction of the spin magnetic moment of one electron with the orbital motion of another;

(j) the spin-spin term comprising a dipole-dipole interaction between the spin magnetic moments of two electrons together with a Fermi-contact-type interaction;

(k) a correction to the Coulomb term (f) analogous to the Darwin term (d).

In the Hamiltonian (Equation 9.75) the mechanical momentum $\boldsymbol{\pi}_i = (\mathbf{p}_i + e\mathbf{A}_i)$ appears in many terms. It is expedient not to expand these terms, since this ensures that the Hamiltonian is gauge invariant. However, it should be noted that when, for example, $\pi_i^2/2m$ is expanded, an orbital Zeeman term appears together with a contribution to the diamagnetism (Section 7.1).

Bibliography

The derivation of the two-electron Hamiltonian from the Dirac equation

Carrington, A., Levy, D. H., and Miller, T. A. (1970), 'Electron resonance of gaseous diatomic molecules', *Adv. Chem. Phys.*, **18**, 149.

McWeeny, R., and Sutcliffe, B. T. (1969), *Methods of Molecular Quantum Mechanics*, Academic Press, New York: Appendix 4.

The Breit interaction

Bethe and Salpeter: pages 170-183.

Rose: pages 225-229.

Reduction of the Breit equation to non-relativistic form

Chraplyvy, Z. V. (1953), 'Reduction of relativistic two-particle wave equations to approximate forms. I and II', *Phys. Rev.*, **91**, 388 and **92**, 1310: Foldy-Wouthuysen-type transformations.

Barker, W. A., and Glover, F. N. (1955), 'Reduction of relativistic two-particle wave equations to approximate forms. III', *Phys. Rev.*, **99**, 317: the Foldy-Wouthuysen-type transformation.

Estabrooks, P. G. (1967), 'Reduction of the two-electron Breit equation', WIS-TCI-264, The University of Wisconsin, Theoretical Chemistry Institute: the method of small components.

The electron g factor

Wesley, J. C., and Rich, A. (1971), 'High-field electron *g*-2 measurement', *Phys. Rev. A*, **4**, 1341: one of the more recent experimental determinations of the *g* factor of the electron.

Levine, M. J., and Wright, J. (1971), 'Sixth-order magnetic moment of the electron', *Phys. Rev. Letters*, **26**, 1351: one of the more recent theoretical evaluations of the *g* factor of the electron.

Sikolov, A. A., Koskutov, Y. M., and Ternov, I. M. (1966), *Quantum Mechanics*, Holt, Rinehart and Winston, New York: a qualitative discussion of the origin of the *g* factor correction is given on pages 354-356.

The many-electron Hamiltonian

Itoh, T. (1965), 'Derivation of non-relativistic Hamiltonian for electrons from quantum electrodynamics', *Rev. Mod. Phys.*, **37**,159.

Present, R. D. (1971), 'Non-additive interactions', *Contemp. Phys.*, **12**, 595: a review of the magnitude of three-body interactions.

CHAPTER TEN

The Molecular Hamiltonian

In extending the many-electron Hamiltonian to a molecule, nuclei must be introduced. This poses serious problems, since nuclear theory is not in as advanced a state as electron theory; although simple models exist to explain nuclear properties, a serious treatment of the topic is beyond the scope of this book. Fortunately, many of the nuclear contributions to the Hamiltonian are much smaller than the corresponding electronic contributions and nuclei can usually be included quite satisfactorily using phenomenological arguments.

Even when a suitable molecular Hamiltonian has been established, its use is by no means trivial and would require another book to deal with it in depth. However, there are several excellent texts which essentially start with the Hamiltonian and treat its application to specific problems in detail. Nevertheless the transformation to molecular coordinates and the use of effective Hamiltonians is outlined briefly here as a general introduction to the problem of spectral analysis.

10.1 The introduction of nuclei

To obtain a molecular Hamiltonian nuclei must be introduced and to do this it is necessary to assume that nuclei can be treated as Dirac particles, that is particles that are described by the Dirac equation and behave in the same way as electrons. The non-relativistic approximation (Equation 9.75) to the many-electron Hamiltonian can then be extended to a molecule by pretending that some of the electrons are nuclei and making appropriate changes in notation. For m, the mass of an electron, we write m_α, the mass of nucleus α. Similarly, $\mathbf{s}_i$ is replaced by $\mathbf{I}_\alpha$, the spin of nucleus α, and $Z_\alpha e$ is used in place of $-e$, where Z_α is the atomic number of nucleus α.

This procedure involves a number of drastic assumptions, but it has been shown to be valid provided that only terms involving (nuclear mass)$^{-1}$ are retained. (The arguments used are quantum electrodynamical and cannot be given here; the reader is referred to the book of Bethe and Salpeter (pages 193-4) for details.) This means that all terms involving (nuclear mass)$^{-2}$ and smaller are neglected, but in fact all these terms are formally of order $mc^2\alpha^4$ to start with and are at least 10^{-6} times smaller than the corresponding electron terms, because the ratio of the proton mass to the electron mass is $1{\cdot}8631 \times 10^3$.

It might be thought that only the nuclear and electron-nuclear terms in the Hamiltonian are to be regarded with any suspicion. However, a note of caution must be sounded about the terms involving just the electrons. It is recalled that in reducing the Dirac and Breit equations to non-relativistic form, the order of magnitude of the terms was estimated on the basis of an electron in the first Bohr orbit of the hydrogen atom. In general higher nuclear charges are involved and for a one-electron atom in which the nucleus has a charge Ze it is no longer appropriate to express the magnitude of the terms in the Hamiltonian as $mc^2\alpha^n$. Instead the terms $c(\boldsymbol{\alpha}.\mathbf{p})$, $e\phi$ and $ce\boldsymbol{\alpha}.\mathbf{A}$ in the Dirac Hamiltonian are of order $mc^2(Z\alpha)$, $mc^2(Z\alpha)^2$ and $mc^2(Z\alpha)^3$, so that $Z\alpha$ should be used as the expansion parameter rather than the fine structure constant itself. Since α is approximately 1/137, these comments only become important for atoms or molecules involving nuclei with large atomic numbers, and then only for the inner electrons, since the outer electrons are effectively screened from the nuclei. It is clear from these remarks that relativistic effects are more important if inner electrons and nuclei with large Z are involved. In such cases the use of an approximate Hamiltonian, which is a power series expansion in $Z\alpha$, is not satisfactory. One way of avoiding this problem is to take as the zero-order problem one electron under the influence of a nucleus and then introduce as perturbations the electron-electron interactions including the Coulomb interactions. The Dirac equation is solved explicitly for a one-electron atom in the next chapter, but the general problem of nuclei with large atomic numbers is too specialized to be considered here.

By assuming that nuclei behave as Dirac particles, it is also assumed that all nuclei have spin $\frac{1}{2}$, but this is not true. Some nuclei have no spin angular momentum while nuclear spins as high as 9/2 are not uncommon; in addition, nuclei with integral spins are bosons and do not obey Fermi-Dirac statistics. However, if the policy is adopted of assuming that nuclei are Dirac particles, but that some of them have anomalous spins, the resulting theory is not in disagreement with experiment; again quantum electrodynamics indicates that this is a valid approach, if only terms involving (nuclear mass)$^{-1}$ are retained. A further complication associated with nuclei which have spins greater than or equal to 1 is that they have quadrupole moments. These are discussed in the next section, which is concerned with effects associated with the finite size of the nucleus; for the present nuclear quadrupole contributions to the Hamiltonian are merely noted and labelled as $\mathcal{H}_Q$.

Associated with a nuclear spin $\mathbf{I}_\alpha$ is a spin magnetic moment:

$$g_\alpha \mu_N \mathbf{I}_\alpha = g_\alpha(e\hbar/2m_p)\mathbf{I}_\alpha, \quad (10.1)$$

where μ_N is the nuclear magneton and differs from the Bohr magneton by involving the mass of the proton m_p instead of the electron mass. If nuclei were true Dirac particles (with, if necessary, anomalous nuclear spins), the spin magnetic moment would be:

$$2(Z_\alpha e\hbar/2m_\alpha)\mathbf{I}_\alpha. \quad (10.2)$$

Even in the case of the proton, where $m_\alpha = m_p$ and $Z_\alpha = 1$, the nuclear g factor g_α is not 2 and as for the electron it is anomalous. However, the anomaly is much larger than for the electron and the proton g factor is in fact 5·5855 even though its spin is $\frac{1}{2}$. In this connection we might note that the magnetic moment of the neutron, which is of the same magnitude as that of the proton, is completely anomalous, since the neutron is chargeless and would thus be expected to have no magnetic moment at all. These anomalous magnetic moments have been explained in terms of the interaction of the proton and neutron with their virtual meson charge clouds, although meson theory is not in as satisfactory a state as quantum electrodynamics. Larger nuclei also have anomalous magnetic moments and some are even negative. Finally, we should note that in practice the ratio (Z_α/m_α) in 10.2 is absorbed into g_α, so that nuclear magnetic moments are given by $g_\alpha \mu_N \mathbf{I}_\alpha$ (Equation 10.1).

We are now in a position to generalize the many-electron Hamiltonian (Equation 9.75) to a molecule by assuming that nuclei are Dirac particles with anomalous magnetic moments and spins. In doing this it is helpful to keep in mind the interpretation of the terms in Equation 9.75 and to consult Equations 9.37 or 9.71 to see which particle is responsible for quantities such as mass and charge in an individual term. The molecular Hamiltonian may then be written as:

$$\mathcal{H} = \mathcal{H}_e + \mathcal{H}_n + \mathcal{H}_{en}, \quad (10.3)$$

where $\mathcal{H}_e$ is the many-electron Hamiltonian (Equation 9.75) and $\mathcal{H}_n$ and $\mathcal{H}_{en}$ are, respectively, the nuclear Hamiltonian and the electron-nuclear interaction Hamiltonian.

Written explicitly the nuclear Hamiltonian is:

$$\mathcal{H}_n = \sum_{\alpha} \{m_\alpha c^2 + Z_\alpha e\phi_\alpha + \pi_\alpha^2/2m_\alpha - g_\alpha \mu_N(\mathbf{I}_\alpha.\mathbf{B}_\alpha) + \sum_{\beta \neq \alpha} Z_\alpha Z_\beta e^2/8\pi\epsilon_0 r_{\alpha\beta}\} + O(mc^2\alpha^4(m/m_p)^2), \tag{10.4}$$

where

$$\boldsymbol{\pi}_\alpha = \mathbf{p}_\alpha - Z_\alpha e\mathbf{A}_\alpha. \tag{10.5}$$

The terms in Equation 10.4 represent, respectively, nuclear rest energy, the energy of interaction between nuclear charge and the external electric potential, nuclear kinetic energy, nuclear Zeeman energy and nuclear-nuclear Coulomb repulsion. The interaction Hamiltonian appears as:

$$\left.\begin{aligned}
\mathcal{H}_{en} = \sum_{i,\alpha} \{ & -Z_\alpha e^2/4\pi\epsilon_0 r_{i\alpha} && \text{(a)}\\
& + (Z_\alpha e^2/8\pi\epsilon_0 m m_\alpha c^2)[\boldsymbol{\pi}_i.(r_{i\alpha}^{-1}\boldsymbol{\pi}_\alpha) + (\boldsymbol{\pi}_i.\mathbf{r}_{i\alpha})r_{i\alpha}^{-3}(\mathbf{r}_{i\alpha}.\boldsymbol{\pi}_\alpha)] && \text{(b)}\\
& - (g\mu_B Z_\alpha e/8\pi\epsilon_0 mc^2 r_{i\alpha}^3)\mathbf{s}_i.(\mathbf{r}_{i\alpha} \wedge \boldsymbol{\pi}_i) && \text{(c)}\\
& - (g\mu_B Z_\alpha e/4\pi\epsilon_0 m_\alpha c^2 r_{i\alpha}^3)\mathbf{s}_i.(\mathbf{r}_{i\alpha} \wedge \boldsymbol{\pi}_\alpha) && \text{(d)}\\
& + (g_\alpha \mu_N e/4\pi\epsilon_0 mc^2 r_{i\alpha}^3)\mathbf{I}_\alpha.(\mathbf{r}_{i\alpha} \wedge \boldsymbol{\pi}_i) && \text{(e)}\\
& - (g g_\alpha \mu_B \mu_N/4\pi\epsilon_0 c^2)[(\mathbf{s}_i.\mathbf{I}_\alpha)/r_{i\alpha}^3 - 3(\mathbf{s}_i.\mathbf{r}_{i\alpha})(\mathbf{I}_\alpha.\mathbf{r}_{i\alpha})/r_{i\alpha}^5 \\
& \qquad - (8\pi/3)\delta(\mathbf{r}_{i\alpha})(\mathbf{s}_i.\mathbf{I}_\alpha)] && \text{(f)}\\
& + (Z_\alpha e^2\hbar^2/8\epsilon_0 m^2 c^2)\delta(\mathbf{r}_{i\alpha})\} && \text{(g)}\\
& + O(mc^2\alpha^4(m/m_p)^2) + \mathcal{H}_Q.
\end{aligned}\right\} \tag{10.6}$$

In this Hamiltonian (a) is the electron-nuclear Coulomb attraction term, while (b) is a retarded orbit-orbit interaction. The electric field due to the nucleus is responsible for the spin-orbit interaction term (c). Both (d) and (e) are spin-other-orbit interactions involving the spin magnetic moment of an electron and the motion of a nucleus and vice versa. The term (f) represents the interaction between the spin magnetic moments of an electron and a nucleus; in addition to the dipole-dipole interaction, which has a classical analogue, there is a term involving a Dirac delta function, which represents the Fermi contact interaction and has no classical analogue. Finally, there is a Darwin-type term (g) and the nuclear quadrupole interaction term $\mathcal{H}_Q$.

No terms involving (nuclear mass)$^{-2}$ have been included in Equations 10.4 or 10.6, since their presence cannot be rigorously justified, but this does not mean that there are no significant terms of this type. Indeed the dipole-dipole interaction between two nuclear spin magnetic moments, that is observed in broad line nuclear magnetic resonance experiments, is of this type and so is the interaction between molecular rotation and nuclear spin magnetic moments, that is seen in molecular beam studies. Terms like this have analogues in the two-electron Hamiltonian and are quite plausible, but their inclusion must be based on phenomenological arguments.

10.2 Finite nuclear size effects

In this section two additional nuclear effects are mentioned, which may be attributed to the fact that nuclei are not point charges but have finite extent; of course, both the angular momentum and the magnetic moment associated with nuclear spin might also be attributable to this.

Apparently electrons can penetrate nuclei and move about inside them. In atoms only s electrons have a finite probability of being at the nucleus, and when such electrons are inside the nucleus the electric potential they experience is not given by $Ze/4\pi\epsilon_0 r$, but is presumably less than this. Consequently the energy of an s electron is shifted upwards because of the finite size of the nucleus, but the magnitude of this effect is dependent on the structure of the nucleus involved. In particular, different isotopes of the same element have different shifts, since the nuclei are different; isotope shifts have been detected experimentally, but the effect is small and has only been observed in the spectra of heavy atoms ($Z > 50$). A more important consequence of this effect is in Mössbauer spectroscopy, where the observed transitions are between different nuclear energy levels. Since the size of a nucleus alters during these transitions, the exact energy change involved depends on the environment of the nuclei and in particular the distribution of the electrons in the molecule. These effects must be treated phenomenologically in view of the uncertainty about nuclear structure.

The other consequence of finite nuclear size is that many nuclei have quadrupole moments. One model that is frequently given takes the nucleus to be a collection of (charged) protons and (uncharged) neutrons. If the position of proton p in nucleus α relative to the centre of charge of nucleus α is $\mathbf{r}_{p\alpha}$, the electrostatic interaction energy between the nucleus and an electron at position $\mathbf{r}_{e\alpha}$ is:

$$-e\phi_e = -\sum_p e^2/4\pi\epsilon_0|\mathbf{r}_{e\alpha} - \mathbf{r}_{p\alpha}|. \qquad (10.7)$$

Normally the electron is well outside the nucleus so that Equation 10.7 may be expanded as a power series in $\mathbf{r}_{p\alpha}$ about $\mathbf{r}_{e\alpha}$. This may be accomplished by noting that:

$$f(\mathbf{r} - \mathbf{a}) = [1 - (\mathbf{a}.\nabla) + \tfrac{1}{2}(\mathbf{a}.\nabla)^2 + \ldots]f(\mathbf{r}). \qquad (10.8)$$

Use of Equations 2.78 and 2.96 then gives:

$$1/|\mathbf{r}_{e\alpha}-\mathbf{r}_{p\alpha}| = 1/r_{e\alpha}+(\mathbf{r}_{p\alpha}.\mathbf{r}_{e\alpha})/r_{e\alpha}^3-r_{p\alpha}^2/2r_{e\alpha}^3+3(\mathbf{r}_{p\alpha}.\mathbf{r}_{e\alpha})^2/2r_{e\alpha}^5+\ldots; \tag{10.9}$$

a term involving $\delta(\mathbf{r}_{e\alpha})$ has been omitted, since it is assumed that $r_{e\alpha} \gg r_{p\alpha}$.

When Equation 10.9 is substituted into Equation 10.7 the first term that arises is $-\sum_p e^2/4\pi\epsilon_0 r_{e\alpha} = -Z_\alpha e^2/4\pi\epsilon_0 r_{e\alpha}$, which is the familiar Coulomb term. The next contribution involves $\sum_p e\mathbf{r}_{p\alpha}$, which is the electric dipole moment of nucleus α, and vanishes, since $\mathbf{r}_{p\alpha}$ is the position of a proton relative to the centre of charge of the nucleus. Although the model used here is a simple one, it is a general result, both experimentally and theoretically, that nuclei have no electric dipole moments. The theoretical demonstration of this fact is based on the assumption that the states of nuclei have definite parity, that is they are either symmetric or antisymmetric with respect to the inversion of all coordinates. The expectation value of the electric dipole moment operator, which is of odd parity, must then vanish.

The third and fourth terms in Equation 10.9 are both quadratic in the components of $\mathbf{r}_{p\alpha}$ and may also be factorized into the product of an electronic term and a term dependent on nuclear structure:

$$-r_{p\alpha}^2/2r_{e\alpha}^3+3(\mathbf{r}_{p\alpha}.\mathbf{r}_{e\alpha})^2/2r_{e\alpha}^5=(3r_{p\alpha j}r_{p\alpha k}-\delta_{jk}r_{p\alpha}^2)(3r_{e\alpha j}r_{e\alpha k}-\delta_{jk}r_{e\alpha}^2)/6r_{e\alpha}^5, \tag{10.10}$$

as may easily be verified; the additional subscripts, j and k, distinguish vector components and, since the repeated subscript summation convention is in operation, $\delta_{jk}\delta_{jk}=3$ (Equation 2.21). The contribution of these terms to Equation 10.7 is thus a scalar product of two second-rank tensors:

$$(1/6)Q_{jk}^{(\alpha)}V_{jk}^{(\alpha)}, \tag{10.11}$$

where

$$Q_{jk}^{(\alpha)} = e\sum_p (3r_{p\alpha j}r_{p\alpha k}-\delta_{jk}r_{p\alpha}^2) \tag{10.12}$$

is a component of the nuclear quadrupole tensor of nucleus α; it has even parity so that its expectation value is non-zero and, since it depends on nuclear structure, it is treated phenomenologically, although its matrix elements do depend on the nuclear spin $\mathbf{I}_\alpha$. The other tensor has components:

$$V_{jk}^{(\alpha)} = -(e/4\pi\epsilon_0 r_{e\alpha}^5)(3r_{e\alpha j}r_{e\alpha k}-\delta_{jk}r_{e\alpha}^2), \tag{10.13}$$

and is known as the electric field gradient tensor, since its components may also be written in the form:

$$V_{jk}^{(\alpha)} = \nabla_j\nabla_k(-e/4\pi\epsilon_0 r_{e\alpha}), \tag{10.14}$$

provided the electron is always taken to be outside the nucleus so that the delta function contained in the right-hand side of Equation 10.14 may be ignored. Equation 10.13 may be extended by summing over all the electrons in the molecule and, with a little modification, the contribution of other nuclei may be included. The total nuclear quadrupole interaction energy $\mathcal{H}_Q$ is then given by a summation of the expression 10.11 over all the nuclei in the molecule.

The largest of the terms that have been neglected in the expansion (Equation 10.9) are cubic in the components of $\mathbf{r}_{p\alpha}$ and are related to the electric octupole moment of the nucleus. However, they vanish, since parity arguments show that nuclei do not have electric octupole moments. Thus the next nonzero contribution to Equation 10.7 involves the electric hexadecapole moment of the nucleus, but at present this is of no experimental significance. In this connection we might note that, of the nuclear magnetic moments, the monopole and quadrupole moments are zero, but that the dipole and octupole moments do not vanish; the dipole moment has already been considered, but, although magnetic octupole moment effects have been detected in iodine, indium and gallium atoms, they are at present at the limit of experimental significance.

10.3 Spectroscopically useful Hamiltonians

The molecular Hamiltonian that has been derived here is expressed in terms of the position vectors of the individual particles. This form has the disadvantage that it does not reflect the basic physical features of molecules. A molecule is essentially a semi-rigid nuclear framework supporting a distribution of electrons which is responsible for the chemical bonds. (In this and the following section molecules are considered, but atoms can always be regarded as special cases.) Because nuclei are much more massive than electrons, nuclear and electronic motion are largely separated from each other and have very different energies; in addition, molecular translation, rotation and vibration also have characteristically different energies from each other. Consequently, it is expedient to write the Hamiltonian of a molecule in such a way that translation, rotation, vibration and electronic motion are separated as much as possible. To do this it is appropriate to change the molecular variables from the position vectors of individual particles to translational, rotational, vibrational and electronic coordinates.

Since the Hamiltonian includes relativistic corrections, it might be thought that a Lorentz transformation should be used in transforming to molecular coordinates. However, the frame of reference in which physical measurements are made is not changed so that the transformation is merely a change of variables within a particular inertial frame.

The transformation may be thought of as taking place in three stages. First, translation is separated by changing the origin of the coordinates from a space-fixed one to the molecular centre of mass keeping the axes parallel to the original space-fixed axes. Without moving the origin, but changing to axes rotating with the molecule, rotation may be separated off. Finally, the electronic motion may be separated most effectively from the vibrations by shifting the origin of the coordinates to the centre of mass of the nuclei without altering the orientation of the axes.

The transformation of the approximate field-free Hamiltonian:

$$\mathcal{H} = \sum_\alpha p_\alpha^2/2m_\alpha + \sum_i p_i^2/2m + V, \tag{10.15}$$

where the potential energy V comprises the Coulomb interactions, is performed in a number of texts with the result that:

$$\mathcal{H} = P^2/2M + \tfrac{1}{2}\mathbf{N}.\boldsymbol{\mu}.\mathbf{N} + \tfrac{1}{2}\sum_r P_r^2 + \sum_i p_i^2/2m + (\sum_i \mathbf{p}_i)^2/2M_N + V, \tag{10.16}$$

where $\mathbf{p}_i$ is now the momentum of electron i in the molecule, that is it is the momentum conjugate to the position $\mathbf{r}_i$ of electron i relative to the new origin at the centre of mass of the nuclei. Other new quantities in Equation 10.16 are the total mass of the molecule $M\,(= \sum_\alpha m_\alpha + \sum_i m)$ and the mass of the nuclei alone $M_N\ (= \sum_\alpha m_\alpha)$. The translational momentum $\mathbf{P}$ is the momentum conjugate to $\mathbf{R}$, the position of the centre of mass of the molecule; in terms of the original space-fixed coordinates this is given by $\mathbf{R} = (\sum_\alpha m_\alpha \mathbf{r}_\alpha + \sum_i m\mathbf{r}_i)/M$. The position $\mathbf{r}_\alpha$ of a nucleus in the molecule may be expressed in terms of normal coordinates Q_r:

$$\mathbf{r}_\alpha = \mathbf{r}_\alpha^0 + \sum_r m_\alpha^{-\frac{1}{2}} \mathbf{l}_{\alpha r} Q_r, \tag{10.17}$$

where $\mathbf{r}_\alpha^0$ is the equilibrium position of nucleus α and the $\mathbf{l}_{\alpha r}$ are constant vectors chosen so that the normal coordinates are orthogonal to each other and to the molecular rotations and translations; the P_r are then the vibrational momenta conjugate to the vibrational coordinates Q_r. The vector $\mathbf{N}$ is the rotational angular momentum of the nuclei and differs from the total angular momentum $\mathbf{J}$ of the molecule by the electronic and vibrational angular momenta, $\mathbf{L}$ and $\mathbf{G}$ respectively:

$$\mathbf{N} = \mathbf{J} - \mathbf{L} - \mathbf{G} = \mathbf{J} - \sum_i \mathbf{r}_i \wedge \mathbf{p}_i - \sum_{r,s} \boldsymbol{\zeta}_{rs} Q_r P_s, \tag{10.18}$$

where the components of $\boldsymbol{\zeta}_{rs}$ are the so-called Coriolis coupling constants:

$$\boldsymbol{\zeta}_{rs} = \sum_\alpha \mathbf{l}_{\alpha r} \wedge \mathbf{l}_{\alpha s}. \tag{10.19}$$

Finally, $\boldsymbol{\mu}$ is the inverse of the instantaneous inertia tensor $\mathbf{I}$, the elements of which are given by:

$$I_{ij} = \epsilon_{ikm}\epsilon_{jlm} \sum_{\alpha} m_\alpha r_{\alpha k} r_{\alpha l} - \sum_{r,s,t} (\zeta_{rs})_j (\zeta_{ts})_i Q_r Q_t. \qquad (10.20)$$

The main problem associated with the derivation of Equation 10.16 is that the rotational momentum $\mathbf{N}$ is not conjugate to any coordinates. However, the components of the total angular momentum $\mathbf{J}$ are conjugate to angles defining the orientation of the molecule. Even so Equation 10.16 is not at all easy to derive and it is usual to perform the transformation in a round about way rather than directly. First, it is easier to transform classical expressions rather than quantum mechanical expressions; after the transformation it is necessary to write the resulting classical Hamiltonian in a quantum mechanical form and this involves choosing the correct order for operators, but techniques exist for doing this. Another simplification is to use the classical Lagrangian instead of the corresponding Hamiltonian, since the Lagrangian is a function of velocities, which are easier to transform than the momenta used in the Hamiltonian; the transitions from velocities to momenta and Lagrangians to Hamiltonians may be performed using the relations to be found in Chapter 3. Finally, we should note that difficulties are encountered in considering linear molecules, which have no nuclear moment of inertia about the molecular axis in the equilibrium position. Consequently, it is not possible to invert the inertia tensor $\mathbf{I}$ to get $\boldsymbol{\mu}$, but this difficulty can also be circumvented.

Of course the Hamiltonian that we have derived is much more complicated than Equation 10.15, since it includes relativistic corrections and external electromagnetic fields. Nevertheless, the full Hamiltonian has been transformed to molecular coordinates. The main feature of the result is that, after negligible terms have been omitted, the leading terms in the transformed Hamiltonian are similar to Equation 10.16 and the correction terms, which were omitted from the approximate Hamiltonian (Equation 10.15), are essentially unchanged in form by the transformation except that they are expressed in terms of molecular coordinates. This is particularly true of the electronic terms, although some so-called mass polarization correction terms appear because the positions of the electrons are now measured relative to a molecule-fixed rather than a space-fixed origin. However, the parts of the Hamiltonian involving nuclear momentum have to be reinterpreted in terms of molecular rotation and vibration. For example, the spin-other-orbit interactions 10.6d between the spin magnetic moments of the electrons and the motion of the nuclei leads to spin-vibration and spin-rotation interaction terms.

Any further discussion of this topic would take us outside the scope of this book and reference must be made to the sources given in the bibliography at the end of the chapter.

10.4 Effective Hamiltonians

To use the molecular Hamiltonian it is necessary to solve the corresponding Schrödinger equation to obtain the eigenvalues and eigenvectors of the molecule under consideration. Exact analytic or numerical solution of the appropriate Schrödinger equation is of course impossible, but even if it were possible such an approach would not be rigorously justifiable. This is because the Hamiltonian itself is based on the Breit equation, which is only correct to order $mc^2\alpha^4$. In addition, the Breit operator is obtained using perturbation theory, so that operators derived from it (those based on the terms appearing in Equation 9.70) should only be used in first order. This means that only an approximate solution of the Schrödinger equation is necessary and the use of perturbation theory is quite in order.

The usual approach is to choose a set of basis functions (approximate eigenfunctions of the Hamiltonian), construct the Hamiltonian matrix and diagonalize it. However, a complete set of basis functions would be infinite and some method of simplification or approximation is necessary. For many problems the majority of the terms in our Hamiltonian are too small to be significant and may be omitted, but the influence of these terms may be detected by a number of high resolution techniques such as magnetic resonance or microwave spectroscopy. The information obtained from these high resolution experiments is not in general concerned with the energy separation between vibronic (vibrational-electronic) states, but with the rotational and spin energy levels associated with a particular vibronic state.

For the Hamiltonian expressed in molecular coordinates it is appropriate to use a separable representation with the basis functions of the form:

$$\psi = \psi_{\mathbf{trans}}\psi_{\mathbf{rot}}\psi_{\mathbf{vib}}\psi_{\mathbf{el}}\psi_{\mathbf{elsp}}\psi_{\mathbf{nucsp}}, \tag{10.21}$$

where $\psi_{\mathbf{trans}}$ is a translational wave function and depends only on the position $\mathbf{R}$ of the centre of mass of the molecule, $\psi_{\mathbf{rot}}$ is a rotational wave function depending on the angles specifying the orientation of the molecule, $\psi_{\mathbf{vib}}$ is a function of the (vibrational) normal coordinates Q_r and the electronic wave function $\psi_{\mathbf{el}}$ depends only on the electronic coordinates $\mathbf{r}_i$ (and possibly on Q_r as well); $\psi_{\mathbf{elsp}}$ and $\psi_{\mathbf{nucsp}}$ are electron and nuclear spin functions, respectively. The translational part of the problem may easily be separated off, since at present there appear to be no significant interactions between translation and other types of molecular motion. However, there are important interactions between electronic, vibrational and rotational motion and, in addition, external electromagnetic fields may be present.

In principle the molecular Schrödinger equation could now be solved by setting up the Hamiltonian matrix and using appropriate techniques to diagonalize it with sufficient accuracy. However, the state of electronic theory is such that suitable electronic basis functions for a molecule are difficult to obtain and their energies are not known very accurately; this is particularly true for the excited electronic states. The same is true to a lesser extent for the vibrational functions, which depend on a knowledge of the vibrational potential function. Consequently, this approach is not suitable for the interpretation of high resolution spectra involving the rotational and spin fine structure of a specific vibronic state.

What is needed is an effective Hamiltonian that is only appropriate to the vibronic state under consideration. Only spin and rotational operators (and perhaps electronic and vibrational angular momentum operators) appear in such a Hamiltonian together with a number of parameters specific to the vibronic state. These parameters are averages of electronic and vibrational operators over the vibronic state and in addition contain contributions to allow for the mixing of vibronic states by coupling terms in the Hamiltonian. The numerical values of these parameters are such that the eigenvalues of the effective Hamiltonian are exactly the same as those that would be obtained for the particular vibronic state using complete diagonalization of the Hamiltonian, if that were possible. Thus an effective Hamiltonian imitates the complete Hamiltonian so far as one particular vibronic state is concerned.

The advantage of effective Hamiltonians is that it is easy to choose the basis functions to be used for their diagonalization and, in particular, the rigid rotor wave functions may be used for the rotational wave functions. Of course there are an infinite number of rotational basis functions and in general the effective Hamiltonian matrix does not factorize but, since interest usually centres on the lower rotational levels of a vibronic state, this Hamiltonian matrix may usually be truncated at a fairly low rotational level without loss of accuracy. In addition, the matrix elements of the operators are well known.

Estimates of the parameters in an effective Hamiltonian may be obtained in two ways. First, analysis of a spectrum using the effective Hamiltonian gives an experimental determination of their values. Secondly, although diagonalization of the complete Hamiltonian is not possible, the parameters may also be estimated theoretically from vibronic wave functions by using perturbation theory. In addition to first-order contributions from the vibronic state under consideration, there are second- and higher-order contributions arising from the mixing of vibronic states. This calculation of the parameters depends largely on the quality of the electronic wave functions used, so that comparison with the experimental values provides a test of the electronic wave functions.

To illustrate these statements, we consider the simple example of a rotating closed-shell spherical-top molecule in its ground electronic state. If vibrations (and translations) are neglected, the approximate Hamiltonian (Equation 10.16) may be written as:

$$\mathcal{H} = \mathcal{H}_{\mathrm{el}} + \mathcal{H}_{\mathrm{rot}} = \mathcal{H}_{\mathrm{el}} + \tfrac{1}{2}BN^2, \tag{10.22}$$

where $\mathcal{H}_{\mathrm{el}}$ is a purely electronic Hamiltonian. For a molecule with spherical symmetry the rotational kinetic energy can be written in terms of one inertial parameter B and, since vibrations are neglected, B is a function of the masses and the equilibrium positions of the nuclei only and commutes with the nuclear rotational angular momentum operator $\mathbf{N}$, which differs from the total angular momentum $\mathbf{J}$ by the electronic angular momentum $\mathbf{L}$:

$$\mathbf{N} = \mathbf{J} - \mathbf{L}. \tag{10.23}$$

The eigenvectors and eigenvalues of the electronic Hamiltonian $\mathcal{H}_{\mathrm{el}}$ are taken to be ψ_n and E_n so that:

$$\mathcal{H}_{\mathrm{el}}\psi_n = E_n\psi_n. \tag{10.24}$$

In this notation the ground electronic state is ψ_0 and it is further assumed that this is non-degenerate so that the expectation value of $\mathbf{L}$ over ψ_0 vanishes. However, the perturbation $\mathcal{H}_{\mathrm{rot}}$ introduces some electronic orbital angular momentum by mixing ψ_0 with excited electronic states. This may be seen by expanding $\mathcal{H}_{\mathrm{rot}}$ using Equation 10.23:

$$\mathcal{H}_{\mathrm{rot}} = \tfrac{1}{2}BJ^2 - B\mathbf{J}.\mathbf{L} + \tfrac{1}{2}BL^2. \tag{10.25}$$

The third term is diagonal in the electronic state and may conveniently be absorbed in $\mathcal{H}_{\mathrm{el}}$, but the second term has matrix elements between different electronic states. The effect of this term may be accounted for by using first-order perturbation theory to obtain an improved wave function ψ_0' for the ground electronic state:

$$\begin{aligned}\psi_0' &= \psi_0 - \sum_{n\neq 0}\langle\psi_n|-B\mathbf{J}.\mathbf{L}|\psi_0\rangle|\psi_n\rangle/(E_n - E_0)\\ &= \psi_0 + BJ_i\sum_{n\neq 0}\langle\psi_n|L_i|\psi_0\rangle|\psi_n\rangle/(E_n - E_0).\end{aligned} \tag{10.26}$$

The expectation value of a component L_j of the orbital angular momentum in the ground electronic state is then given to first order by:

$$\begin{aligned}\langle L_j\rangle &= \langle\psi_0'|L_j|\psi_0'\rangle\\ &= 2BJ_i\sum_{n\neq 0}\langle\psi_0|L_j|\psi_n\rangle\langle\psi_n|L_i|\psi_0\rangle/(E_n - E_0).\end{aligned} \tag{10.27}$$

This expression for $\langle L_j\rangle$ still contains the operator J_i, but the introduction of orbital angular momentum into the ground state is due to molecular rotation and Equation 10.27 still has to be averaged over the rotational states.

The total Hamiltonian is from Equations 10.22 and 10.25:

$$\mathcal{H} = \mathcal{H}_{\text{el}} + \tfrac{1}{2}BJ^2 - B\mathbf{J}.\mathbf{L}, \tag{10.28}$$

where the third term in Equation 10.25 has been absorbed in $\mathcal{H}_{\text{el}}$. This may be changed into an effective Hamiltonian operating only within the electronic ground state, if the operator **L** in Equation 10.28 is replaced by its expectation value in that state. Thus substitution of Equation 10.27 gives the effective Hamiltonian:

$$\tilde{\mathcal{H}} = E_0 + \tfrac{1}{2}BJ^2 - B^2 J_i J_j \sum_{n\neq 0} \langle\psi_0|L_i|\psi_n\rangle\langle\psi_n|L_j|\psi_0\rangle/(E_n - E_0). \tag{10.29}$$

When it is recalled that the molecule under consideration has spherical symmetry, Equation 10.29 can be simplified further:

$$\tilde{\mathcal{H}} = E_0 + \tfrac{1}{2}B'J^2, \tag{10.30}$$

where B' is:

$$B' = B - B^2 \sum_{n\neq 0} |\langle\psi_0|L_x|\psi_n\rangle|^2/(E_n - E_0). \tag{10.31}$$

Thus to consider the rotational levels of the ground electronic state it is only necessary to use an effective Hamiltonian (Equation 10.30) operating entirely within this state. The influence of the excited electronic states is accounted for by the use of an effective rotational constant B', which differs from B, the rotational constant of the nuclei alone, by a term that can be interpreted as the contribution of the electrons to the molecular rotational constant.

Another example of the use of effective Hamiltonians concerns the g factors of free radicals. Experimental measurements show that for radicals, which contain one unpaired electron and which are in electronic states devoid of electronic orbital angular momentum, the g factors often differ significantly from the free electron value. This occurs because spin-orbit coupling can mix electronic states and consequently introduce some orbital angular momentum into the state under consideration. This may be allowed for by using an (effective) spin Hamiltonian in which the unpaired electron g factor is permitted to differ from the free electron value. The magnitude of the discrepancy then gives information about the way in which spin-orbit coupling operates in the particular molecule.

These are simple examples and in general the situation is more complicated. Depending on the problem under consideration there can be many small terms in the Hamiltonian that are responsible for significant mixing of vibronic states. In addition, second-order perturbation theory is often not accurate enough and higher-order perturbations must be included. This can be rather complicated, but the procedure may be simplified by using so-called degenerate perturbation theory; until recently the technique of degenerate perturbation theory suffered from the disadvantage that high-order terms were not Hermitian, but this fault has now been remedied.

At present, the reduction of the molecular Hamiltonian to an effective Hamiltonian operating within a single vibronic state has not been given for a general molecule. Indeed such a reduction would probably be much too complicated to be of any practical use. Instead each new problem has to be treated on its merits, but it must be emphasized that great care must be taken to ensure that all possible contributions to a parameter in the effective Hamiltonian have been considered. This is particularly important if experimental and calculated values of the parameter are being compared as a test of the quality of an electronic wave function.

It is not the purpose of this book to go into the details of spectral analysis, so again the reader is referred to the bibliography for further information.

Bibliography

The introduction of nuclei

Bethe and Salpeter: the justification for treating nuclei as Dirac particles with anomalous spins and magnetic moments is given on pages 193-194.

Ramsey, N. F. (1953), *Nuclear Moments*, Wiley, New York.

Shore, B. W., and Menzel, D. H. (1968), *Principles of Atomic Spectra*, Wiley, New York: nuclear effects are discussed on pages 313-319.

Spectroscopically useful Hamiltonians

Howard, B. J., and Moss, R. E. (1970), 'The molecular Hamiltonian I. Nonlinear molecules', *Molec. Phys.*, **19**, 433: the transformation of the Hamiltonian (Equation 10.3) from space-fixed to molecule-fixed coordinates.

Howard, B. J., and Moss, R. E. (1971), 'The molecular Hamiltonian II. Linear molecules', *Molec. Phys.*, **20**, 147: the transformation to molecular coordinates for linear molecules, which present special problems.

Wilson, E., Bright, Jr., Decius, J. C., and Cross, P. C. (1955), *Molecular Vibrations*, McGraw-Hill, New York: Chapter 11 describes the transformation to molecular coordinates for a nuclear Hamiltonian.

Degenerate perturbation theory

Bloch, C. (1958), 'Sur la théorie des perturbations des états liés', *Nucl. Phys.*, **6**, 329.

Messiah, A. (1961), *Quantum Mechanics*, North-Holland, Amsterdam: pages 712-720.

Pryce, M. H. L. (1950), 'A modified perturbation procedure for a problem in paramagnetism', *Proc. Phys. Soc. A*, **63**, 25.

Soliverez, C. E. (1969), 'An effective Hamiltonian and time-independent perturbation theory', *J. Phys. C.*, **2**, 2161: the effective Hamiltonian derived here is Hermitian to all orders.

Examples of effective Hamiltonians and their use

Brown, J. M. (1972), 'Electron resonance of gaseous free radicals', *MTP International Reviews of Physical Chemistry (Series 1)*, **4**, 235.

Carrington, A., Levy, D. H., and Miller, T. A. (1970), 'Electron resonance of gaseous diatomic molecules', *Adv. Chem. Phys.*, **18**, 149.

Carrington, A., and McLachlan, A. D. (1967), *Introduction to Magnetic Resonance*, Harper and Row, New York: spin Hamiltonians are discussed in Chapter 9.

Eshbach, J. R., and Strandberg, M. W. P. (1952), 'Rotational magnetic moments of $^1\Sigma$ molecules', *Phys. Rev.*, **85**, 24.

Van Vleck, J. H. (1951), 'The coupling of angular momentum vectors in molecules', *Rev. Mod. Phys.*, **23**, 213: Appendix 1.

CHAPTER ELEVEN

The Hydrogen Atom

In non-relativistic theory the hydrogen atom is one of the few cases for which the Schrödinger equation may be solved exactly. For this to be possible the proton and electron are assumed to be spinless, and the only role that the proton plays is to provide an attractive Coulombic potential; the finite mass of the proton may, of course, be taken into account by using a reduced mass for the electron.

If the proton is regarded as merely an infinitely massive source of electrostatic potential, the Dirac equation for the electron may also be solved exactly. This chapter is devoted to its solution using this model for a one-electron atom, although the more general case where the nucleus has charge Ze is considered, since the results demonstrate that relativistic effects are increasingly important for heavy atoms. It is noted that the effect of the finite mass of the nucleus could be largely accounted for by using a reduced mass for the electron, although unlike the non-relativistic case this procedure is not exact. Even with the simplification of an infinite mass for the nucleus, the mathematics involved is tedious though straightforward and for this reason only an outline will be given. However, the reader should by now have acquired some facility in the manipulations involved in evaluating commutators, for example, and no difficulties should be encountered.

11.1 Non-relativistic theory for a one-electron atom

The solutions of the non-relativistic Schrödinger equation for a one-electron atom are summarized here, since they provide a starting point for the solution of the relativistic problem and a number of their properties will be used later.

In addition, this brief review facilitates comparison of the relativistic and non-relativistic solutions.

The non-relativistic Hamiltonian is:

$$\mathcal{H}_0 = p^2/2m - Ze^2/4\pi\epsilon_0 r, \tag{11.1}$$

where **r** is the position of the electron relative to the nucleus and the second term represents the potential energy of the electron in the electric field of the nucleus. This Hamiltonian commutes with the components of the orbital angular momentum operator $\hbar\mathbf{l}$ $(= \mathbf{r} \wedge \mathbf{p})$, so that $\mathcal{H}_0$, l^2 and l_z are separately constants of motion with the eigenvalues E_0, $l(l+1)$ and m_l, respectively.

If the quantum number n $(= 1, 2, \ldots)$ is introduced the eigenvalues of the Hamiltonian are:

$$E_0 = -me^4Z^2/8h^2\epsilon_0^2 n^2 = -mc^2Z^2\alpha^2/2n^2, \tag{11.2}$$

where α $(= e^2/2h\epsilon_0 c)$ is the fine structure constant (Equation 8.85). Using spherical polar coordinates (r, θ, ϕ) for the position of the electron relative to the nucleus the corresponding eigenfunctions may be written as the product of radial and angular parts:

$$\psi_{nlm_l} = R_{nl}(r)Y_{lm_l}(\theta, \phi); \tag{11.3}$$

for a given value of n the quantum number l is restricted to the values $0, 1, \ldots, (n-1)$, while m_l can only take the values $l, (l-1), \ldots, -l$. All solutions with the same value of n have the same energy, so that the energy levels exhibit a degeneracy of n^2.

The radial functions $R_{nl}(r)$ may be expressed as:

$$R_{nl}(r) = \left[\left(\frac{2Z}{na_0}\right)^3 \frac{(n-l-1)!}{2n[(n+l)!]^3}\right]^{\frac{1}{2}} \exp(-\rho/n)(2\rho/n)^l L_{n+l}^{2l+1}(2\rho/n), \tag{11.4}$$

where

$$L_{n+l}^{2l+1}(2\rho/n) = \sum_{k=0}^{n-l-1} (-1)^{k+1}[(n+l)!]^2(2\rho/n)^k/(n-l-1-k)!(2l+1+k)!k! \tag{11.5}$$

is an associated Laguerre polynomial, ρ is a dimensionless quantity:

$$\rho = Zr/a_0 \tag{11.6}$$

and a_0 $(= 4\pi\epsilon_0\hbar^2/me^2)$ is the Bohr radius. In general these radial functions exhibit nodes, that is they vanish for certain values of r; in particular, the function $R_{nl}(r)$ has $(n-l-1)$ nodes.

The angular functions $Y_{lm_l}(\theta, \phi)$ are of more interest, since they will be used to construct the solutions of the Dirac equation. Specifically they are spherical harmonics and may be written as:

$$Y_{lm_l}(\theta, \phi) = [(2l+1)(l-m_l)!/4\pi(l+m_l)!]^{\frac{1}{2}} P_{lm_l}(\cos\theta) \exp(im_l\phi), \quad (11.7)$$

where

$$P_{lm_l}(\cos\theta) = \frac{(-1)^{l+m_l}}{2^l l!}(\sin\theta)^{m_l}\frac{\mathrm{d}^{l+m_l}}{(\mathrm{d}\cos\theta)^{l+m_l}}(\sin\theta)^{2l} \quad (11.8)$$

is an associated Legendre polynomial. These angular functions also have nodes, the number being given by the quantum number l. However, the important thing to note is that they are eigenfunctions of the operators l^2 and l_z:

$$l^2 Y_{lm_l} = l(l+1)Y_{lm_l}; \qquad l_z Y_{lm_l} = m_l Y_{lm_l}. \quad (11.9)$$

In addition, their behaviour under the operators l_x and l_y may be expressed in the following manner:

$$\begin{aligned} l_+ Y_{lm_l} &= (l_x + il_y)Y_{lm_l} = [(l-m_l)(l+m_l+1)]^{\frac{1}{2}} Y_{l,m_l+1}; \\ l_- Y_{lm_l} &= (l_x - il_y)Y_{lm_l} = [(l+m_l)(l-m_l+1)]^{\frac{1}{2}} Y_{l,m_l-1}. \end{aligned} \quad (11.10)$$

The probability density of the electron in an atomic orbital is given by the square of the wave function (Equation 11.3). Since the radial and angular parts of the wave function have $(n-l-1)$ and l nodes, respectively, the probability density exhibits $(n-1)$ nodes. This fact raises the problem of how an electron in an atomic orbital crosses these nodes. This question can be dismissed as being improper, since the uncertainty principle does not permit an electron to be located at a node anyway. Alternatively, a node may be regarded as being a mathematical concept of no physical significance. However, it will be seen later that in relativistic theory the electron probability densities are in fact nodeless.

This section is concluded by noting certain expectation values, which will be of use later. For an electron in an atomic orbital represented by the wave function ψ_{nlm_l}, the expectation values for the operators r^{-1}, r^{-2}, r^{-3} and $\delta(\mathbf{r})$ are given by:

$$\begin{aligned} &\langle r^{-1}\rangle = Z/a_0n^2; \qquad \langle r^{-2}\rangle = Z^2/a_0^2n^3(l+\tfrac{1}{2}); \\ &\langle r^{-3}\rangle = Z^3/a_0^3n^3l(l+\tfrac{1}{2})(l+1); \qquad \langle\delta(\mathbf{r})\rangle = \delta_{l0}Z^3/a_0^3n^3\pi; \end{aligned} \quad (11.11)$$

for an s orbital ($l=0$) the expectation value of r^{-3} is infinite and this fact causes difficulties in the next section. In addition, the expectation value of p^4 may be determined by first using Equations 11.1 and 11.2 to show that:

$$p^2 = -m^2c^2\alpha^2(Z^2/n^2 - 2Za_0/r). \tag{11.12}$$

Thus:

$$\langle p^4 \rangle = m^4c^4\alpha^4[Z^4/n^4 - 4Z^3a_0\langle r^{-1}\rangle/n^2 + 4Z^2a_0^2\langle r^{-2}\rangle]$$
$$= (m^4c^4Z^4\alpha^4/n^3)[-3/n + 4/(l+\tfrac{1}{2})], \tag{11.13}$$

where Equations 11.11 have been used.

11.2 The non-relativistic approximation of the Dirac equation

In Chapter 8 the non-relativistic approximation to the Dirac equation was developed. The approximate Hamiltonian may be written as:

$$\mathcal{H} = mc^2 + \mathcal{H}_0 + \mathcal{H}_1 + O(mc^2\alpha^6), \tag{11.14}$$

where $\mathcal{H}_0$ is given by Equation 11.1 and $\mathcal{H}_1$ is to be regarded as a perturbation.

The explicit form of $\mathcal{H}_1$ may be obtained by substituting the appropriate electromagnetic potentials into Equation 8.142. For a one-electron atom the electric potential experienced by the electron is $Ze/4\pi\epsilon_0 r$, but the vector potential vanishes. Consequently, the electric field at the electron is $Ze\mathbf{r}/4\pi\epsilon_0 r^3$, while the magnetic field is zero. It is easy to show that the divergence of the electric field is:

$$\mathbf{\nabla}.\mathbf{E} = Ze\delta(\mathbf{r})/\epsilon_0 \tag{11.15}$$

and that:

$$\mathbf{p}\wedge\mathbf{E} - \mathbf{E}\wedge\mathbf{p} = -2\mathbf{E}\wedge\mathbf{p} = -(Ze\hbar\mathbf{l}/2\pi\epsilon_0 r^3), \tag{11.16}$$

where $\hbar\mathbf{l}$ is the orbital angular momentum operator. Thus:

$$\mathcal{H}_1 = -p^4/8m^3c^2 + (Ze^2\hbar^2/8\pi\epsilon_0 m^2c^2)[r^{-3}\mathbf{s}.\mathbf{l} + \pi\delta(\mathbf{r})]$$
$$= -p^4/8m^3c^2 + \tfrac{1}{2}mc^2Z\alpha^4a_0^3[r^{-3}\mathbf{s}.\mathbf{l} + \pi\delta(\mathbf{r})], \tag{11.17}$$

a result which may be confirmed by referring to the approximate Hamiltonian (Equation 9.71) for two electrons and allowing the momentum and spin of one of the electrons to vanish; the three terms in Equation 11.17 are, respectively, a relativistic correction to the kinetic energy, a spin-orbit coupling term and a Darwin-type correction term.

The perturbation $\mathcal{H}_1$ contains the spin-operator $\mathbf{s}$, so that the basis functions (Equation 11.3) are not satisfactory as they stand. Since the components of $\mathbf{s}$ commute with $\mathcal{H}_0$, this may be remedied by multiplying each of the orbital basis functions by a two-component spin function characterized by the spin quantum numbers $s\,(=\tfrac{1}{2})$ and $m_s\,(=\pm\tfrac{1}{2})$. Such a spin function is an eigenfunction of the operators s^2 and s_z with eigenvalues $s(s+1)\,(=3/4)$ and m_s, respectively, and the behaviour of the functions under the operators s_x and s_y is given by equations analogous to Equations 11.10 for the corresponding orbital angular momentum operators. However, perturbation theory using these new basis functions cannot be used directly, since the operator $\mathbf{s.l}$ in $\mathcal{H}_1$ (Equation 11.17) has non-zero off-diagonal matrix elements between levels which are degenerate in zeroth order; the other operators in $\mathcal{H}_1$ only connect levels with different n. This problem may be overcome by choosing new basis functions, which are linear combinations of the old, but which are diagonal in the operator $\mathbf{s.l}$. One way of doing this is to diagonalize the matrix of $\mathbf{s.l}$ for a given value of n, but a different approach is used here.

The root of the problem is that the basis functions are characterized by the quantum numbers m_l and m_s, but that the perturbation $\mathcal{H}_1$ does not commute with the components of the operators $\mathbf{l}$ and $\mathbf{s}$. On the other hand l^2 and s^2 do commute with $\mathcal{H}_1$, so that l and s are still good quantum numbers. However, the components of the total angular momentum operator:

$$\mathbf{j} = \mathbf{l}+\mathbf{s} = \mathbf{l}+\tfrac{1}{2}\boldsymbol{\sigma}, \tag{11.18}$$

also commute with $\mathcal{H}_1$ as one might expect from the results of Section 8.4. Thus, instead of using simultaneous eigenfunctions of l^2, s^2, l_z and s_z as a basis, a new set of commuting operators may be used to construct new basis functions; explicitly these operators may be chosen to be j^2, l^2, s^2 and j_z, which have the eigenvalues $j(j+1)$, $l(l+1)$, $s(s+1)$ and m_j. The way in which the two sets of basis functions are related and in particular the identity of the new set need not concern us yet and is considered in detail in the next section. The advantage of the new set is that they must also be eigenfunctions of the operator $\mathbf{s.l}$ with eigenvalues $\frac{1}{2}[j(j+1)-l(l+1)-s(s+1)]$, since from Equation 11.18:

$$\mathbf{s.l} = \tfrac{1}{2}(j^2-l^2-s^2); \tag{11.19}$$

this relationship also shows that $\mathbf{s.l}$ commutes with j^2, l^2, s^2 and $\mathcal{H}$.

Of course, the new quantum numbers j and m_j are restricted. For a given value of j the quantum number m_j can only adopt one of the $(2j+1)$ values j, $(j-1), \ldots, -j$. More importantly, reference to Equation 11.18 shows that, since $s=\frac{1}{2}$, the values of the quantum number j can only be $(l+\frac{1}{2})$ or $(l-\frac{1}{2})$ for a given value of l. In addition, since j is related to the eigenvalue of the operator j^2, it must be positive, so that for $l=0$ the only value that j can take is $+\frac{1}{2}$, the value $-\frac{1}{2}$ not being allowed.

The specific value of j for a given value of l indicates whether the electron spin is oriented parallel or antiparallel to the total angular momentum. Another way of distinguishing these possibilities is to quote the eigenvalue of the operator **s**.**l** or the related operator:

$$K = \boldsymbol{\sigma}.\mathbf{l} + 1 = 2\mathbf{s}.\mathbf{l} + 1, \quad (11.20)$$

whose eigenvalues are given the symbol k. This new quantum number conveys no new information, but may be used instead of l or j. The reason for its introduction here is that its use facilitates the solution of the Dirac equation later in this chapter.

The restrictions on the values of k may easily be determined. From the definition of the operator K (Equation 11.20):

$$\begin{aligned} K^2 &= 1 + 2(\boldsymbol{\sigma}.\mathbf{l}) + (\boldsymbol{\sigma}.\mathbf{l})^2 \\ &= 1 + 2(\boldsymbol{\sigma}.\mathbf{l}) + l^2 + i\boldsymbol{\sigma}.(\mathbf{l}\wedge\mathbf{l}) \\ &= 1 + (\boldsymbol{\sigma}.\mathbf{l}) + l^2, \end{aligned} \quad (11.21)$$

where use has been made of the relation 8.40 and the fact that $\mathbf{l}\wedge\mathbf{l}$ is just $i\mathbf{l}$. Substitution for $(\boldsymbol{\sigma}.\mathbf{l})$ using Equation 11.20 now gives the operator equation:

$$l^2 = K^2 - K, \quad (11.22)$$

so that:

$$l(l+1) = k(k-1). \quad (11.23)$$

A relationship between j and k may be obtained from Equation 11.19 by substituting for **s**.**l** (Equation 11.20), l^2 (Equation 11.22) and s^2 (= 3/4). The resulting operator equation:

$$j^2 = K^2 - \tfrac{1}{4} \quad (11.24)$$

shows that:

$$j(j+1) = (k+\tfrac{1}{2})(k-\tfrac{1}{2}). \quad (11.25)$$

Solution of the Equations 11.23 and 11.25, together with a knowledge of the allowed values of j and l gives the following possible combinations of quantum numbers; if k is positive:

$$l = k-1; \quad j = l+\tfrac{1}{2} = k-\tfrac{1}{2}; \quad k = +1, +2, \ldots, +n; \quad (11.26a)$$

while for negative k:

$$l = -k; \quad j = l-\tfrac{1}{2} = -k-\tfrac{1}{2}; \quad k = -1, -2, \ldots, -(n-1). \quad (11.26b)$$

Thus, the sign of k indicates whether the electron spin is parallel or anti-parallel to the total angular momentum; k cannot be zero, since j and l must both be positive. Again it is noted that k is an additional quantum number that gives no extra information but leads to simplifications later.

The new set of basis functions, characterized by the quantum numbers j, l,

$s\,(=\frac{1}{2})$, k and m_j, can now be used straightforwardly in a first-order perturbation treatment, since $\mathcal{H}_1$ now has no matrix elements between degenerate basis functions. The contribution from the first term in the perturbation (Equation 11.17) is, from Equation 11.13:

$$-\langle p^4\rangle/8m^3c^2 = (mc^2Z^4\alpha^4/2n^3)[3/4n - 1/(l+\tfrac{1}{2})], \qquad (11.27)$$

while the spin-orbit coupling term gives:

$$\tfrac{1}{2}mc^2Z\alpha^4a_0^3\langle r^{-3}\mathbf{s}.\mathbf{l}\rangle = mc^2Z^4\alpha^4[j(j+1)-l(l+1)-s(s+1)]/4n^3l(l+\tfrac{1}{2})(l+1), \qquad (11.28)$$

where Equations 11.11 and 11.19 have been used. For $l = 0$ the expression 11.28 is strictly indeterminate, since $j = s = \frac{1}{2}$ and both the denominator and numerator vanish. However, agreement with experiment is obtained if it is taken to be zero, a choice which is confirmed by the exact solution of the Dirac equation. Finally, the Darwin-type term makes a contribution:

$$\tfrac{1}{2}mc^2Z\alpha^4a_0^3\pi\langle\delta(\mathbf{r})\rangle = mc^2Z^4\alpha^4\delta_{l0}/2n^3, \qquad (11.29)$$

where again Equations 11.11 have been used. Although the spin-orbit term vanishes for $l = 0$, this is compensated by the Darwin-type term, which only contributes when l is zero.

To obtain the total first-order correction to the energy provided by the perturbation $\mathcal{H}_1$ the last three equations must be added. In doing this, use may be made of the facts that $s = \frac{1}{2}$ and that j can only take the values $(l+\frac{1}{2})$ and $(l-\frac{1}{2})$. The result for both these possibilities is given by the same expression:

$$-(mc^2Z^4\alpha^4/2n^3)[1/(j+\tfrac{1}{2}) - 3/4n]. \qquad (11.30)$$

For a given value of n the maximum value of j is $(n-\frac{1}{2})$ so that 11.30 is negative and the inclusion of relativistic corrections always leads to a lowering of the energy.

The eigenvalues of the approximate Hamiltonian (Equation 11.14) are thus:

$$E = mc^2\{1 - Z^2\alpha^2/2n^2 - (Z^4\alpha^4/2n^3)[1/(j+\tfrac{1}{2}) - 3/4n] + O(Z^6\alpha^6)\}. \qquad (11.31)$$

In the non-relativistic theory all levels with the same value of n are degenerate, but this degeneracy is lifted when the relativistic corrections are included. However, for a given n the energy of a level only depends on the quantum number j and not on l. If only the spin-orbit coupling had been included, the energy would have been a function of l as well, but the relativistic kinetic energy correction ensures that levels characterized by the same values of n and j, but different values of l, are degenerate. This degeneracy persists even if the Dirac equation for a one-electron atom is solved exactly, as will be seen in Section 11.7.

If the notation used for a level is nl_j the first few levels are:

$$1s_{\frac{1}{2}}(k=+1)$$

$$[2s_{\frac{1}{2}}(k=+1),\ 2p_{\frac{1}{2}}(k=-1)],\ 2p_{3/2}(k=+2)$$

$$[3s_{\frac{1}{2}}(k=+1),\ 3p_{\frac{1}{2}}(k=-1)],\ [3p_{3/2}(k=+2),\ 3d_{3/2}(k=-2)],\ 3d_{5/2}(k=+3)$$

$$\ldots, \tag{11.32}$$

where degenerate levels are enclosed in square parentheses. Of course, each of these levels has a further $(2j+1)$-fold degeneracy corresponding to the possible values of m_j, but this degeneracy is only removed by an external magnetic field. Finally, it should be noted that, although the Dirac theory predicts that, for example, the $2s_{\frac{1}{2}}$ and $2p_{\frac{1}{2}}$ levels are degenerate, experiment shows that in the hydrogen atom the $2s_{\frac{1}{2}}$ level has a slightly higher energy than the $2p_{\frac{1}{2}}$ level. This effect, which is called the Lamb shift, may be explained by quantum electrodynamics and is considered briefly in Section 11.9.

11.3 The simultaneous eigenfunctions of j^2, j_z, l^2, s^2 and K

In the preceding section use was made of basis functions which are eigenfunctions of the operators j^2, l^2, s^2, K and j_z, but they were not related explicitly to the original basis functions which are eigenfunctions of l^2, s^2, l_z and s_z. Their identity must now be considered and a number of additional properties established, since they are to be used in the exact solution of the Dirac equation.

Since they are linear combinations of the original basis functions, they must be two-component functions, which can be written as:

$$\psi = \begin{pmatrix} \psi_1 \\ \psi_2 \end{pmatrix}, \tag{11.33}$$

where ψ_1 and ψ_2 are functions of space coordinates only; in this form ψ is automatically an eigenfunction of s^2 with eigenvalue $s(s+1) = 3/4$. It is also an eigenfunction of l^2 with eigenvalue $l(l+1)$, if the two components ψ_1 and ψ_2 are both chosen to be a product of a radial function and spherical harmonics of the same degree:

$$\psi_1 = f_1(r)Y_{lm_l}(\theta,\phi); \qquad \psi_2 = f_2(r)Y_{lm_l'}(\theta,\phi). \tag{11.34}$$

The relationship between m_l and m_l' may be established by using the condition that ψ is to be an eigenfunction of j_z:

$$j_z\begin{pmatrix}\psi_1\\ \psi_2\end{pmatrix} = \begin{pmatrix} l_z+\frac{1}{2}, & 0\\ 0, & l_z-\frac{1}{2}\end{pmatrix}\begin{pmatrix}\psi_1\\ \psi_2\end{pmatrix} = \begin{pmatrix}(l_z+\frac{1}{2})\psi_1\\ (l_z-\frac{1}{2})\psi_2\end{pmatrix} = m_j\begin{pmatrix}\psi_1\\ \psi_2\end{pmatrix}, \tag{11.35}$$

where Equation 11.18 and the matrix representation (Equations 1.36) of σ_z have been used. Reference to Equations 11.9 and 11.34 shows that $m_l' = m_l + 1$ and that $m_j = m_l + \frac{1}{2}$.

The next stage is to show that the radial functions, $f_1(r)$ and $f_2(r)$, can only differ by a constant. To do this use is made of the operator K (Equation 11.20), which can be written as a two-by-two matrix with the aid of the matrix representations of the components of $\boldsymbol{\sigma}$:

$$K\psi = \begin{pmatrix} 1+l_z, & l_x - il_y\\ l_x+il_y, & 1-l_z\end{pmatrix}\begin{pmatrix}\psi_1\\ \psi_2\end{pmatrix} = k\begin{pmatrix}\psi_1\\ \psi_2\end{pmatrix}. \tag{11.36}$$

Expansion of this relation gives two equations connecting ψ_1 and ψ_2, but only one is needed:

$$(1+l_z)\psi_1 + (l_x - il_y)\psi_2 = k\psi_1. \tag{11.37}$$

On substitution for ψ_1 and ψ_2 (Equations 11.34) and use of Equations 11.9 and 11.10 this gives:

$$(k - m_l - 1)f_1 = [(l+m_l+1)(l-m_l)]^{\frac{1}{2}}f_2. \tag{11.38}$$

Thus f_1 and f_2 only differ by a constant and ψ can be written as:

$$\psi = f(r)\chi, \tag{11.39}$$

where

$$\chi = \begin{pmatrix} c_1Y_{lm_l}\\ c_2Y_{l,m_l+1}\end{pmatrix}. \tag{11.40}$$

The constants c_1 and c_2 may easily be obtained, since their ratio is given by Equation 11.38 and, if $f(r)$ is normalized, normalization of ψ requires that $|c_1|^2 + |c_2|^2 = 1$. In quoting the results two possibilities may be recognized depending on whether k is greater than or less than zero; these two cases are distinguished by a superscript + or −, respectively. For k positive (Equations 11.26a):

$$\psi = f(r)\chi^{+}_{jm_j} = f(r)\begin{pmatrix} c_1^+ Y_{j-\frac{1}{2},m_j-\frac{1}{2}} \\ c_2^+ Y_{j-\frac{1}{2},m_j+\frac{1}{2}} \end{pmatrix}, \qquad (11.41)$$

where

$$c_1^+ = [(j+m_j)/2j]^{\frac{1}{2}}; \qquad c_2^+ = [(j-m_j)/2j]^{\frac{1}{2}}, \qquad (11.42)$$

while for k negative (Equations 11.26b):

$$\psi = f(r)\chi^{-}_{jm_j} = f(r)\begin{pmatrix} c_1^- Y_{j+\frac{1}{2},m_j-\frac{1}{2}} \\ c_2^- Y_{j+\frac{1}{2},m_j+\frac{1}{2}} \end{pmatrix}, \qquad (11.43)$$

where

$$c_1^- = [(j-m_j+1)/2(j+1)]^{\frac{1}{2}}; \quad c_2^- = -[(j+m_j+1)/2(j+1)]^{\frac{1}{2}}. \qquad (11.44)$$

It is also easy to confirm that these two-component angular functions $\chi^{\pm}_{jm_j}$ behave in the expected manner under the various operators:

$$\begin{aligned} j^2\chi^{\pm}_{jm_j} &= j(j+1)\chi^{\pm}_{jm_j}; \qquad j_z\chi^{\pm}_{jm_j} = m_j\chi^{\pm}_{jm_j}; \\ j_+\chi^{\pm}_{jm_j} &= [(j-m_j)(j+m_j+1)]^{\frac{1}{2}}\chi^{\pm}_{j,m_j+1}; \\ j_-\chi^{\pm}_{jm_j} &= [(j+m_j)(j-m_j+1)]^{\frac{1}{2}}\chi^{\pm}_{j,m_j-1}; \\ K\chi^{\pm}_{jm_j} &= k\chi^{\pm}_{jm_j} = \pm(j+\tfrac{1}{2})\chi^{\pm}_{jm_j}. \end{aligned} \qquad (11.45)$$

There is one other property of these functions which is needed later, and this concerns their behaviour under the operator $(\boldsymbol{\sigma}.\mathbf{r})/r$, which is the radial component of $\boldsymbol{\sigma}$. It is easy to show that this operator commutes with $\mathbf{j}$, but anticommutes with K. Consequently, the result of operating on $\chi^{\pm}_{jm_j}$ with $(\boldsymbol{\sigma}.\mathbf{r})/r$ is a function which is also an eigenfunction of j^2 and j_z with the same quantum numbers j and m_j, and which is an eigenfunction of K but with an eigenvalue differing by a factor of -1 from that of $\chi^{\pm}_{jm_j}$. Thus:

$$[(\boldsymbol{\sigma}.\mathbf{r})/r]\chi^{\pm}_{jm_j} = a^{\pm}_{jm_j}\chi^{\mp}_{jm_j}, \qquad (11.46)$$

where $a^{\pm}_{jm_j}$ is a constant that must now be determined.

It may first be shown that $a^{\pm}_{jm_j}$ is independent of m_j, since from Equations 11.45 and 11.46:

$$j_+[(\boldsymbol{\sigma}.\mathbf{r})/r]\chi^{\pm}_{jm_j} = [(j-m_j)(j+m_j+1)]^{\frac{1}{2}}a^{\pm}_{jm_j}\chi^{\mp}_{j,m_j+1}, \qquad (11.47)$$

while

$$[(\boldsymbol{\sigma}.\mathbf{r})/r]j_+\chi^{\pm}_{jm_j} = [(j-m_j)(j+m_j+1)]^{\frac{1}{2}}a^{\pm}_{j,m_j+1}\chi^{\mp}_{j,m_j+1}. \qquad (11.48)$$

Now, since $(\boldsymbol{\sigma}.\mathbf{r})/r$ and j_+ commute, $a^{\pm}_{jm_j}$ is the same as $a^{\pm}_{j,m_j+1}$ and in future the eigenvalue of j_z may be omitted.

To evaluate $a_j^{\pm}$ the special case when $\theta = 0$ is considered; the operator $(\boldsymbol{\sigma}.\mathbf{r})/r$ is then just σ_z, while for a spherical harmonic:

$$Y_{lm_l}(\theta = 0) = [(2l+1)/4\pi]^{\frac{1}{2}} \delta_{m_l 0}, \tag{11.49}$$

so that from Equations 11.41 to 11.44:

$$\chi^{\pm}_{jm_j}(\theta = 0) = [(j+\tfrac{1}{2})/4\pi]^{\frac{1}{2}} \begin{pmatrix} \delta_{m_j,+\frac{1}{2}} \\ \pm\,\delta_{m_j,-\frac{1}{2}} \end{pmatrix}. \tag{11.50}$$

Thus for $\theta = 0$:

$$[(\boldsymbol{\sigma}.\mathbf{r})/r]\chi^{\pm}_{jm_j} = \sigma_z \chi^{\pm}_{jm_j} = [(j+\tfrac{1}{2})/4\pi]^{\frac{1}{2}} \begin{pmatrix} 1 & 0 \\ 0 & -1 \end{pmatrix} \begin{pmatrix} \delta_{m_j,+\frac{1}{2}} \\ \pm\,\delta_{m_j,-\frac{1}{2}} \end{pmatrix}$$

$$= [(j+\tfrac{1}{2})/4\pi]^{\frac{1}{2}} \begin{pmatrix} \delta_{m_j,+\frac{1}{2}} \\ \mp\,\delta_{m_j,-\frac{1}{2}} \end{pmatrix} = \chi^{\mp}_{jm_j}, \tag{11.51}$$

so that $a_j^{\pm} = 1$. The required result is therefore:

$$[(\boldsymbol{\sigma}.\mathbf{r})/r]\chi^{\pm}_{jm_j} = \chi^{\mp}_{jm_j}, \tag{11.52}$$

since, although it has been determined for the special case $\theta = 0$, $a_j^{\pm}$ is a constant and independent of θ, as may be seen by noting that the square of $(\boldsymbol{\sigma}.\mathbf{r})/r$ is just 1 by Equation 8.40.

11.4 Commutation relations for the Dirac Hamiltonian

In a one-electron atom the electron experiences no magnetic potential, so that the Dirac Hamiltonian is:

$$\mathcal{H} = \beta mc^2 + c\boldsymbol{\alpha}.\mathbf{p} - e\phi, \tag{11.53}$$

where the electric potential ϕ is $Ze/4\pi\epsilon_0 r$. Since ϕ commutes with the orbital angular momentum operator $\hbar\mathbf{l}$, the results of Section 8.4 for a free electron still hold and $\mathcal{H}$ does not commute with $\mathbf{l}$ or $\boldsymbol{\sigma}$. However, $\mathcal{H}$ does commute with the total angular momentum $\mathbf{j}$ (Equation 11.18) and consequently the eigenfunctions of the Dirac Hamiltonian are also eigenfunctions of the operators j^2 and j_z.

Although the Dirac Hamiltonian also commutes with σ^2, it differs from the non-relativistic case in that it does not commute with l^2 or $\boldsymbol{\sigma}.\mathbf{l}$. Nevertheless, the new operator:

$$K = \beta(\boldsymbol{\sigma}.\mathbf{l} + 1) \tag{11.54}$$

does commute with $\mathcal{H}$, j^2 and j_z, and may also be used to classify the eigenfunctions of the Dirac Hamiltonian. It is similar to the operator (Equation 11.20) used in the non-relativistic theory and is denoted by the same symbol, but since K now possesses a four-by-four matrix representation no confusion should arise. It may be noted that the operator $\beta(\boldsymbol{\sigma}.\mathbf{l})$ does not commute with the Dirac Hamiltonian and the non-relativistic operator was chosen to be $(\boldsymbol{\sigma}.\mathbf{l}+1)$ rather than $\boldsymbol{\sigma}.\mathbf{l}$ so that the same eigenfunction k can be used in both the exact and the approximate theory.

Since $\mathcal{H}$, j^2, j_z and K all commute with each other, the solutions of the Dirac equation are simultaneous eigenfunctions of these operators. However, an eigenfunction Ψ must now have four components and may be written in terms of two two-component functions, ψ_+ and ψ_-:

$$\Psi = \begin{pmatrix} \psi_+ \\ \psi_- \end{pmatrix}. \tag{11.55}$$

Since Ψ is to be an eigenfunction of j^2, j_z and K (Equation 11.54) with eigenvalues $j(j+1)$, m_j and k, respectively, ψ_+ and ψ_- possess the following properties:

$$\begin{aligned} j^2\psi_\pm &= j(j+1)\psi_\pm; \\ j_z\psi_\pm &= m_j\psi_\pm; \\ (\boldsymbol{\sigma}.\mathbf{l}+1)\psi_\pm &= \pm k\psi_\pm . \end{aligned} \tag{11.56}$$

The last of these three relations is obtained by using the explicit form of the Dirac matrix β; here the operator $(\boldsymbol{\sigma}.\mathbf{l}+1)$ has a two-by-two matrix representation and is just the operator (Equation 11.20) used earlier. Comparison of Equations 11.56 and 11.45 now shows that the two-component functions ψ_+ and ψ_- may be taken to be the functions 11.41 and 11.43 of the previous section; again two possibilities arise depending on whether k is positive or negative. Explicitly:

$$\Psi = \begin{pmatrix} \psi_+ \\ \psi_- \end{pmatrix} = \begin{pmatrix} f(r)\chi^{\pm}_{jm_j} \\ g(r)\chi^{\mp}_{jm_j} \end{pmatrix}, \tag{11.57}$$

where the upper signs refer to $k = j + \frac{1}{2}$ and the lower to $k = -(j+\frac{1}{2})$.

Finally, we may record that the result of $(\boldsymbol{\sigma}.\mathbf{r})/r$ operating on these four-component functions is:

$$[(\boldsymbol{\sigma}.\mathbf{r})/r]\begin{pmatrix} f(r)\chi^{\pm}_{jm_j} \\ g(r)\chi^{\mp}_{jm_j} \end{pmatrix} = \begin{pmatrix} f(r)\chi^{\mp}_{jm_j} \\ g(r)\chi^{\pm}_{jm_j} \end{pmatrix}. \tag{11.58}$$

This result may easily be derived by using Equation 11.52 and recalling that the four-by-four matrix representation of $\boldsymbol{\sigma}$ is just the direct product of the two-by-two representation with the two-by-two unit matrix.

11.5 The Dirac equation in polar coordinates

The components of the solutions of the Dirac equation have now been expressed as the products of radial and angular functions. To proceed further it is necessary to express the Dirac Hamiltonian itself in polar coordinates. Substitution of Equation 11.57 into the Dirac equation will then give equations for the radial functions $f(r)$ and $g(r)$.

To perform the transformation to polar coordinates, we use the fact that, from Equation 2.38:

$$\mathbf{r}_\wedge(\mathbf{r}_\wedge \boldsymbol{\nabla}) = \mathbf{r}(\mathbf{r}.\boldsymbol{\nabla}) - r^2\boldsymbol{\nabla}. \tag{11.59}$$

Rearrangement of this equation shows that the operator $\boldsymbol{\nabla}$ may be split into a radial and an angular part:

$$\boldsymbol{\nabla} = (1/r^2)[\mathbf{r}(\mathbf{r}.\boldsymbol{\nabla}) - \mathbf{r}_\wedge(\mathbf{r}_\wedge \boldsymbol{\nabla})]. \tag{11.60}$$

The radial part involves $(1/r)(\mathbf{r}.\boldsymbol{\nabla})$, which may readily be shown to be just $(\partial/\partial r)$, while the angular part contains $(\mathbf{r}_\wedge \boldsymbol{\nabla})$, which is related to the orbital angular momentum operator $\hbar\mathbf{l}$. Thus, the operator $(\boldsymbol{\alpha}.\mathbf{p})$ is given by:

$$\boldsymbol{\alpha}.\mathbf{p} = -i\hbar\boldsymbol{\alpha}.\boldsymbol{\nabla} = -(i\hbar/r)(\boldsymbol{\alpha}.\mathbf{r})(\partial/\partial r) - (\hbar/r^2)\boldsymbol{\alpha}.(\mathbf{r}_\wedge \mathbf{l}). \tag{11.61}$$

The way in which the angular functions behave under operators involving $\boldsymbol{\sigma}$ is already known and Equation 11.61 can be rewritten in terms of $\boldsymbol{\sigma}$ by recalling that $\boldsymbol{\alpha} = \rho\boldsymbol{\sigma}$ (Equations 8.19). In addition, the relationship 8.40 shows that:

$$(\boldsymbol{\sigma}.\mathbf{r})(\boldsymbol{\sigma}.\mathbf{l}) = i\boldsymbol{\sigma}.(\mathbf{r}_\wedge \mathbf{l}), \tag{11.62}$$

where the fact that $\mathbf{r}.\mathbf{l}$ vanishes has been used. Finally, multiplication of both sides of Equation 11.54 by β followed by rearrangement gives:

$$\boldsymbol{\sigma}.\mathbf{l} = \beta K - 1, \tag{11.63}$$

so that Equation 11.61 becomes:

$$\boldsymbol{\alpha}.\mathbf{p} = -(i\hbar/r)\rho(\boldsymbol{\sigma}.\mathbf{r})[\partial/\partial r - (1/r)(\beta K - 1)]. \tag{11.64}$$

Substitution of Equation 11.64 into Equation 11.53 now gives the Dirac Hamiltonian in polar coordinates:

$$\mathcal{H} = \beta mc^2 - Ze^2/4\pi\epsilon_0 r - (i\hbar c/r)\rho(\boldsymbol{\sigma}.\mathbf{r})[\partial/\partial r - (1/r)(\beta K - 1)]. \tag{11.65}$$

In seeking solutions of the equation $\mathcal{H}\Psi = E\Psi$, substitution of Equation 11.57 for Ψ and use of the explicit forms of the Dirac matrices β (Equation 8.23) and ρ (Equation 8.25) yields a pair of coupled two-component equations. The way in which the two-component angular functions behave under the operators $(\boldsymbol{\sigma}.\mathbf{r})/r$ and $(\boldsymbol{\sigma}.\mathbf{l} + 1)$ is given in Equations 11.58 and 11.56, so that:

$$\left.\begin{aligned}[(mc^2 - Ze^2/4\pi\epsilon_0 r - E)f - (i\hbar c/r)(k+1)g - i\hbar c(\partial g/\partial r)]\chi^{\pm}_{jm_j} &= 0\\ [(-mc^2 - Ze^2/4\pi\epsilon_0 r - E)g + (i\hbar c/r)(k-1)f - i\hbar c(\partial f/\partial r)]\chi^{\mp}_{jm_j} &= 0\end{aligned}\right\} \quad (11.66)$$

The angular functions are now just multiplicative factors and may be dropped.

The resulting radial equations are simplified by using two new functions:

$$F = rf; \qquad G = irg. \qquad (11.67)$$

If, in addition, the fine structure constant α and the energy parameter ϵ $(= E/mc^2)$ are introduced, then after division throughout by $\hbar c$ the equations become:

$$\left.\begin{aligned}[(mc/\hbar)(1-\epsilon) - Z\alpha/r]F - (\partial G/\partial r) - kG/r &= 0\\ [-(mc/\hbar)(1+\epsilon) - Z\alpha/r]G + (\partial F/\partial r) - kF/r &= 0\end{aligned}\right\} \quad (11.68)$$

These equations may be checked by considering their non-relativistic limit, when the coefficient of G in the second equation is of the order of $-2mc/\hbar$ so that:

$$G \simeq \hbar(\partial F/\partial r - kF/r)/2mc. \qquad (11.69)$$

Substitution of this into the first of Equations 11.68 now gives:

$$\partial^2 F/\partial r^2 - k(k-1)F/r^2 - (2mc/\hbar)[(mc/\hbar)(1-\epsilon) - Z\alpha/r]F = 0, \quad (11.70)$$

an equation which may be identified with the usual non-relativistic radial equation by replacing F by rR, identifying $k(k-1)$ with $l(l+1)$ (Equation 11.23) and allowing for the different energy origins.

11.6 Solution of the radial equations

To solve the radial equations (11.68) series solutions are sought, but in doing this two conditions may be imposed. First, we may restrict ourselves to bound states, so that ϵ $(= E/mc^2)$ is less than unity. Secondly, for bound states it must be possible to normalize the eigenfunctions.

A starting point for the solution is provided by solving the radial equations for the special case when r tends to infinity and they reduce to:

$$\left.\begin{aligned}(\partial G/\partial r) &\simeq (mc/\hbar)(1-\epsilon)F\\(\partial F/\partial r) &\simeq (mc/\hbar)(1+\epsilon)G\end{aligned}\right\}. \tag{11.71}$$

These equations may be uncoupled by substituting one in the other to give two second-order equations:

$$\begin{aligned}(\partial^2 F/\partial r^2) &\simeq (mc/\hbar)^2(1-\epsilon^2)F;\\(\partial^2 G/\partial r^2) &\simeq (mc/\hbar)^2(1-\epsilon^2)G,\end{aligned} \tag{11.72}$$

the normalizable solutions of which are:

$$\begin{aligned}F &\simeq c_F \exp\,[-(mc/\hbar)(1-\epsilon^2)^{\frac{1}{2}}r];\\G &\simeq c_G \exp\,[-(mc/\hbar)(1-\epsilon^2)^{\frac{1}{2}}r];\end{aligned} \tag{11.73}$$

the original equations (11.71) show that the constants c_F and c_G are related by:

$$c_F/c_G = -\,[(1+\epsilon)/(1-\epsilon)]^{\frac{1}{2}}. \tag{11.74}$$

For bound states it is clear from the definition of ϵ that F and G are both real and that $c_F \gg c_G$, so that the components of ψ_+ and ψ_- may again be thought of as the large and small components, respectively.

The solutions (Equation 11.73) for very large r suggest that the substitutions:

$$\begin{aligned}F &= (1+\epsilon)^{\frac{1}{2}} \exp\,(-\rho/2)(u+v);\\G &= (1-\epsilon)^{\frac{1}{2}} \exp\,(-\rho/2)(u-v)\end{aligned} \tag{11.75}$$

may be appropriate, where u and v are new functions of r which, because of Equation 11.74, are subject to the condition that $v \gg u$ when r tends to infinity. In Equations 11.75 the new dimensionless variable:

$$\rho = (2mc/\hbar)(1-\epsilon^2)^{\frac{1}{2}}r \tag{11.76}$$

has been introduced; no confusion should occur between this and the Dirac matrix which has the same symbol. Substitution of Equations 11.75 into Equations 11.68 followed by rearrangement and cancellation of the exponential terms gives equations in $(u+v)$ and $(u-v)$, but by addition and subtraction a pair of coupled equations in u and v themselves are obtained:

$$\left.\begin{aligned}\partial u/\partial\rho &= [\rho - Z\alpha\epsilon(1-\epsilon^2)^{-\frac{1}{2}}](u/\rho) + [k - Z\alpha(1-\epsilon^2)^{-\frac{1}{2}}](v/\rho)\\\partial v/\partial\rho &= [Z\alpha\epsilon(1-\epsilon^2)^{-\frac{1}{2}}](v/\rho) + [k + Z\alpha(1-\epsilon^2)^{-\frac{1}{2}}](u/\rho)\end{aligned}\right\}. \tag{11.77}$$

We now investigate the possibility that u and v may be written as ascending power series in ρ:

$$u = \rho^\lambda \sum_{\nu=0}^{\infty} a_\nu \rho^\nu; \qquad v = \rho^\lambda \sum_{\nu=0}^{\infty} b_\nu \rho^\nu, \tag{11.78}$$

where one (or both) of a_0 and b_0 is non-zero. Substituting in Equations 11.77 and equating the coefficients of $\rho^{\lambda+\nu-1}$ on the two sides of the equations gives:

$$\left.\begin{aligned}(\lambda+\nu)a_\nu &= a_{\nu-1} - Z\alpha\epsilon(1-\epsilon^2)^{-\frac{1}{2}}a_\nu + [k - Z\alpha(1-\epsilon^2)^{-\frac{1}{2}}]b_\nu \\ (\lambda+\nu)b_\nu &= Z\alpha\epsilon(1-\epsilon^2)^{-\frac{1}{2}}b_\nu + [k + Z\alpha(1-\epsilon^2)^{-\frac{1}{2}}]a_\nu\end{aligned}\right\}. \tag{11.79}$$

For the special case $\nu = 0$, the coefficient $a_{\nu-1}$ is zero by definition and a pair of coupled equations in a_0 and b_0 results:

$$\begin{pmatrix}\lambda + Z\alpha\epsilon(1-\epsilon^2)^{-\frac{1}{2}}, & -[k - Z\alpha(1-\epsilon^2)^{-\frac{1}{2}}] \\ -[k + Z\alpha(1-\epsilon^2)^{-\frac{1}{2}}], & \lambda - Z\alpha\epsilon(1-\epsilon^2)^{-\frac{1}{2}}\end{pmatrix}\begin{pmatrix}a_0 \\ b_0\end{pmatrix} = 0. \tag{11.80}$$

For non-trivial solutions of these equations both a_0 and b_0 must in general be non-vanishing, so that the series u and v both start at the same power of ρ. To determine this power, that is to determine the value of λ in Equations 11.78, use can be made of the fact that for non-trivial solutions the determinant of the two-by-two matrix in Equation 11.80 must vanish. Expansion of the determinant gives a quadratic equation in λ, the solutions of which are:

$$\lambda = \pm(k^2 - Z^2\alpha^2)^{\frac{1}{2}}. \tag{11.81}$$

A decision may be made between the two possibilities for λ by invoking the condition that the eigenfunctions of the Dirac Hamiltonian must be normalizable for bound states. Since the angular functions are already normalized, this normalization condition reduces to one for the radial functions:

$$\int(|f|^2 + |g|^2)r^2\mathrm{d}r = \int(F^2 + G^2)\mathrm{d}r = 1. \tag{11.82}$$

Here use has been made of Equations 11.57 and 11.67 and the fact that F and G are both real; the extra factor of r^2 on the left-hand side of the equation comes from the volume element in polar coordinates. Now as ρ approaches zero $(F^2 + G^2)$ behaves as $\rho^{2\lambda}$ and for the normalization (Equation 11.82) to be possible λ must be greater than $-\frac{1}{2}$. Since the minimum value that k^2 may adopt is $+1$, the positive solution for λ must be taken. (For $k^2 = 1$ and $Z \geqslant 119$ the negative solution cannot be excluded, but this possibility may safely be ignored until superheavy elements have been made.) Thus for atoms with low atomic number λ is approximately $|k|$.

To investigate the coefficients a_ν and b_ν further it is convenient to define a new dimensionless quantity:

$$n' = -\lambda + Z\alpha\epsilon(1-\epsilon^2)^{-\frac{1}{2}}. \tag{11.83}$$

From the second of Equations 11.79 the ratio a_ν/b_ν is seen to be:

$$a_\nu/b_\nu = -(n'-\nu)/[k+Z\alpha(1-\epsilon^2)^{-\frac{1}{2}}] = +(n'-\nu)a_0/n'b_0, \quad (11.84)$$

while use of this result with the first equation gives, after some manipulation, the recurrence relation:

$$a_\nu = -(n'-\nu)a_{\nu-1}/\nu(2\lambda+\nu). \quad (11.85)$$

This equation leads to the general expression for a_ν:

$$a_\nu = \frac{(-1)^\nu(n'-\nu)(n'-\nu+1)\dots(n'-1)a_0}{\nu!(2\lambda+\nu)(2\lambda+\nu-1)\dots(2\lambda+1)} \quad (11.86)$$

and a similar expression for b_ν may be obtained using Equation 11.84:

$$b_\nu = \frac{(-1)^\nu(n'-\nu+1)(n'-\nu+2)\dots n'b_0}{\nu!(2\lambda+\nu)(2\lambda+\nu-1)\dots(2\lambda+1)}. \quad (11.87)$$

The ratio of a_0 and b_0 is given by Equation 11.84, but their actual magnitude is determined by the normalization condition (Equation 11.82). However, although λ has been chosen so that F and G do not diverge as r approaches zero, their behaviour when r becomes infinite has yet to be considered. In particular the series u and v must not diverge faster than $\exp(-\rho/2)$ converges as r increases. For large values of ν both $a_\nu/a_{\nu-1}$ and $b_\nu/b_{\nu-1}$ are approximately $1/\nu$, so that the series u and v both behave as $\exp(\rho)$ and F and G diverge as $\exp(\rho/2)$. This behaviour prevents the normalization that is necessary for bound states, but may be avoided by insisting that both the series u and v terminate at a finite value of ν. Reference to Equations 11.86 and 11.87 shows that this only happens if n' is a positive integer. The series u then terminates at $\nu = n'$, while v terminates at $\nu = n'+1$.

The case when $n' = 0$ requires special consideration, since, although v terminates at $\nu = 1$, Equation 11.86 shows that the series u only terminates if a_0 itself vanishes and $u = 0$. From Equation 11.84:

$$a_0/b_0 = -n'/[k+Z\alpha(1-\epsilon^2)^{-\frac{1}{2}}], \quad (11.88)$$

so that this possibility is acceptable provided the denominator does not vanish as well as the numerator n'. Now for $n' = 0$:

$$\lambda = (k^2 - Z^2\alpha^2)^{\frac{1}{2}} = Z\alpha\epsilon(1-\epsilon^2)^{-\frac{1}{2}}, \quad (11.89)$$

where Equations 11.81 and 11.83 have been used. Solution of this equation for k gives:

$$k = \pm Z\alpha(1-\epsilon^2)^{-\frac{1}{2}}; \quad (11.90)$$

for the negative solution the denominator in Equation 11.88 vanishes and a_0/b_0 is indeterminate, but the positive solution is permissible. Thus there are solutions for $n' = 0$, but they are only allowed if the quantum number k is positive.

To complete the solution of the radial equations for bound states it is only necessary to normalize the radial functions, that is to determine the magnitudes of a_0 and b_0. This is not a trivial problem, since F and G involve an exponential and, because λ is non-integral, fractional powers of ρ. Nevertheless, normalization is possible using gamma functions, which have to be evaluated numerically and may be obtained from tables; the inclusion of the details here would serve little purpose.

11.7 The energy levels

Although the eigenfunctions of the Dirac Hamiltonian have now been determined, the eigenvalues $E\,(= \epsilon mc^2)$ have yet to be found. Now from Equation 11.83:

$$\epsilon(1-\epsilon^2)^{-\frac{1}{2}} = (n'+\lambda)/Z\alpha, \tag{11.91}$$

where λ is $(k^2 - Z^2\alpha^2)^{\frac{1}{2}}$ and n' is a positive integer or zero. Solution of this equation gives:

$$E = mc^2\epsilon = mc^2[1 + (Z\alpha/(n'+\lambda))^2]^{-\frac{1}{2}}. \tag{11.92}$$

At this stage it is convenient to introduce:

$$n = n' + |k|. \tag{11.93}$$

Since $n' = 0, 1, 2, \ldots$ and the minimum value of $|k|$ is $+1$, the new quantum number n can be any positive integer but not zero. In addition, for a given value of n the maximum value that $|k|$ can take is n, although $k = -n$ is not allowed, since this corresponds to $n' = 0$ for which k cannot be negative. Thus, the possible values of k for a given n are $1, 2, \ldots, n$ if k is positive and $-1, -2, \ldots, -(n-1)$ if k is negative. When it is recalled that k is $\pm(j+\frac{1}{2})$, it may be seen that j, k, and n can have just the same values as in the approximate two-component theory (Equations 11.26) and in particular the new quantum number n may be identified with the non-relativistic principal quantum number n. In the four-component theory l is no longer a good quantum number, although the large and small components, ψ_+ and ψ_-, are separately eigenfunctions of l^2 as may be seen from Equations 11.41, 11.43 and 11.57. It was for this reason that the quantum number k was introduced earlier. Nevertheless, it is convenient to use the notation 11.32 for the different levels, the l value for the large components being employed.

Substituting for n' and λ in Equation 11.92 gives finally for the eigenvalues:

$$E = mc^2\{1 + [Z\alpha/(n - |k| + (k^2 - Z^2\alpha^2)^{\frac{1}{2}})]^2\}^{-\frac{1}{2}}. \qquad (11.94)$$

This may be expanded as a power series in $(Z\alpha)^2$:

$$E = mc^2[1 - Z^2\alpha^2/2n^2 - (Z^4\alpha^4/2n^3)(1/|k| - 3/4n) + O(Z^6\alpha^6)]; \qquad (11.95)$$

since $|k| = (j + \frac{1}{2})$, this is the same result as that obtained earlier (Equation 11.31). However, as Z increases such an expansion becomes less satisfactory, since the expansion parameter gets larger, and the exact solution (Equation 11.94) should be used.

As for the approximate theory, the exact energy (Equation 11.94) only depends on n and $|k|$, so that degeneracies arise as shown in 11.32. These degeneracies occur because in a one-electron atom the potential experienced by the electron is Coulombic; in many-electron atoms this is no longer true and the degeneracies are removed. However, even in one-electron atoms some of the degeneracies are lifted by the Lamb shift, which is the subject of Section 11.9.

11.8 Comparison of Dirac and non-relativistic atomic orbitals

The eigenfunctions of the Dirac Hamiltonian, which may be said to represent Dirac orbitals, are four-component functions in contrast to non-relativistic atomic orbital functions, which have one component, and the two-component functions of the approximate theory.

It has already been seen that for bound states the two components of a Dirac eigenfunction involving the radial function F are the large components, while those involving G are the small components. The ratio of the amplitudes of the small and large components thus has the approximate magnitude:

$$G/F \simeq [(1 - \epsilon)/(1 + \epsilon)]^{\frac{1}{2}}, \qquad (11.96)$$

where Equations 11.75 have been used. Substitution for ϵ and expansion of this ratio as a power series in $Z\alpha$ then gives:

$$G/F \simeq Z\alpha/2n; \qquad (11.97)$$

classically this is just the ratio of the Bohr velocity of the electron to the velocity of light. This shows that for light nuclei the small components are indeed not very important, but for large Z their influence is not negligible. For instance for $Z = 80$ and $n = 1$ the ratio of the small components to the large components is approximately 0.3. Even when this is squared to give the ratio of the contributions to the probability density, it is seen that the small components are responsible for about ten per cent.

To compare and contrast relativistic and non-relativistic atomic orbitals, approximate Dirac eigenfunctions for $n = 1$ and $n = 2$ are used. These are given in the table and are correct to order $Z\alpha$. They are normalized to this order provided each is multiplied by the factor $\pi^{-\frac{1}{2}}(Z/a_0)^{3/2}$, where a_0 is the Bohr radius; in addition, the dimensionless parameter ρ (Equation 11.76) is to this approximation just Zr/a_0, the same quantity as that employed in the non-relativistic theory (Equation 11.6). Although these functions become less accurate as Z increases, they exhibit the same qualitative features as the exact functions, and are easier to handle; the results obtained may be generalized to higher values of n without difficulty.

The probability density distribution $|\Psi|^2$ is the real interest, but radial parts of the individual components of Ψ are considered first; the angular parts involve spherical harmonics and do not need special consideration. For each four-component function there are two radial functions, one for the large components $f(r)$ and one for the small components $g(r)$. For f there are n' or $(n' - 1)$ nodes in the radial function depending on whether k is positive or negative, respectively. If the value of l appropriate to the large components is used, we may say that f has $(n - l - 1)$ nodes. This result is the same as that for non-relativistic orbitals and indeed, to the order of accuracy used in the table, the radial functions of the large components are the same as those in the non-relativistic theory. For g the number of radial nodes is n' regardless of the sign of k; this means that when k is positive the large and small components have the same number of radial nodes as each other, and in the special case when $n' = 0$ and k can only be positive f and g are both nodeless. The important thing to note is that, although the large and small components may have the same number of radial nodes, a node in g never occurs at the same value of ρ as a node in the corresponding radial function f.

When $|\Psi|^2$ is considered it is immediately apparent from the table that it may be factorized into a radial and an angular part. This is because, although $|\psi_+|^2$ and $|\psi_-|^2$ have different radial dependences, their angular behaviour is similar and the ratio of $|\psi_+|^2$ to $|\psi_-|^2$ is independent of θ and ϕ.

The radial part of $|\Psi|^2$ is thus a linear combination of $|f|^2$ and $|g|^2$. Since the large components predominate, the radial dependence is very like that of the corresponding non-relativistic orbitals, but unlike these the Dirac orbitals have no radial nodes. Although f has nodes at the same places as the corresponding non-relativistic radial functions, g never vanishes at these points. As an example the $2s_{\frac{1}{2}}$ orbital may be considered. At $\rho = 2$ both f and the non-relativistic $2s$ radial function possess a node, but the node in g occurs at

Table 11.1. Eigenfunctions of the Dirac Hamiltonian for a one-electron atom correct to order $Z\alpha$. Each function should be multiplied by a factor $\pi^{-\frac{1}{2}}(Z/a_0)^{3/2}$, where a_0 is the Bohr radius; ρ is the dimensionless parameter Zr/a_0.

$1s_{\frac{1}{2}}$ $(n = 1, j = \frac{1}{2}, k = +1, n' = 0)$

$m_j = +\frac{1}{2}$

$$\begin{pmatrix} e^{-\rho} \\ 0 \\ \frac{1}{2}iZ\alpha e^{-\rho}\cos\theta \\ \frac{1}{2}iZ\alpha e^{-\rho}\sin\theta e^{i\phi} \end{pmatrix}$$

$m_j = -\frac{1}{2}$

$$\begin{pmatrix} 0 \\ e^{-\rho} \\ \frac{1}{2}iZ\alpha e^{-\rho}\sin\theta e^{-i\phi} \\ -\frac{1}{2}iZ\alpha e^{-\rho}\cos\theta \end{pmatrix}$$

$2s_{\frac{1}{2}}$ $(n = 2, j = \frac{1}{2}, k = +1, n' = 1)$

$m_j = +\frac{1}{2}$

$$\frac{1}{2\sqrt{2}}\begin{pmatrix} e^{-\rho/2}(1-\rho/2) \\ 0 \\ \frac{1}{2}iZ\alpha e^{-\rho/2}(1-\rho/4)\cos\theta \\ \frac{1}{2}iZ\alpha e^{-\rho/2}(1-\rho/4)\sin\theta e^{i\phi} \end{pmatrix}$$

$m_j = -\frac{1}{2}$

$$\frac{1}{2\sqrt{2}}\begin{pmatrix} 0 \\ e^{-\rho/2}(1-\rho/2) \\ \frac{1}{2}iZ\alpha e^{-\rho/2}(1-\rho/4)\sin\theta e^{-i\phi} \\ -\frac{1}{2}iZ\alpha e^{-\rho/2}(1-\rho/4)\cos\theta \end{pmatrix}$$

$2p_{\frac{1}{2}}$ $(n = 2, j = \frac{1}{2}, k = -1, n' = 1)$

$m_j = +\frac{1}{2}$

$$\frac{1}{4\sqrt{6}}\begin{pmatrix} e^{-\rho/2}\rho\cos\theta \\ e^{-\rho/2}\sin\theta e^{i\phi} \\ -(3/2)iZ\alpha e^{-\rho/2}(1-\rho/6) \\ 0 \end{pmatrix}$$

$m_j = -\frac{1}{2}$

$$\frac{1}{4\sqrt{6}}\begin{pmatrix} e^{-\rho/2}\rho\sin\theta e^{-i\phi} \\ -e^{-\rho/2}\rho\cos\theta \\ 0 \\ -(3/2)iZ\alpha e^{-\rho/2}(1-\rho/6) \end{pmatrix}$$

$2p_{3/2}(n=2, j=3/2, k=+2, n'=0)$

$$m_j = +3/2 \qquad \frac{1}{8}\begin{pmatrix} e^{-\rho/2}\rho \sin\theta e^{i\phi} \\ 0 \\ (1/4)iZ\alpha e^{-\rho/2}\rho \sin\theta \cos\theta e^{i\phi} \\ (1/4)iZ\alpha e^{-\rho/2}\rho \sin^2\theta e^{2i\phi} \end{pmatrix}$$

$$m_j = -3/2 \qquad \frac{1}{8}\begin{pmatrix} 0 \\ e^{-\rho/2}\rho \sin\theta e^{-i\phi} \\ (1/4)iZ\alpha e^{-\rho/2}\rho \sin^2\theta e^{-2i\phi} \\ -(1/4)iZ\alpha e^{-\rho/2}\rho \sin\theta \cos\theta e^{-i\phi} \end{pmatrix}$$

$$m_j = +\tfrac{1}{2} \qquad \frac{1}{4\sqrt{3}}\begin{pmatrix} e^{-\rho/2}\rho \cos\theta \\ -\frac{1}{2}e^{-\rho/2}\rho \sin\theta e^{i\phi} \\ -(1/8)iZ\alpha e^{-\rho/2}\rho(1-3\cos^2\theta) \\ (3/8)iZ\alpha e^{-\rho/2}\rho \sin\theta \cos\theta e^{i\phi} \end{pmatrix}$$

$$m_j = -\tfrac{1}{2} \qquad \frac{1}{4\sqrt{3}}\begin{pmatrix} \frac{1}{2}e^{-\rho/2}\rho \sin\theta e^{-i\phi} \\ e^{-\rho/2}\rho \cos\theta \\ (3/8)iZ\alpha e^{-\rho/2}\rho \sin\theta \cos\theta e^{-i\phi} \\ (1/8)iZ\alpha e^{-\rho/2}\rho(1-3\cos^2\theta) \end{pmatrix}$$

$\rho = 4$. In fact $|g|^2$ at $\rho = 2$ is of the order of $(Z/a_0)^3 Z^2\alpha^2 \exp(-2)/8$ and, although this is very small, it ensures that the $2s_{\frac{1}{2}}$ orbital is radially nodeless.

Another qualitative feature of the radial part of Dirac orbitals is that the probability charge distribution is slightly contracted compared with the corresponding non-relativistic orbitals. This cannot be seen from the approximate functions in the table, since the effect is of higher order than $Z\alpha$. However, it might be supposed that, since the inclusion of relativistic effects leads to a slight lowering of the electron's energy, this corresponds to the electron being on average closer to the nucleus. This naive argument also suggests that the contraction should be more marked for $|k| = 1$ and for low values of n; detailed consideration of the exact eigenfunctions shows that these statements are correct.

The angular dependence of Dirac orbitals has yet to be considered. Unlike the radial part it is not influenced by the small components, since $|\psi_+|^2$ and $|\psi_-|^2$ are both proportional to the same angular function. We will see that, in contrast to non-relativistic orbitals, there are no angular nodes, but this is to

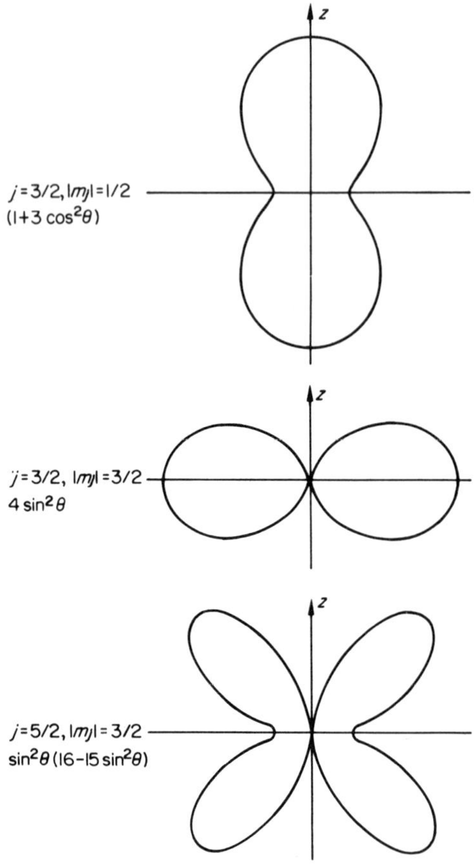

Fig. 11.1. The angular distribution for selected Dirac orbitals. Each is cylindrically symmetric about the z axis.

be regarded as being the result of a correct handling of spin using two-component wavefunctions as in Section 11.3.

The first thing to note is that the probability density distribution associated with a Dirac orbital is independent of the polar coordinate ϕ, so that Dirac orbitals are all cylindrically symmetric about the z axis. In addition, the angular dependence of $|\Psi|^2$ depends only on the quantum numbers j and m_j; for example, when $j = \frac{1}{2}$ the orbital is spherically symmetric as may be seen explicitly for $1s_{\frac{1}{2}}$, $2s_{\frac{1}{2}}$ and $2p_{\frac{1}{2}}$ by reference to the table.

When the value of j differs from $\frac{1}{2}$, the angular dependence is more complicated, but two limiting cases may be distinguished. For $|m_j| = \frac{1}{2}$ the angular dependence resembles a dumb-bell with its axis along the z axis; for $j = 3/2$ the angular part of $|\Psi|^2$ is proportional to $(1 + 3\cos^2\theta)$ and is plotted in Fig. 11.1. At the opposite extreme, when $|m_j|$ takes its maximum value of j, the distribution is toroidal and an example of this case is also given; the specific case shown is also for $j = 3/2$, the angular dependence being proportional to $4\sin^2\theta$. For intermediate values of m_j the orbitals are also toroidal, but with a number of lobes, the example in the figure being for $j = 5/2$ and $m_j = 3/2$, for which $|\Psi|^2$ is proportional to $\sin^2\theta(16 - 15\sin^2\theta)$.

Although the first angular function shown in the figure is not unlike that of a non-relativistic p_z orbital, it has no node perpendicular to the z axis. The other angular functions are also nodeless, since they are cylindrically symmetric about the z axis. The lack of angular nodes is a general property of Dirac orbitals.

We have already seen that relativistic orbitals have no radial nodes, so that it may be concluded that except at the nucleus itself they have no nodes at all; there are no radial nodes because the eigenfunctions of the Dirac Hamiltonian have four components and the small components prevent them occuring, while the lack of angular nodes may be attributed to the fact that the electron has spin $\frac{1}{2}$.

This is in sharp contrast to non-relativistic orbitals, which in general have both angular and radial nodes, a feature which raises the question of how an electron gets from one part of an orbital to another, if there is a node in between. However, this problem no longer arises if relativistic atomic orbitals are used, since they are nodeless.

11.9 The Lamb shift

Both the exact (Equation 11.94) and approximate (Equation 11.95) eigenvalues of the Dirac Hamiltonian for a one-electron atom show that levels characterized by the same values of the quantum numbers n and j are degenerate. However, this degeneracy is in fact removed by the Lamb shift, which is of quantum electrodynamic origin. The largest manifestation of this effect occurs for the $n = 2$ levels in the hydrogen atom and is illustrated in Fig. 11.2, which shows the successive effects on the non-relativistic energy of including relativistic effects and the Lamb shift; the energy separations are given in terms of frequency (1 GHz $= 10^9$ Hz $\equiv 0.03336$ cm^{-1}) and are to be compared with the non-relativistic energy lowering of 8.22×10^5 GHz relative to the rest energy of the electron.

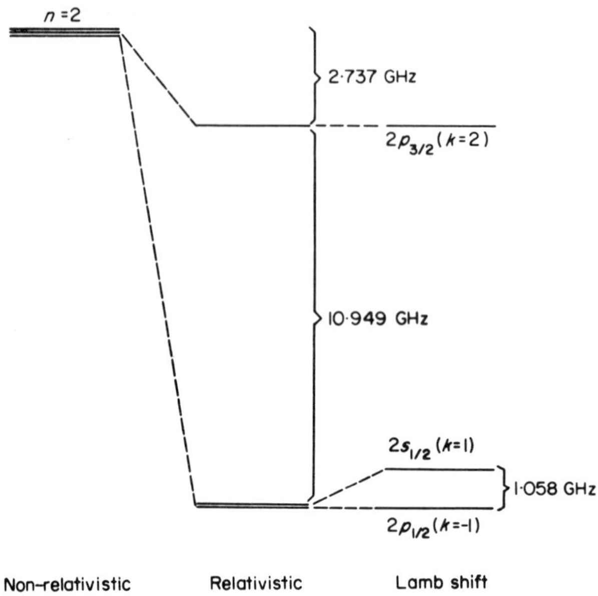

Fig. 11.2 The effect of relativistic corrections and the Lamb shift on the energies of the $n = 2$ levels in the hydrogen atom.

Although the Lamb shift is a quantum electrodynamic effect and cannot be explained fully here, an attempt will be made to indicate its origin, since at least two authors have wrongly attributed it to the Zitterbewegung of the electron. As we have seen, the concept of Zitterbewegung arises when the non-relativistic approximation to the Dirac Hamiltonian is used (see Section 8.10). The Darwin-type term then accounts for the correction to the electrostatic interaction energy between the nucleus and the electron due to the smearing out of the electron's charge that is associated with the Zitterbewegung; this correction term was included in the approximate theory of Section 11.2. However, if the Dirac equation itself is solved this effect is automatically accounted for and mention of the Zitterbewegung is out of place and misleading. Moreover, the relativistic corrections to the energy are of order $mc^2Z^4\alpha^4$ and in particular the Darwin-type correction term in the approximate theory may be seen to be of this order by reference to Equation 11.29. The Lamb shift, on the other hand, is smaller than this by one order in α.

The confusion has arisen because an estimate of the magnitude of the Lamb shift may be obtained using arguments similar to those used in Section 8.10 to explain the Darwin term on the basis of the Zitterbewegung. There it is shown that, if the electron has a highly oscillatory motion superimposed on its mean position $\mathbf{r}$, then the correction to an electrostatic interaction energy $-e\phi(\mathbf{r})$ is given approximately by:

$$-e\langle\delta^2\rangle\nabla^2\phi(\mathbf{r})/6, \tag{11.98}$$

where $\phi(\mathbf{r})$ is the scalar potential and $\boldsymbol{\delta}$ is the instantaneous deviation of the electron from its mean position. For the Zitterbewegung the time average $\langle\delta^2\rangle$ is of the order of $(\hbar/mc)^2$.

The Lamb shift may be thought of as arising from an additional smearing out of the electron's charge that is superimposed on that due to the Zitterbewegung. This arises from the quantization of the electromagnetic field, which is the subject of the next chapter. Nevertheless, we may anticipate this discussion to the extent of recognizing that the quantized field has zero-point energy in the same way that a quantized harmonic oscillator does. This result is not unexpected, since the classical field may be described in terms of a collection of classical harmonic oscillators (Section 6.7). Consequently, even a field-free vacuum contains electromagnetic energy in the form of so-called 'vacuum fluctuations', which interact with the electron and cause it to undergo a rapid random motion. Detailed calculations, which cannot be presented here, show that for an electron in an atom the mean square deviation of this random motion is:

$$\langle\delta^2\rangle = (2\alpha/\pi)(\hbar/mc)^2\ln(2n^2/\alpha^2), \tag{11.99}$$

which is smaller than that due to the Zitterbewegung by one order in α.

If the result (Equation 11.99) is accepted, the crude model of Section 8.10 may be used to estimate the magnitude of the Lamb shift. In addition to $\langle\delta^2\rangle$ the expression 11.98 involves $\nabla^2\phi(\mathbf{r})$ and in a one-electron atom this is just $-Ze\delta(\mathbf{r})/\epsilon_0$ as may be seen from Equation 11.15. Thus the Lamb shift is given by the operator:

$$(e^2Z\alpha/3\pi\epsilon_0)(\hbar/mc)^2\delta(\mathbf{r})\ln(2n^2/\alpha^2). \tag{11.100}$$

The expectation value for an electron in a non-relativistic s orbital with principal quantum number n is from Equations 11.11 just:

$$mc^2Z^4\alpha^5(4/3\pi n^3)\ln(2n^2/\alpha^2), \tag{11.101}$$

while in other atomic orbitals it vanishes. Although the model used is crude, more sophisticated approaches give the same result for non-relativistic orbitals.

For the hydrogen atom 11.101 gives 1.040 GHz for the Lamb shift of the $2s$ level; the $2p$ levels are unaffected. This is to be compared with 1.058 GHz, the experimental separation of $2s_{\frac{1}{2}}$ and $2p_{\frac{1}{2}}$. The discrepancy has been explained to within 1 MHz by relativistic, higher-order quantum electrodynamic and other effects. These corrections also produce small shifts in levels other than s levels. Lamb shifts have also been observed in the hydrogen atom for $n>2$ and in other one-electron atoms and they have all been explained quantitatively. They also occur in many-electron systems, but are more complicated; electron-electron interactions are of course present and full calculations have only been performed for the simplest species.

11.10 More complicated systems

So far this chapter has been devoted to the relativistic treatment of one-electron atoms in which the nucleus is assumed to be stationary and spinless. It has already been remarked that the finite mass of the nucleus may largely be accounted for by using a reduced mass for the electron and, except for a few remarks about the influence of nuclear spin, this section is concerned with many-electron atoms.

The presence of a nuclear spin magnetic moment in a one-electron atom may be effectively allowed for by using perturbation theory, since the corrections introduced are relatively small. If non-relativistic atomic orbitals are employed as spatial basis functions, it is appropriate to use a perturbation based on the approximate Hamiltonian of chapter 10. In particular the nuclear hyperfine interaction arises from the Fermi contact term (10.6f); the dipole-dipole term averages to zero for an atom because of its spherical symmetry. However, it is only permissible to use first-order perturbation theory for the Fermi contact interaction, since it has its origin in the Breit operator. Indeed, for the hydrogen atom the estimated hyperfine splitting is infinite if second-order perturbation theory is used, although the cause of this divergence has also been attributed to treating the proton as a point particle.

An alternative approach is to employ Dirac atomic orbitals but, although it is tempting to use, for example, the Fermi contact operator as a perturbation, this is not permissible, since it is a non-relativistic operator. If it is used, even first-order perturbation theory gives infinite answers, since Dirac orbitals with $j=\frac{1}{2}$ have weak singularities at the nucleus as the exact radial functions of Section 11.6 show. The correct procedure is to substitute the vector potential due to the nuclear spin magnetic moment into the Dirac Hamiltonian. This gives:

$$(g_\alpha \mu_N e/4\pi\epsilon_0 c)\boldsymbol{\alpha}.(\mathbf{I}_\alpha \wedge \mathbf{r})/r^3 \tag{11.102}$$

as the perturbation, where $\boldsymbol{\alpha}$ is the Dirac operator. Exact diagonalization of this operator using all Dirac orbitals with the same n as basis functions gives the same results as the non-relativistic theory except for a correction of relative order $(Z\alpha)^2$. For example, the correction factor for the $1s_{\frac{1}{2}}$ level is $(1 + 3(Z\alpha)^2/2 + \ldots)$; correction factors for other levels have been tabulated in the literature, but except for large Z they are relatively unimportant.

For many-electron atoms perturbation theory must again be used to estimate the effect of small terms in the Hamiltonian, such as those involving nuclear spin. However, the choice arises as to what electronic basis functions to use. The usual approach is to obtain eigenfunctions of the non-relativistic Hamiltonian, which consists of the kinetic energy of the electrons and all the Coulomb interactions. Even these can only be obtained approximately and in general consist of a linear combination of determinants involving one-electron functions that have been optimized using a self-consistent field criterion, the starting point for the iterative procedure being eigenfunctions for one-electron atoms. For inner electrons and heavy nuclei the relativistic corrections are relatively large and their calculation by perturbation theory is not very satisfactory. This is becoming a serious drawback, since attention is now being focused on inner electrons by the use of experimental techniques such as X-ray photoelectron spectroscopy. This situation has led to the use of the Dirac Hamiltonian for the individual electrons, since allowance for the relativistic corrections involving individual electrons is then automatic; in particular, account is taken of such things as spin-orbit coupling and the relativistic corrections to the kinetic energies of the electrons. An appropriate procedure is described briefly here; for the details reference must be made to the research literature, since the mathematical techniques used are outside the scope of this book.

The approximate Hamiltonian for a many-electron atom may be taken as:

$$\mathcal{H} = \sum_i \mathcal{H}_i + \sum_{i \neq j} e^2/4\pi\epsilon_0 r_{ij}, \qquad (11.103)$$

where $\mathcal{H}_i$ is the Dirac Hamiltonian for electron i and includes the Coulomb interaction between the electron and the nucleus. The term representing the Coulomb interaction between the electrons is not even approximately Lorentz invariant, but the terms that have been omitted from the full Hamiltonian may be treated as perturbations on the eigenfunctions of 11.103. Indeed, to describe the interaction between two electrons better the Breit operator 9.41 may be added, but as we have emphasized repeatedly it is only appropriate to use it in a first-order perturbation treatment. Other perturbations that could be included are those involving the nuclear spin magnetic moment and quadrupole moment; electron spin is of course included automatically in the Hamiltonian 11.103. The problem then is to find eigenfunctions of

11.103 that can be used as basis functions for the perturbation treatment.

Now the eigenfunctions of the operator $\sum_i \mathcal{H}_i$ in 11.103 are just products of the eigenfunctions of the individual Dirac Hamiltonians $\mathcal{H}_i$; these provide suitable starting functions for an iterative procedure. Of course, these functions are not unique since permutation of the electrons gives other eigenfunctions of $\sum_i \mathcal{H}_i$ with the same eigenvalue so that a linear combination of these products is taken. The Pauli exclusion principle restricts the possible linear combinations to those which are antisymmetric with respect to interchange of any pair of electrons and this linear combination may conveniently be written as a determinant, the elements being (four-component) Dirac functions. For closed shell configurations only one determinant is needed, but for open shell configurations a linear combination of determinants is necessary, although the number involved may be reduced by choosing the combinations to be eigenfunctions of the total angular momentum operator $\sum_i \mathbf{j}_i$.

When the Coulomb interaction between electrons is introduced the exact eigenfunctions of the Hamiltonian can no longer be written as the product of one-electron functions, since a definite energy cannot be associated with an individual particle and the probability of an electron being at a particular point in space depends on the position of the other electrons. However, it is not practicable to obtain exact eigenfunctions of the Hamiltonian (Equation 11.103) and an approximation must be made. The usual policy is to assume that each electron experiences an average repulsive potential due to the other electrons in addition to the attractive potential of the nucleus. With this simplification the potential energy of an electron is a function only of its distance from the nucleus, although the electrostatic potential is no longer Coulombic. This approximation is not too unrealistic since for Dirac atomic orbitals closed shells of electrons have spherically symmetric charge distributions and this fact does not depend on the assumption of a Coulomb potential.

Thus for a many-electron atom the eigenfunctions of the Hamiltonian are assumed to be products of one-electron functions, which are themselves eigenfunctions of the operator $\mathcal{H}_i + V(r_i)$. Since a central field has been assumed, the angular dependence of the components of the one-electron functions is the same as for a one-electron atom and may be factored out. However, since the potential is no longer Coulombic, the resulting radial equations cannot be solved analytically and numerical techniques must be used. Of course the potential energy $V(r_i)$ is an average over all the electrons other than i and a knowledge of their radial distributions is necessary for it to be calculated.

This means that an iterative procedure must be adopted, the Dirac orbitals for a one-electron atom being used to calculate initial potential energies $V(r_i)$.

The improved one-electron functions so obtained are then used to obtain better potentials and the process repeated until a self-consistent field is achieved. This procedure is quite laborious and in practice the whole calculation may be accomplished in one step using the variation method; this has the added advantage that the exchange interaction is included, if determinental wave functions are used.

In Section 11.7 it was seen that for one-electron atoms eigenfunctions with the same values of n and j are degenerate. These degeneracies of one-electron functions do not occur for many-electron atoms, since the potential experienced by an electron is no longer Coulombic. The energy of an electron then depends on how effectively it is screened from the nucleus by the other electrons. The lower the value of l associated with the large components of a one-electron function the lower is its energy; the same situation occurs in the non-relativistic theory.

The wave functions obtained using this self-consistent field approach are only approximate, but they may be improved by using techniques, such as configuration interaction, which are well established in the non-relativistic theory of many-electron atoms.

The relativistic theory for atoms is thus similar to the non-relativistic theory, although the details must be modified to allow for the fact that relativistic eigenfunctions have four components; the amount of labour involved is greater and the mathematical techniques employed are more sophisticated. In addition, it is only comparatively recently that relativistic calculations of atomic structure along these lines have been attempted in any great number and at present they are by no means routine. For these reasons the details are not given here and reference must be made to the specialist literature.

Finally, a few comments may be made about calculations on molecules using relativistic atomic orbitals. Since molecules do not possess spherical symmetry, a central-field approximation is not possible and the wavefunctions cannot be factorized into radial and angular parts. A popular approach in non-relativistic theory is to construct one-electron molecular orbitals, which are linear combinations of a limited basis set of atomic functions, the variation principle being used to calculate the coefficients or other adjustable parameters. In principle there is no reason why this should not be done using relativistic atomic orbitals as a basis. However, although this approach has been used for atoms themselves, at the present time little attention appears to have been directed towards the use of relativistic atomic orbitals to describe molecular electronic structure. Recently attempts have been made to solve the non-relativistic self-consistent field equations for simple molecules by using numerical techniques and in principle this approach could also be used in the relativistic theory.

Bibliography

Solution of the Dirac equation for the hydrogen atom

Bethe and Salpeter: pages 63-70.

Corinaldesi, E., and Strocchi, F. (1963), *Relativistic Wave Mechanics,* North-Holland, Amsterdam: part II, Chapters 8 and 9.

Dirac, P. A. M. (1958), *Quantum Mechanics,* Clarendon Press, Oxford: pages 269-273.

Rose: pages 157-181.

Sakurai: pages 122-131.

Comparison of Dirac and non-relativistic atomic orbitals

Bethe and Salpeter: pages 70-71.

Powell, R. E. (1968), 'Relativistic quantum chemistry. The electrons and the nodes', *J. Chem. Fduc.,* **45**, 558.

Szabo, A. (1969), 'Contour diagrams for relativistic orbitals', *J. Chem. Educ.,* **46**, 678.

The Lamb shift

Series, G. W. (1957), *The Spectrum of Atomic Hydrogen,* Clarendon Press, Oxford: pages 37-65.

Sikolov, A. A., Koskutov, Y. M., and Ternov, I. M. (1966), *Quantum Mechanics,* Holt, Rinehart and Winston, New York: pages 350-354.

The introduction of nuclear spin

Breit, G. (1930), 'Possible effects of nuclear spin on X-ray terms', *Phys. Rev.,* **35**, 1447: relativistic corrections to hyperfine splittings.

Rose: relativistic corrections to nuclear hyperfine splittings are discussed on pages 188-191.

Velenik, A., Živković, T., de Jeu, W. H., and Murrell, J. N. (1970), 'The hydrogen atom in the presence of the Fermi contact interaction', *Molec. Phys.,* **18**, 693: a demonstration that the Fermi contact operator taken to second order with non-relativistic atomic orbitals leads to an infinite hyperfine splitting.

Many-electron atoms

Grant, I. P. (1970), 'Relativistic calculation of atomic structures', *Adv. Phys.,* **19**, 747: a review of relativistic calculations on atoms.

CHAPTER TWELVE

Quantum Field Theory

The classical theory of fields was developed in Chapter 6. In particular, the electromagnetic field was considered and it is the quantization of this field that is the subject of the present chapter, although it should be noted that other fields may be quantized.

It is, of course, well known that electromagnetic radiation has a quantum nature. Indeed, quantum theory originated with Planck postulating that radiation of angular frequency ω can only have an energy that is an integral multiple of $\hbar\omega$, where $\hbar$ was a new fundamental constant; this was necessary to avoid the so-called ultraviolet catastrophe for black-body radiation. A little later Einstein explained the photoelectric effect by proposing that electromagnetic radiation consists of massless particles, each with energy $\hbar\omega$.

So far we have been concerned with quantized matter moving under the influence of electromagnetic fields and the semiclassical approach, which treats this radiation classically, has proved quite adequate. However, this procedure is not very satisfactory when it comes to determining what effect the quantized matter can have on the radiation; here we are thinking particularly of time-dependent processes such as the absorption and emission of radiation by molecules. Besides, it is inconsistent to treat one part of an interacting system quantum-mechanically and the other part classically; for example, it is possible to design experiments which in principle would violate Heisenberg's uncertainty principle by observing quantized particles with classical radiation.

The last chapter is concerned with the interaction of radiation with matter. Here, electromagnetic radiation in free space is discussed.

12.1 Quantization of the electromagnetic field

Before considering the electromagnetic field, the more familiar particle quantization is reviewed, since it provides a guide for the quantization of field variables. Classically the Hamiltonian of a particle may be expressed in terms of its position **r** and its momentum **p** conjugate to **r**. Quantization may then be accomplished by making **r** and **p** time-independent operators and in particular by identifying **p** with the operator $-i\hbar(\partial/\partial\mathbf{r})$. This procedure is equivalent to imposing the condition that the operators **p** and **r** satisfy the commutation relations:

$$[p_i, r_j] = -i\hbar\delta_{ij}; \quad [p_i, p_j] = [r_i, r_j] = 0. \qquad (i, j = x, y, z). \tag{12.1}$$

The resulting energy operator is then used in the time-dependent Schrödinger equation, $\mathcal{H}\Psi = i\hbar\partial\Psi/\partial t$, where the wave function Ψ is a function of both space and time coordinates.

This suggests that to quantize the electromagnetic field its energy should be expressed in terms of conjugate pairs of field variables; that is, for each pair one variable should be the momentum conjugate to the other. Quantization may then be achieved by interpreting these variables as operators, insisting that they satisfy commutation relations analogous to Equations 12.1 and allowing them to operate on appropriate wave functions.

In Chapter 6 it was seen that the gauge of the electromagnetic potentials can always be chosen so that at all points in space the scalar potential ϕ vanishes and the vector potential **A** satisfies the Coulomb gauge ($\nabla.\mathbf{A} = 0$). Thus, the field is determined by three field variables; these are the components of the vector potential, which is given by Equation 6.65:

$$\mathbf{A}(\mathbf{r}, t) = \sum_{\mathbf{k},\lambda} [q_{\mathbf{k}\lambda}(t)\mathbf{A}_{\mathbf{k}\lambda}(\mathbf{r}) + q^*_{\mathbf{k}\lambda}(t)\mathbf{A}^*_{\mathbf{k}\lambda}(\mathbf{r})], \tag{12.2}$$

where

$$q_{\mathbf{k}\lambda}(t) = q_{\mathbf{k}\lambda}(0)\exp(-i\omega_{\mathbf{k}}t) \tag{12.3}$$

and

$$\mathbf{A}_{\mathbf{k}\lambda}(\mathbf{r}) = (\epsilon_0 V)^{-\frac{1}{2}}\boldsymbol{\epsilon}_{\mathbf{k}\lambda}\exp(i\mathbf{k}.\mathbf{r}). \tag{12.4}$$

Here the wave vector **k** and the angular frequency $\omega_{\mathbf{k}}$ are related by $\omega_{\mathbf{k}} = kc$, and V, the volume under consideration, may be made arbitrarily large. In addition, for each value of **k** there are two independent polarization vectors $\boldsymbol{\epsilon}_{\mathbf{k}\lambda}$ and these are distinguished by the subscript $\lambda(=1, 2)$. The two possibilities for $\boldsymbol{\epsilon}_{\mathbf{k}\lambda}$ are orthogonal to each other and because of the Coulomb gauge condition they are both orthogonal to **k**:

$$\boldsymbol{\epsilon}_{\mathbf{k}\lambda}.\boldsymbol{\epsilon}_{\mathbf{k}\lambda'} = \delta_{\lambda\lambda'}; \qquad \boldsymbol{\epsilon}_{\mathbf{k}\lambda}.\mathbf{k} = 0. \tag{12.5}$$

The energy of the field is then given by the classical Hamiltonian (Equation 6.83):

$$\mathcal{H} = \sum_{\mathbf{k},\lambda} \omega_{\mathbf{k}}^2 (q_{\mathbf{k}\lambda}^* q_{\mathbf{k}\lambda} + q_{\mathbf{k}\lambda} q_{\mathbf{k}\lambda}^*). \tag{12.6}$$

Although this Hamiltonian is expressed in terms of pairs of variables, $q_{\mathbf{k}\lambda}$ and $q_{\mathbf{k}\lambda}^*$, they do not form conjugate pairs and are not appropriate for quantization. This situation was remedied in Section 6.7, where the new real variables:

$$Q_{\mathbf{k}\lambda} = q_{\mathbf{k}\lambda} + q_{\mathbf{k}\lambda}^*; \qquad P_{\mathbf{k}\lambda} = -i\omega_{\mathbf{k}}(q_{\mathbf{k}\lambda} - q_{\mathbf{k}\lambda}^*) \tag{12.7}$$

were introduced, the inverse relation being:

$$q_{\mathbf{k}\lambda} = (\omega_{\mathbf{k}} Q_{\mathbf{k}\lambda} + iP_{\mathbf{k}\lambda})/2\omega_{\mathbf{k}}. \tag{12.8}$$

There it was shown that $P_{\mathbf{k}\lambda}$ may be regarded as the momentum conjugate to $Q_{\mathbf{k}\lambda}$ and that in terms of these new variables the field energy is:

$$\mathcal{H} = \sum_{\mathbf{k},\lambda} \tfrac{1}{2}(P_{\mathbf{k}\lambda}^2 + \omega_{\mathbf{k}}^2 Q_{\mathbf{k}\lambda}^2). \tag{12.9}$$

Consequently, we consider the theory based on quantization of the conjugate pairs, $P_{\mathbf{k}\lambda}$ and $Q_{\mathbf{k}\lambda}$. The final justification for this policy is, of course, that the resulting theory works.

To get from the classical theory to the quantum theory the $Q_{\mathbf{k}\lambda}$ and $P_{\mathbf{k}\lambda}$ are regarded as operators satisfying commutation relations analogous to Equations 12.1. Operators corresponding to different values of $\mathbf{k}$ or λ are independent of each other, so that the appropriate commutation relations are:

$$\begin{aligned} [P_{\mathbf{k}\lambda}, Q_{\mathbf{k}'\lambda'}] &= -i\hbar\delta_{\mathbf{k}\mathbf{k}'}\delta_{\lambda\lambda'}; \\ [P_{\mathbf{k}\lambda}, P_{\mathbf{k}'\lambda'}] &= [Q_{\mathbf{k}\lambda}, Q_{\mathbf{k}'\lambda'}] = 0. \end{aligned} \tag{12.10}$$

The Hamiltonian (Equation 12.9) is now to be regarded as an operator and is to be used in the Schrödinger equation $\mathcal{H}\Psi = i\hbar\partial\Psi/\partial t$.

For stationary states the Schrödinger equation reduces to $\mathcal{H}\Psi = E\Psi$ and this may be solved using the separation of variables method with the result that:

$$E = \sum_{\mathbf{k},\lambda} E_{\mathbf{k}\lambda}; \qquad \Psi = \prod_{\mathbf{k},\lambda} \psi_{\mathbf{k}\lambda}, \tag{12.11}$$

where $E_{\mathbf{k}\lambda}$ and $\psi_{\mathbf{k}\lambda}$ are the solutions of the separated equation:

$$\mathcal{H}_{\mathbf{k}\lambda}\psi_{\mathbf{k}\lambda} = E_{\mathbf{k}\lambda}\psi_{\mathbf{k}\lambda} \tag{12.12}$$

with

$$\mathcal{H}_{\mathbf{k}\lambda} = \tfrac{1}{2}(P_{\mathbf{k}\lambda}^2 + \omega_{\mathbf{k}}^2 Q_{\mathbf{k}\lambda}^2). \tag{12.13}$$

We note that after quantization the field can still be regarded as a collection of radiation oscillators, since Equation 12.13 is just the Hamiltonian for a one-dimensional harmonic oscillator. The solution of Equation 12.12 is considered in the next section, but before then other aspects of quantization are considered, since the operator forms of the vector potential and the electric and magnetic fields are needed later.

Although $q_{\mathbf{k}\lambda}$ and its complex conjugate $q^*_{\mathbf{k}\lambda}$ cannot be used directly in the quantization procedure, they are now to be regarded as operators defined by the operator form of Equation 12.8. Using this definition their commutation relations may easily be determined from Equations 12.10:

$$\begin{aligned}[q_{\mathbf{k}\lambda}, q^*_{\mathbf{k}'\lambda'}] &= (\hbar/2\omega_{\mathbf{k}})\delta_{\mathbf{k}\mathbf{k}'}\delta_{\lambda\lambda'};\\ [q_{\mathbf{k}\lambda}, q_{\mathbf{k}'\lambda'}] &= [q^*_{\mathbf{k}\lambda}, q^*_{\mathbf{k}'\lambda'}] = 0.\end{aligned} \tag{12.14}$$

We may now investigate the validity of quantizing the vector potential (Equation 12.2) by interpreting $q_{\mathbf{k}\lambda}$ and $q^*_{\mathbf{k}\lambda}$ as time-independent operators. This procedure must be consistent with the rest of the quantization scheme and to confirm this the Hamiltonian must now be derived from the quantized potential.

To do this the operator forms of the electric and magnetic fields are needed. The magnetic field operator is readily shown to be:

$$\mathbf{B} = \nabla_\wedge \mathbf{A} = \sum_{\mathbf{k},\lambda} i(q_{\mathbf{k}\lambda}\mathbf{k}_\wedge \mathbf{A}_{\mathbf{k}\lambda} - q^*_{\mathbf{k}\lambda}\mathbf{k}_\wedge \mathbf{A}^*_{\mathbf{k}\lambda}), \tag{12.15}$$

but a problem arises for the electric field operator, since it is given by $\mathbf{E} = -\,\partial\mathbf{A}/\partial t$ and $q_{\mathbf{k}\lambda}$ and $q^*_{\mathbf{k}\lambda}$ are both time-independent operators. The origin of this difficulty is that we are using the Schrödinger picture in which operators are time independent and the wave functions evolve with time. The alternative formalism involves time-dependent operators and time-independent wave functions. This is the Heisenberg picture, but it may be made to coincide with the Schrödinger picture at $t = 0$. Heisenberg's equation of motion (Equation 1.24) shows that:

$$\mathrm{d}q_{\mathbf{k}\lambda}/\mathrm{d}t = (i/\hbar)[\mathcal{H}, q_{\mathbf{k}\lambda}] = -i\omega_{\mathbf{k}} q_{\mathbf{k}\lambda}, \tag{12.16}$$

a result that may easily be obtained from Equations 12.8 to 12.10; this is analogous to the classical result and may be used in the Schrödinger picture at time $t = 0$. The electric field operator is then:

$$\mathbf{E} = -\,\partial\mathbf{A}/\partial t = \sum_{\mathbf{k},\lambda} i\omega_{\mathbf{k}}(q_{\mathbf{k}\lambda}\mathbf{A}_{\mathbf{k}\lambda} - q^*_{\mathbf{k}\lambda}\mathbf{A}^*_{\mathbf{k}\lambda}). \tag{12.17}$$

Equations 12.15 and 12.17 for the electric and magnetic field operators may now be used to determine the Hamiltonian from Equation 6.46:

$$\mathcal{H} = \tfrac{1}{2}\epsilon_0 \int (c^2 B^2 + E^2)\mathrm{d}\tau. \tag{12.18}$$

The manipulations involved are the same as those in the classical derivation of Section 6.7 except that it is now important to preserve the order of $q_{\mathbf{k}\lambda}$ and $q^*_{\mathbf{k}\lambda}$, since they do not commute. However, this was anticipated in the classical treatment so that Equation 6.83 is also valid quantum-mechanically:

$$\mathcal{H} = \sum_{\mathbf{k},\lambda} \omega_{\mathbf{k}}^2 (q^*_{\mathbf{k}\lambda} q_{\mathbf{k}\lambda} + q_{\mathbf{k}\lambda} q^*_{\mathbf{k}\lambda}). \tag{12.19}$$

When this Hamiltonian is expressed in terms of the operators $P_{\mathbf{k}\lambda}$ and $Q_{\mathbf{k}\lambda}$ using Equation 12.8, we recover:

$$\mathcal{H} = \sum_{\mathbf{k},\lambda} \tfrac{1}{2}(P^2_{\mathbf{k}\lambda} + \omega_{\mathbf{k}}^2 Q^2_{\mathbf{k}\lambda}), \tag{12.20}$$

so that the quantization procedure outlined is self-consistent.

The classical derivation of the field momentum in Chapter 6 also needs no modification after quantization, since the order of $q_{\mathbf{k}\lambda}$ and $q^*_{\mathbf{k}\lambda}$ was preserved; in addition Equations 12.14 now justify the policy used there of ignoring the order of, for example, $q_{\mathbf{k}\lambda}$ and $q_{\mathbf{k}'\lambda'}$. Thus:

$$\mathbf{G} = \sum_{\mathbf{k},\lambda} \mathbf{G}_{\mathbf{k}\lambda}, \tag{12.21}$$

where

$$\begin{aligned} \mathbf{G}_{\mathbf{k}\lambda} &= \omega_{\mathbf{k}} \mathbf{k} (q^*_{\mathbf{k}\lambda} q_{\mathbf{k}\lambda} + q_{\mathbf{k}\lambda} q^*_{\mathbf{k}\lambda}) \\ &= \tfrac{1}{2}(\mathbf{k}/\omega_{\mathbf{k}})(P^2_{\mathbf{k}\lambda} + \omega_{\mathbf{k}}^2 Q^2_{\mathbf{k}\lambda}) \\ &= (\mathbf{k}/\omega_{\mathbf{k}})\mathcal{H}_{\mathbf{k}\lambda}. \end{aligned} \tag{12.22}$$

Finally, we may show that Maxwell's equations for a free field may be recovered after quantization. From Equations 12.15 to 12.17 the first pair are obtained:

$$\nabla.\mathbf{B} = 0; \qquad \nabla_\wedge \mathbf{E} = -(\partial \mathbf{B}/\partial t), \tag{12.23}$$

but these are merely a consequence of the definition of the electric and magnetic fields in terms of the potentials. The divergence of the electric field is easily shown to vanish, since in the gauge chosen ϕ vanishes and $\mathbf{A}$ satisfies the Coulomb condition ($\nabla.\mathbf{A} = 0$), so that:

$$\nabla.\mathbf{E} = -\partial(\nabla.\mathbf{A})/\partial t = 0. \tag{12.24}$$

The remaining equation is more difficult to obtain. It is first noted that, from Equations 12.16 and 12.17:

$$(\partial \mathbf{E}/\partial t) = \sum_{\mathbf{k},\lambda} \omega_{\mathbf{k}}^2 (q_{\mathbf{k}\lambda}\mathbf{A}_{\mathbf{k}\lambda} + q^*_{\mathbf{k}\lambda}\mathbf{A}^*_{\mathbf{k}\lambda}), \tag{12.25}$$

while the curl of Equation 12.15 gives:

$$\nabla_\wedge \mathbf{B} = \sum_{\mathbf{k},\lambda} k^2 (q_{\mathbf{k}\lambda}\mathbf{A}_{\mathbf{k}\lambda} + q^*_{\mathbf{k}\lambda}\mathbf{A}^*_{\mathbf{k}\lambda}), \tag{12.26}$$

where

$$\nabla_{\wedge}(\mathbf{k}_{\wedge}\mathbf{A}_{\mathbf{k}\lambda}) = \mathbf{k}(\nabla.\mathbf{A}_{\mathbf{k}\lambda}) - (\mathbf{k}.\nabla)\mathbf{A}_{\mathbf{k}\lambda} = -ik^2\,\mathbf{A}_{\mathbf{k}\lambda} \tag{12.27}$$

has been used. Since $\omega_{\mathbf{k}} = ck$, comparison of Equations 12.25 and 12.26 shows that:

$$\nabla_{\wedge}\mathbf{B} = c^{-2}(\partial\mathbf{E}/\partial t), \tag{12.28}$$

which is the required equation.

12.2 Solution of the one-dimensional harmonic oscillator equation

For simplicity the subscripts in the one-dimensional harmonic oscillator equation may be omitted. Equation 12.12 is then $\mathcal{H}\psi = E\psi$ and the Hamiltonian is given by Equation 12.20 as $\mathcal{H} = \frac{1}{2}(P^2 + \omega^2 Q^2)$. The solutions of this equation could be obtained by replacing P by $-i\hbar(\partial/\partial Q)$ and solving the resulting differential equation for ψ in the usual manner. However, for the electromagnetic field it is not possible to give a simple physical meaning to the operator Q or to wave functions that are functions of Q. As it happens interpretations of Q and $\psi(Q)$ are not needed and we choose to solve the harmonic oscillator equation in a different way. Furthermore, this alternative approach is much more instructive in the present context.

Now that quantization has been accomplished the operators q and q^* may again be used. However, it is more convenient to define the dimensionless operator:

$$a = (2\omega/\hbar)^{\frac{1}{2}}q, \tag{12.29}$$

for which the commutation relations are:

$$[a, a^*] = 1; \qquad [a, a] = [a^*, a^*] = 0, \tag{12.30}$$

as may be seen by reference to Equations 12.14. The Hamiltonian now becomes:

$$\mathcal{H} = \tfrac{1}{2}\hbar\omega(a^*a + aa^*), \tag{12.31}$$

or using Equations 12.30:

$$\mathcal{H} = \hbar\omega(a^*a + \tfrac{1}{2}) = \hbar\omega(N + \tfrac{1}{2}). \tag{12.32}$$

The new operator $N = a^*a$ is of particular interest, since unlike a and a^* it is Hermitian. If n is an eigenvalue of N and the corresponding normalized eigenfunction is $|n\rangle$, we may write:

$$N|n\rangle = n|n\rangle \tag{12.33}$$

and, since N is Hermitian, n must be real. From Equation 12.32 it is seen that the $|n\rangle$ are also eigenfunctions of $\mathcal{H}$ with eigenvalues $\hbar\omega(n+\frac{1}{2})$.

To investigate the properties of N and its eigenvalues, the commutation relations of N with a and a^* are first determined using Equations 12.30:

$$\begin{aligned}[a, N] &= [a, a^*a] = [a, a^*]a = a; \\ [a^*, N] &= [a^*, a^*a] = a^*[a^*, a] = -a^*.\end{aligned} \qquad (12.34]$$

Now, operating on $|n\rangle$ with the operator Na^* and using these commutation relations:

$$Na^*|n\rangle = (a^*N + a^*)|n\rangle = a^*(N+1)|n\rangle = (n+1)a^*|n\rangle, \quad (12.35)$$

so that $a^*|n\rangle$ is also an eigenfunction of N and, since the corresponding eigenvalue is $(n+1)$, it must be proportional to $|n+1\rangle$:

$$a^*|n\rangle = c_+|n+1\rangle, \quad (12.36)$$

where c_+ is a constant. Similarly, it may be shown that $a|n\rangle$ is an eigenfunction of N with eigenvalue $(n-1)$:

$$Na|n\rangle = (aN - a)|n\rangle = a(N-1)|n\rangle = (n-1)a|n\rangle, \quad (12.37)$$

so that:

$$a|n\rangle = c_-|n-1\rangle. \quad (12.38)$$

The proportionality constants c_+ and c_- have been introduced because the eigenfunctions $a^*|n\rangle$ and $a|n\rangle$ are not necessarily normalized, but they may be evaluated by considering:

$$|c_+|^2 = \langle n|aa^*|n\rangle = \langle n|N+1|n\rangle = n+1 \quad (12.39)$$

and

$$|c_-|^2 = \langle n|a^*a|n\rangle = \langle n|N|n\rangle = n. \quad (12.40)$$

The phases of c_+ and c_- cannot be determined and it is normal practice to take them to be zero, so that c_+ and c_- are $(n+1)^{\frac{1}{2}}$ and $n^{\frac{1}{2}}$, respectively:

$$a^*|n\rangle = (n+1)^{\frac{1}{2}}|n+1\rangle; \qquad a|n\rangle = n^{\frac{1}{2}}|n-1\rangle. \quad (12.41)$$

The next stage in this development is to find the possible values of n. From Equation 12.40 it may be seen that n must be positive or zero, since $|c_-|^2$ is obtained by taking a wave function, multiplying it by its complex conjugate and integrating over the appropriate space. If n were non-integer it would be possible to apply the operator a repeatedly to $|n\rangle$ until an eigenfunction with a negative value of n was obtained. However, it is possible for n to be an integer, since Equations 12.41 show that this inconsistency does not then arise:

$$a|n\rangle = n^{\frac{1}{2}}|n-1\rangle; \ldots; \quad a|1\rangle = |0\rangle; \quad a|0\rangle = 0; \quad (12.42)$$

further application of a can only give zero. We conclude that n is a positive integer or zero.

The eigenvalues of the Hamiltonian (Equation 12.32) are thus:

$$E = \hbar\omega(n + \tfrac{1}{2}); \qquad n = 0, 1, 2, \ldots, \tag{12.43}$$

a result in accord with that of the more traditional approach. In addition, we may note that the $|n\rangle$ are also eigenfunctions of the momentum operator, which from Equations 12.22 and 12.29 is:

$$\begin{aligned} \mathbf{G} &= \tfrac{1}{2}\hbar\mathbf{k}(a^*a + aa^*) \\ &= \hbar\mathbf{k}(N + \tfrac{1}{2}), \end{aligned} \tag{12.44}$$

so that:

$$\mathbf{G}|n\rangle = \hbar\mathbf{k}(n + \tfrac{1}{2})|n\rangle. \tag{12.45}$$

The eigenvectors $|n\rangle$ may all be expressed in terms of $|0\rangle$, since repeated use of Equations 12.41 shows that:

$$|n\rangle = (n!)^{-\frac{1}{2}}(a^*)^n|0\rangle. \tag{12.46}$$

For our purposes in the present chapter there is no need to enquire more closely into the nature of $|n\rangle$. If, however, the one-dimensional harmonic oscillator wave functions are required in terms of the coordinate Q, this approach may easily be extended. Provided $|0\rangle$ is known as a function of Q, other eigenfunctions may be generated using Equation 12.46 with a^* replaced by the operator $(\omega/2\hbar)^{\frac{1}{2}}[Q - (\hbar/\omega)\partial/\partial Q]$. The explicit form of $|0\rangle$ is readily obtained from the equation $a|0\rangle = 0$, which is equivalent to the differential equation:

$$[\partial/\partial Q + (\omega/\hbar)Q]|0\rangle = 0. \tag{12.47}$$

The normalized solution of Equation 12.47 is just:

$$|0\rangle = (\omega/\hbar\pi)^{\frac{1}{4}}\exp(-\omega Q^2/2\hbar) \tag{12.48}$$

and this too is in agreement with the results of other approaches.

12.3 Creation and annihilation operators

In the previous section the energy of the eigenfunction $|n\rangle$ for an individual radiation oscillator was seen to be $\hbar\omega(n + \frac{1}{2})$. The quantum number n may be regarded as the number of quanta of energy $\hbar\omega$ in the oscillator in addition to the zero-point energy of $\frac{1}{2}\hbar\omega$. For this reason n is known as the occupation number and, since it is the eigenvalue of N, this operator is called the number operator.

Now a^* operating on $|n\rangle$ increases the number of quanta by one to give, within a numerical factor, the eigenfunction $|n+1\rangle$. Consequently, a^* is called a creation operator. In the same way a is an annihilation operator, since it reduces the occupation number by one. In this terminology it is the quanta of energy that are created or annihilated.

So far we have confined ourselves to just one radiation oscillator, although in reality there are an infinite number each characterized by a wave vector **k** and a polarization λ. For every oscillator we can define a creation and annihilation operator and these are distinguished by the subscripts k and λ. From Equations 12.14 and 12.29 these may be seen to satisfy the commutation relations:

$$[a_{\mathbf{k}\lambda}, a^*_{\mathbf{k}'\lambda'}] = \delta_{\mathbf{k}\mathbf{k}'}\delta_{\lambda\lambda'}; \qquad [a_{\mathbf{k}\lambda}, a_{\mathbf{k}'\lambda'}] = [a^*_{\mathbf{k}\lambda}, a^*_{\mathbf{k}'\lambda'}] = 0. \quad (12.49)$$

In addition, a number operator $N_{\mathbf{k}\lambda}$ and a Hamiltonian $\mathcal{H}_{\mathbf{k}\lambda}$ may be defined for each oscillator and the corresponding eigenfunctions may be designated by $|n_{\mathbf{k}\lambda}\rangle$.

The electromagnetic field as a whole is described (Equations 12.11) by an eigenfunction which is the product of the eigenfunctions for the individual radiation oscillators:

$$\prod_{\mathbf{k},\lambda} |n_{\mathbf{k}\lambda}\rangle = |n_{\mathbf{k}_1\lambda_1}\rangle|n_{\mathbf{k}_2\lambda_2}\rangle \ldots |n_{\mathbf{k}_i\lambda_i}\rangle \ldots,$$
$$= |n_{\mathbf{k}_1\lambda_1}, n_{\mathbf{k}_2\lambda_2}, \ldots, n_{\mathbf{k}_i\lambda_i}, \ldots\rangle. \quad (12.50)$$

An operator specific to one wave vector and polarization only acts on the corresponding part of the eigenfunction, so that, for example, use of $N_{\mathbf{k}_i\lambda_i}$ tells us how many quanta of type $\mathbf{k}_i$, λ_i are present:

$$N_{\mathbf{k}_i\lambda_i}|n_{\mathbf{k}_1\lambda_1}, n_{\mathbf{k}_2\lambda_2}, \ldots, n_{\mathbf{k}_i\lambda_i}, \ldots\rangle$$
$$= n_{\mathbf{k}_i\lambda_i}|n_{\mathbf{k}_1\lambda_1}, n_{\mathbf{k}_2\lambda_2}, \ldots, n_{\mathbf{k}_i\lambda_i}, \ldots\rangle. \quad (12.51)$$

Thus the total energy of the field is given by:

$$E = \sum_{\mathbf{k},\lambda} \hbar\omega_{\mathbf{k}}(n_{\mathbf{k}\lambda} + \tfrac{1}{2}), \quad (12.52)$$

as would be expected.

In specifying a state of the field it is only necessary to mention those quantum numbers $n_{\mathbf{k}\lambda}$ that differ from zero, although all the possibilities for **k** and λ must be remembered in performing a summation such as in Equation 12.52 because of the zero-point energy. Thus, if there are no quanta present, $|0_{\mathbf{k}_1\lambda_1}, 0_{\mathbf{k}_2\lambda_2}, \ldots, 0_{\mathbf{k}_i\lambda_i}, \ldots\rangle$ may be abbreviated to $|0\rangle$ and this is known as the vacuum state.

Any state of the radiation field may be constructed from the vacuum state by using the appropriate creation operators. For a state of the field with just one quantum of type **k**, λ:

$$|1_{\mathbf{k}\lambda}\rangle = a^*_{\mathbf{k}\lambda}|0\rangle, \tag{12.53}$$

while for two quanta of the same type:

$$|2_{\mathbf{k}\lambda}\rangle = 2^{-\frac{1}{2}}(a^*_{\mathbf{k}\lambda})^2|0\rangle, \tag{12.54}$$

as may be seen by reference to Equation 12.46. For two quanta of different types the factor $2^{-\frac{1}{2}}$ does not occur:

$$|1_{\mathbf{k}\lambda}, 1_{\mathbf{k}'\lambda'}\rangle = a^*_{\mathbf{k}\lambda}a^*_{\mathbf{k}'\lambda'}|0\rangle = a^*_{\mathbf{k}'\lambda'}a^*_{\mathbf{k}\lambda}|0\rangle. \tag{12.55}$$

Finally, we note that, since all creation operators commute with one another (Equations 12.49), it does not matter in which order the quanta are created.

12.4 Photons

In the last section it was seen that the energy of an electromagnetic field is the sum of the quanta of energy associated with the individual radiation oscillators. These quanta may be identified with particles or photons and it is the purpose of this section to show that this identification is consistent with photons being massless particles which obey Bose-Einstein statistics.

If we consider just one photon with wave vector **k** and polarization λ, we know that it has energy $\hbar\omega_{\mathbf{k}}$ and, from Equation 12.45, momentum $\hbar\mathbf{k}$. Now Equation 4.41 relates the energy E and momentum $\mathbf{p}$ of a particle:

$$E = c(m^2c^2 + p^2)^{\frac{1}{2}}, \tag{12.56}$$

so that the mass of the photon is given by:

$$m = (E^2 - c^2p^2)^{\frac{1}{2}}/c^2 = \hbar(\omega_{\mathbf{k}}^2 - c^2k^2)^{\frac{1}{2}}/c^2. \tag{12.57}$$

Since $\omega_{\mathbf{k}} = ck$, Equation 12.57 shows that the photon has no mass.

In this formalism there can be any number of photons in the same state, that is having a particular wave vector, polarization and energy. In addition, in creating photons from the vacuum state the order of creation is immaterial so that the state of the radiation field is symmetric with respect to interchange of any two photons; this fact depends on the commutation of all creation operators. Particles with these properties satisfy Bose-Einstein statistics, so that photons are bosons. These remarks apply to any type of particle or quantum of energy, for which the creation operators all commute; for example vibrational quanta in a molecular oscillator may be thought of as vibrational 'particles' or vibrons. Now Bose-Einstein statistics are obeyed by particles with zero or integer spin, so that we might enquire whether the photon possesses a spin. In fact it may be shown to have spin one and this property is related to the fact that two polarizations are possible for a given wave vector (see bibliography).

Thus the classical wave theory for electromagnetic fields becomes a particle theory on quantization. If there are a great number of photons, the occupation numbers $n_{\mathbf{k}\lambda}$ become very large and may be regarded as continuous variables; this is the classical limit. This situation is to be contrasted with classical particle theory, which on quantization becomes a wave theory. Both radiation and matter demonstrate a wave-particle duality, but the roles are reversed.

12.5 Zero-point energy and vacuum fluctuations

Up to now a discussion of the zero-point energy of the radiation field has been omitted. For the vacuum state there are no photons, but according to Equation 12.52 there is an energy $\sum_{\mathbf{k},\lambda} \frac{1}{2}\hbar\omega_{\mathbf{k}}$ associated with the radiation field and this energy is infinite, since the sum is over all possible wave vectors. From Equation 12.45 the field momentum of the vacuum state may be seen to be $\sum_{\mathbf{k},\lambda} \frac{1}{2}\hbar\mathbf{k}$, but no problem arises here, since for every wave vector $\mathbf{k}$ in the summation there is also a wave vector $-\mathbf{k}$, so that the field momentum vanishes.

Nevertheless, it is disturbing that the vacuum state has an infinite energy. This problem may conveniently be ignored by taking the energy of the vacuum state to be the origin of the energy scale. This may be achieved formally by starting with the classical expression:

$$\mathcal{H} = \sum_{\mathbf{k},\lambda} \omega_{\mathbf{k}}^2 q_{\mathbf{k}\lambda}^* q_{\mathbf{k}\lambda}, \tag{12.58}$$

which on quantization gives:

$$\mathcal{H} = \sum_{\mathbf{k},\lambda} \hbar\omega_{\mathbf{k}} N_{\mathbf{k}\lambda}. \tag{12.59}$$

In addition, we must impose the rule that in products, such as those involved in Equation 12.18, creation operators must always be placed on the left of annihilation operators, since, if they are not, it is possible to have a creation operator acting directly on the vacuum state and this leads to infinite quantities. This restriction may seem a little contrived, but a self-consistent theory can be developed; in effect the zero-point energy of the radiation field is consistently ignored. However, this situation is not very satisfying and we consider the zero-point energy more closely.

The analogy between a radiation oscillator and a molecular vibration has been emphasized on a number of occasions. A vibration also has a zero-point energy and this may be regarded as a manifestation of Heisenberg's uncertainty principle, which may be stated as:

$$(\Delta A)(\Delta B) \geqslant |\langle [A, B] \rangle / 2i|; \tag{12.60}$$

the product of the uncertainties in two observables is greater than or equal to a quantity which depends on the expectation value of the commutator of their operators. For a vibration the vibrational coordinate and the vibrational momentum do not commute, so that they cannot both be zero and there must always be some energy associated with the vibration.

In the case of a radiation field it is instructive to consider the commutators of the number operator $N_{\mathbf{k}\lambda}$ $(= a^*_{\mathbf{k}\lambda} a_{\mathbf{k}\lambda})$ with the electric and magnetic field operators (Equations 12.17 and 12.15). Using Equations 12.29 and 12.49 these may readily be obtained:

$$\begin{aligned} [N_{\mathbf{k}\lambda}, \mathbf{E}] &= -i(\hbar\omega_{\mathbf{k}}/2)^{\frac{1}{2}} (a_{\mathbf{k}\lambda}\mathbf{A}_{\mathbf{k}\lambda} + a^*_{\mathbf{k}\lambda}\mathbf{A}^*_{\mathbf{k}\lambda}); \\ [N_{\mathbf{k}\lambda}, \mathbf{B}] &= -i(\hbar/2\omega_{\mathbf{k}})^{\frac{1}{2}} (a_{\mathbf{k}\lambda}\mathbf{k}_\wedge \mathbf{A}_{\mathbf{k}\lambda} + a^*_{\mathbf{k}\lambda}\mathbf{k}_\wedge \mathbf{A}^*_{\mathbf{k}\lambda}). \end{aligned} \tag{12.61}$$

Neither of these commutators vanish, so that if the number of photons is approximately fixed **E** and **B** are uncertain. For the vacuum state there are no photons at all so that the electric and magnetic fields are very uncertain.

The expectation values of the operators **E** and **B** in the vacuum state are both zero:

$$\langle 0|\mathbf{E}|0\rangle = \langle 0|\mathbf{B}|0\rangle = 0, \tag{12.62}$$

since these operators are linear in creation and annihilation operators. However, the expectation values of E^2 and B^2 do not vanish, since they both involve the operator $a_{\mathbf{k}\lambda}a^*_{\mathbf{k}\lambda}$, for which the expectation value in the vacuum state is unity. For example it may readily be shown that:

$$\begin{aligned} \langle 0|E^2|0\rangle &= \sum_{\mathbf{k},\lambda} \tfrac{1}{2}\hbar\omega_{\mathbf{k}} \mathbf{A}_{\mathbf{k}\lambda}.\mathbf{A}^*_{\mathbf{k}\lambda} \\ &= \sum_{\mathbf{k},\lambda} \hbar\omega_{\mathbf{k}}/2\epsilon_0 V, \end{aligned} \tag{12.63}$$

where Equation 12.4 has been used. The sum (Equation 12.63) is infinite even if the volume V is allowed to become infinite, since the number density (Equation 6.74) for k is proportional to V. Similarly we may obtain:

$$\langle 0|B^2|0\rangle = \sum_{\mathbf{k},\lambda} \hbar\omega_{\mathbf{k}}/2c^2\epsilon_0 V \tag{12.64}$$

and by substituting this and Equation 12.63 into Equation 12.18 and integrating over the volume V the zero-point energy, $\sum_{\mathbf{k},\lambda} \frac{1}{2}\hbar\omega_{\mathbf{k}}$, may be recovered.

Thus, in the vacuum state the electric and magnetic fields fluctuate wildly, although their expectation values vanish. These are the so-called vacuum fluctuations. However, it is not at present clear whether the vacuum fluctuations and the zero-point energy are one and the same thing. As we have already remarked the latter may be formally removed from the theory, but the former may not be avoided so easily. Indeed, the vacuum fluctuations do lead to observable effects. For example, they have already been invoked (Section 11.9) in a crude explanation of the major part of the Lamb shift. However, further discussion of this topic is not possible here, since treating the electron as a point charge leads to infinities and the circumvention of this difficulty involves the more sophisticated techniques of mass and charge renormalization.

12.6 Fermions and second quantization

The theory that has been developed in this chapter is appropriate to photons, but it may readily be adapted to describe other particles that satisfy Bose-Einstein statistics. It is based essentially on the commutation relations (Equation 12.30) for the creation and annihilation operators:

$$[a, a^*] = 1; \qquad [a, a] = [a^*, a^*] = 0. \tag{12.65}$$

These relations lead automatically to the possibility of having any number of particles in the same state and, in addition, they are responsible for many-particle states being symmetric with respect to interchange of two particles; these are the properties characteristic of bosons.

However, this formalism is not adequate to describe fermions and in particular electrons. These obey Fermi-Dirac statistics, which require that there be no more than one particle in a given state and that many-particle states be antisymmetric with respect to interchange of two particles. Thus, the Pauli exclusion principle must somehow be incorporated into the theory. To conclude this chapter the way this may be done is described, but a detailed discussion would be out of place here.

Creation and annihilation operators are again used, but these are given the new symbols b_r^* and b_r, respectively, to distinguish them from the boson operators; the label r indicates the one-particle state to which the operators apply. However, these operators are required to satisfy the anticommutation relations:

$$[b_r, b_s^*]_+ = \delta_{rs}; \qquad [b_r, b_s]_+ = [b_r^*, b_s^*]_+ = 0, \tag{12.66}$$

where

$$[A, B]_+ = AB + BA. \tag{12.67}$$

As for bosons we may define a vacuum state $|0\rangle$ in which there are no particles present. A single-particle state may then be constructed by operating on the vacuum state with a creation operator:

$$|1_r\rangle = b_r^*|0\rangle. \tag{12.68}$$

Further application of a creation operator gives a two-particle state:

$$b_s^* b_r^*|0\rangle = b_s^*|1_r\rangle = |1_s, 1_r\rangle, \tag{12.69}$$

where $r \neq s$. However, it is not possible to have two particles in the same state, since from Equation 12.66:

$$b_r^* b_r^*|0\rangle = \tfrac{1}{2}[b_r^*, b_r^*]_+|0\rangle = 0. \tag{12.70}$$

In addition, the anticommutation relations dictate that the two-particle state (Equation 12.69) is antisymmetric with respect to interchange of the particles:

$$|1_r, 1_s\rangle = b_r^* b_s^*|0\rangle = -b_s^* b_r^*|0\rangle = -|1_s, 1_r\rangle. \tag{12.71}$$

This is just what is required by the Pauli exclusion principle.

By analogy with the number operator for bosons we may define the new operator:

$$N_r = b_r^* b_r \tag{12.72}$$

and it may easily be confirmed that this is a fermion number operator:

$$N_r|0\rangle = b_r^* b_r|0\rangle = 0;$$

$$N_r|1_r\rangle = N_r b_r^*|0\rangle = b_r^* b_r b_r^*|0\rangle$$

$$= b_r^*(1 - b_r^* b_r)|0\rangle = b_r^*|0\rangle = |1_r\rangle. \tag{12.73}$$

The operator N_r is said to be idempotent, since its square is equal to itself:

$$N_r^2 = b_r^* b_r b_r^* b_r = b_r^*(1 - b_r^* b_r)b_r = b_r^* b_r = N_r, \tag{12.74}$$

where we have used the fact (Equations 12.66) that $b_r b_r$ must vanish. Thus, Equation 12.74 shows that:

$$N_r^2 - N_r = N_r(N_r - 1) = 0, \tag{12.75}$$

so that the eigenvalues of N_r are zero or one, that is, a state r may be unoccupied or occupied by just one fermion; this is consistent with our earlier conclusions.

The creation and annihilation operators, a^* and a, for bosons may be expressed in terms of a variable Q and its conjugate momentum P, and indeed the theory was developed via P and Q. However, the fermion operators b_r^* and b_r have been introduced directly, but this is necessary, since they cannot be associated with a variable and its conjugate momentum. This situation is a reflection on the fact that there is no classically measurable field corresponding to quantized fermions.

The use of creation and annihilation operators to describe fermions, and in particular electrons, implies that they are to be treated as particles. Yet the usual quantum theory uses a wave description for fermions and we might enquire how the formalism developed above may be employed. For bosons the classical wave theory becomes a particle theory on quantization. In the same way the quantum-mechanical wave function for a fermion may be treated as a field variable and quantization then gives a particle theory for which creation and annihilation operators are appropriate. This procedure is known as second quantization for obvious reasons. Its advantage is that it provides a convenient way of describing many-particle systems and in particular it may be used to describe situations in which the number of particles is not conserved. For example, in Section 9.5 a crude model was given for the origin of the electron's g factor anomaly. This involved the creation of a virtual electron-positron pair and in the Dirac theory of Chapter 8 this is described as the excitation of an electron from the filled negative energy states; in this theory the number of electrons is conserved, but at the expense of having to describe the electron by a many-particle theory. Using second quantization the necessity for particle number conservation disappears. In addition, the theory based on second quantization may be used to prove the exclusion principle of Pauli. Needless to say these topics cannot be considered further.

Bibliography

Quantization of the electromagnetic field

Grandy: Chapter 9.
Hameka: Section 6-2.
Sakurai: pages 20-36.
Ziman, J. M. (1969), *Elements of Advanced Quantum Theory*, Cambridge University Press, Cambridge: pages 1-5 describe the solution of the harmonic oscillator equation using creation and annihilation operators.

Photon spin

Grandy: pages 138-140.

Sakurai: pages 30-31.

Vacuum fluctuations

Grandy: pages 141-143.
Sakurai: pages 32-36.

Second quantization

Sakurai: pages 28-29 and 143-156.
Schiff, L. I. (1968), *Quantum Mechanics*, McGraw-Hill, New York: pages 498-508.

CHAPTER THIRTEEN

The Interaction of Radiation and Matter

The usual approach to the absorption and emission of radiation by matter is to use the semiclassical theory in which, although the matter is treated quantum-mechanically, the radiation is described classically. For the more elementary applications it is possible to obtain the correct results using this theory, although it is often necessary to use indirect arguments. In addition, there are phenomena which cannot be described by the semiclassical theory.

However, the quantum theory of radiation that was developed in the previous chapter provides an appropriate basis for describing the interaction of radiation with quantized matter. Not only does the resulting theory avoid the problems of the semiclassical approach, but it is more generally applicable and provides a conceptually simpler picture of the processes involved. The semiclassical theory is given in many of the more elementary texts and for this reason it is not detailed here, although a comparison of the two approaches is given in Section 13.5.

The Hamiltonian for electromagnetic radiation in free space was found in the previous chapter. In the first section of this chapter the interaction Hamiltonian is considered, but before this may be done we must decide what sort of Hamiltonian to use for the description of the quantized matter. In much of what follows the interaction of the radiation with a single electron is discussed, but, since the results are to be generalized to a molecule, it is appropriate to use a non-relativistic approximation to the Hamiltonian. Thus, the theory to be developed is not a relativistic one. However, many of the same principles apply if the Dirac Hamiltonian for an electron is used and the differences are pointed out at the appropriate place in the development. As it happens for the phenomena considered here the results are identical, to the accuracy employed.

13.1 The interaction Hamiltonian

To illustrate the principles involved the interaction of radiation with a single electron is considered, but we may anticipate the generalization of the results to molecules by assuming that the electron has discrete energy levels. The Hamiltonian for the system may be written as the sum of three parts, one for the field, one for the electron and one for the interaction between the two:

$$\mathcal{H} = \mathcal{H}_f + \mathcal{H}_e + \mathcal{H}_{int}. \tag{13.1}$$

The quantized field Hamiltonian $\mathcal{H}_f$ was obtained in the last chapter and, as has already been remarked, the non-relativistic approximation is used for the electron's Hamiltonian $\mathcal{H}_e$. The interaction Hamiltonian $\mathcal{H}_{int}$ must now be discussed.

The non-relativistic approximation to the Hamiltonian of an electron in an electromagnetic field was derived in Chapter 8. This is a semiclassical Hamiltonian in the sense that the radiation field was treated classically. From this Hamiltonian (Equation 8.142) we may select those terms involving the electromagnetic potentials to obtain the semiclassical interaction Hamiltonian:

$$\mathcal{H}_{int} = -e\phi + (e/2m)(\mathbf{p.A} + \mathbf{A.p}) + (e^2/2m)A^2 + (e\hbar/m)(\mathbf{s.B}), \tag{13.2}$$

where the symbols have the usual meaning and the term $\pi^2/2m$ has been expanded (Section 7.1). There are, of course, smaller terms that could be added to this, but they are negligible as far as the time-dependent processes considered in this chapter are concerned. When it is recalled that the gauge chosen for the potentials is such that ϕ and $\nabla.\mathbf{A}$ vanish at every point in space, Equation 13.2 reduces to:

$$\mathcal{H}_{int} = (e/m)(\mathbf{A.p}) + (e^2/2m)A^2 + (e\hbar/m)(\mathbf{s.B}). \tag{13.3}$$

The semiclassical interaction Hamiltonian (Equation 13.3) may be put in a fully quantized form merely by interpreting the vector potential $\mathbf{A}$ as the linear combination of annihilation and creation operators obtained in the last chapter:

$$\mathbf{A} = \sum_{\mathbf{k},\lambda} (\hbar/2\omega_{\mathbf{k}})^{\frac{1}{2}} (a_{\mathbf{k}\lambda}\mathbf{A}_{\mathbf{k}\lambda} + a^*_{\mathbf{k}\lambda}\mathbf{A}^*_{\mathbf{k}\lambda}), \tag{13.4}$$

where the $\mathbf{A}_{\mathbf{k}\lambda}$ are given by Equation 12.4. The magnetic field $\mathbf{B}$ is now also an operator and is given by:

$$\mathbf{B} = \sum_{\mathbf{k},\lambda} i(\hbar/2\omega_{\mathbf{k}})^{\frac{1}{2}} (a_{\mathbf{k}\lambda}\mathbf{k}\wedge\mathbf{A}_{\mathbf{k}\lambda} - a^*_{\mathbf{k}\lambda}\mathbf{k}\wedge\mathbf{A}^*_{\mathbf{k}\lambda}). \tag{13.5}$$

In describing the interaction of molecules with electromagnetic radiation the interaction Hamiltonian is relatively small and will be taken as a perturbation in the subsequent treatment. The total Hamiltonian (Equation 13.1) may thus be rewritten as $\mathcal{H}_0 + \mathcal{H}_{int}$. The unperturbed Hamiltonian $\mathcal{H}_0$ is the sum of the field and electron Hamiltonians and because it contains no interaction terms its eigenfunctions may be written as a product of the eigenfunctions of $\mathcal{H}_f$ and $\mathcal{H}_e$ taken separately. In talking about the state of the system we then mean a combined state of electron and photons.

The perturbation $\mathcal{H}_{int}$ contains creation and annihilation operators, so that it connects states with differing numbers of photons. The terms involving **A.p** and **s.B** in Equation 13.3 are linear in these operators so that they have matrix elements between states differing by one photon; it should already be apparent that these terms are involved in single-photon absorption and emission phenomena. The remaining term in Equation 13.3 is proportional to A^2 and is quadratic in creation and annihilation operators so that it connects states differing by zero or two photons. It will be seen that, although this term can be important in scattering processes, where the number of photons is conserved, it is not involved in two-photon absorption or emission and it is the **A.p** term taken to second order that is responsible.

The interaction Hamiltonian (Equation 13.3) was obtained from the non-relativistic approximation to the Dirac Hamiltonian for the electron and we might inquire what would happen if the Dirac Hamiltonian itself was used. In this case the perturbation representing the interaction of the electron with quantized radiation is:

$$\mathcal{H}_{int} = ec\boldsymbol{\alpha}.\mathbf{A}. \qquad (13.6)$$

This differs from Equation 13.3 in that there is no term in A^2 so that it is not obvious that the two interaction Hamiltonians can ever lead to the same results for some phenomena. This is a reflection of the fact that the Dirac theory of the electron is really a many-particle theory, since it is necessary to assume that all the negative energy states are occupied by electrons (Section 8.6). For a correct treatment this has to be taken into account, since the Dirac operator $\boldsymbol{\alpha}$ connects positive and negative energy states. However, we will not go into the details, since for the generalization to molecules it is more convenient to use Equation 13.3 as the interaction Hamiltonian.

13.2 Time-dependent perturbation theory

The interaction Hamiltonian is a relatively small perturbation on molecular systems and the phenomena to be discussed in this chapter are described well by time-dependent perturbation theory. This section is devoted to deriving those results of this theory that are needed later. This material could have been relegated to an appendix, but, although the first-order theory is probably familiar, it is necessary to go to second order to describe some of the processes discussed later. The theory developed is quite general, but it should be remembered that for our purposes a state describes a system consisting of radiation as well as a molecule.

Approximate solutions are required for the time-dependent Schrödinger equation:

$$\mathcal{H}\Psi(t) = i\hbar(\partial\Psi(t)/\partial t), \tag{13.7}$$

where $\mathcal{H} = \mathcal{H}_0 + V$ and it is assumed that the eigenfunctions of $\mathcal{H}_0$ are known:

$$\mathcal{H}_0\psi_n = E_n\psi_n. \tag{13.8}$$

The perturbation V is a time-independent operator, since the Schrödinger picture is being used, but nevertheless it can cause the state of the system to change with time.

The solutions $\Psi(t)$ of Equation 13.7 may be written as a linear combination of the eigenfunctions ψ_n of $\mathcal{H}_0$:

$$\Psi(t) = \sum_n c_n(t)\psi_n \exp(-iE_nt/\hbar). \tag{13.9}$$

The coefficients $c_n(t)$ may be determined by substituting this into Equation 13.7, reorganizing, multiplying on the left by $\psi_k^* \exp(iE_kt/\hbar)$ and integrating over all space. The result is:

$$\dot{c}_k(t) = -(i/\hbar)\sum_n V_{kn}c_n(t)\exp[i(E_k - E_n)t/\hbar], \tag{13.10}$$

where $\dot{c}_k$ is the derivative of c_k with respect to time and V_{kn} is the time-independent quantity:

$$V_{kn} = \langle k|V|n\rangle = \int \psi_k^* V \psi_n \mathrm{d}\tau. \tag{13.11}$$

Let us consider transitions from an initial state i. For the boundary conditions it is assumed that initially, that is at $t = -\infty$, only state i is populated, so that:

$$c_n(-\infty) = \delta_{ni}. \tag{13.12}$$

An approximate solution of the coupled differential equations (13.10) may now be obtained by substituting this condition and integrating:

$$c_k(t) = \delta_{ki} - (i/\hbar)V_{ki}\int_{-\infty}^{t} \exp[i(E_k - E_i)t'/\hbar]\,\mathrm{d}t'. \tag{13.13}$$

The definite integral in Equation 13.13 cannot be evaluated at its lower limit. This could be avoided by taking our initial condition to be $c_n(0) = \delta_{ni}$ instead of Equation 13.12 so that the integration would be from 0 to t (assumed positive). However, this condition corresponds to a sudden and artificial turning on of the perturbation at $t = 0$. Not only is this physically unreasonable, but also it introduces spurious terms in second order. To circumvent these difficulties the initial condition (Equation 13.12) is retained, but a mathematical trick is used. The time-independent perturbation V is replaced by $V \exp(\eta t)$, where η is real and positive. This has the effect of turning the perturbation on gradually as t increases from $-\infty$. The integration may then be performed and the original problem recovered by taking the limit as η goes to zero at the end of the calculation.

With the replacement of V by $V \exp(\eta t)$ Equation 13.10 becomes:

$$\dot{c}_k(t) = -(i/\hbar) \sum_n V_{kn} c_n(t) \exp\left[i(E_k - E_n)t/\hbar + \eta t\right], \qquad (13.14)$$

the approximate solution being:

$$c_k(t) = \delta_{ki} + V_{ki} \exp\left[i(E_k - E_i)t/\hbar + \eta t\right] / (E_i - E_k + i\hbar\eta). \qquad (13.15)$$

Although this is only an approximate solution, it encourages us to enquire if the exact solution of Equation 13.14 may be obtained by merely replacing the V_{ki} by some new quantities R_{ki}:

$$c_k(t) = \delta_{ki} + R_{ki} \exp\left[i(E_k - E_i)t/\hbar + \eta t\right] / (E_i - E_k + i\hbar\eta). \qquad (13.16)$$

If this proves successful the identity of the R_{ki} has then to be determined.

Substitution of Equation 13.16 into Equation 13.14 followed by multiplication throughout by $\exp\left[-i(E_k - E_i)t/\hbar - \eta t\right]$ results in the equation:

$$R_{ki} = V_{ki} + \sum_n V_{kn} R_{ni} \exp(\eta t)/(E_i - E_n + i\hbar\eta). \qquad (13.17)$$

This may be solved iteratively for R_{ki} and so long as the perturbation is small compared with $(E_i - E_n)$ the resulting series converges rapidly:

$$\begin{aligned} R_{ki} = {} & V_{ki} + \sum_n V_{kn} V_{ni} \exp(\eta t)/(E_i - E_n + i\hbar\eta) \\ & + \sum_{n,m} V_{kn} V_{nm} V_{mi} \exp(2\eta t)/(E_i - E_n + i\hbar\eta)(E_i - E_m + i\hbar\eta) + \ldots; \end{aligned} \qquad (13.18)$$

although these R_{ki} are time-dependent this dependence is removed when η is eventually allowed to become zero. Thus, Equation 13.16 is a solution of Equation 13.14, the R_{ki} being given by the series 13.18.

In practice we are interested in the transition from the initial state i to a final state f. The probability of the system being in state f at time t is just $|c_f(t)|^2$ and from Equation 13.16 this is given by:

$$|c_f(t)|^2 = |R_{fi}|^2 \exp(2\eta t)/[(E_i - E_f)^2 + \hbar^2\eta^2]. \tag{13.19}$$

The transition rate w_{fi} from i to f is obtained by taking the time derivative of this probability:

$$\begin{aligned} w_{fi} &= \mathrm{d}|c_f(t)|^2/\mathrm{d}t \\ &= 2\exp(2\eta t)|R_{fi}|^2\{\eta/[(E_i - E_f)^2 + \hbar^2\eta^2]\}. \end{aligned} \tag{13.20}$$

The limit as η tends to zero may now be taken. To do this we note that the term in curly parentheses in Equation 13.20 is related to a representation of a Dirac delta function:

$$\underset{\eta \to 0}{Lt} [\eta/\pi(x^2 + \eta^2)] = \delta(x). \tag{13.21}$$

If $x \neq 0$ the limit tends to zero, while for $x = 0$ the limit is infinite. In addition, it may easily be shown that:

$$\int_{-\infty}^{\infty} [\eta/\pi(x^2 + \eta^2)]\,\mathrm{d}x = 1. \tag{13.22}$$

Thus, Equation 13.20 may be replaced by:

$$w_{fi} = (2\pi/\hbar)|R_{fi}|^2\delta(E_f - E_i); \tag{13.23}$$

here, use has been made of the fact (Equation 2.88) that $\delta(cx) = c^{-1}\delta(x)$, where c is a constant. In this connection we might note that $\delta(E_f - E_i)$ has the dimensions $(\text{energy})^{-1}$, so that w_{fi} has the correct dimensions of $(\text{time})^{-1}$. Equation 13.23 is to be used in conjunction with Equation 13.18, which in the limit as η tends to zero becomes:

$$\begin{aligned} R_{fi} &= V_{fi} + \sum_n V_{fn}V_{ni}/(E_i - E_n) \\ &\quad + \sum_{n,m} V_{fn}V_{nm}V_{mi}/(E_i - E_n)(E_i - E_m) + \ldots; \end{aligned} \tag{13.24}$$

the R_{fi} are called the reaction matrix elements.

Since it involves a Dirac delta function, Equation 13.23 is an operator equation and an appropriate integration must be carried out; the details depend on the problem being discussed and are given explicitly for the applications considered later. However, it is apparent from Equation 13.23 that energy is conserved in a transition.

There are difficulties in this approach to time-dependent perturbation theory. For example, in the expansion 13.24 the terms for $n = i$ and $m = i$

must be omitted from the summations if infinities are to be avoided. Other deficiencies of this theory are discussed in Section 13.8.

13.3 Matrix elements of the interaction Hamiltonian

In this section the matrix elements of the individual terms in the perturbation (Equation 13.3) are considered, since they are needed in the application of Equations 13.23 and 13.24. For convenience, it is assumed that only one type of photon is present, the generalization to all types being made later. The state of the unperturbed system is then given by an eigenfunction of $\mathcal{H}_0\,(=\mathcal{H}_e+\mathcal{H}_f)$:

$$|A, n_{\mathbf{k}\lambda}\rangle = |A\rangle|n_{\mathbf{k}\lambda}\rangle, \tag{13.25}$$

where $|A\rangle$ and $|n_{\mathbf{k}\lambda}\rangle$ are eigenfunctions of $\mathcal{H}_e$ and $\mathcal{H}_f$, respectively.

The perturbation $(e/m)(\mathbf{A}.\mathbf{p})$ is linear in creation and annihilation operators, so that its only non-zero matrix elements are of the type:

$$\langle B, n_{\mathbf{k}\lambda}-1|(e/m)(\mathbf{A}.\mathbf{p})|A, n_{\mathbf{k}\lambda}\rangle$$

$$= (e/m)(\hbar n_{\mathbf{k}\lambda}/2\omega_{\mathbf{k}}V\epsilon_0)^{\frac{1}{2}}\,\boldsymbol{\varepsilon}_{\mathbf{k}\lambda}.\langle B|\exp(i\mathbf{k}.\mathbf{r})\mathbf{p}|A\rangle \tag{13.26}$$

and

$$\langle B, n_{\mathbf{k}\lambda}+1|(e/m)(\mathbf{A}.\mathbf{p})|A, n_{\mathbf{k}\lambda}\rangle$$

$$= (e/m)(\hbar(n_{\mathbf{k}\lambda}+1)/2\omega_{\mathbf{k}}V\epsilon_0)^{\frac{1}{2}}\,\boldsymbol{\varepsilon}_{\mathbf{k}\lambda}.\langle B|\exp(-i\mathbf{k}.\mathbf{r})\mathbf{p}|A\rangle; \tag{13.27}$$

in evaluating the radiation part of these matrix elements use has been made of Equations 13.4, 12.4 and 12.41. The first of these represents the absorption of a photon by the electron accompanied by a transition from state A to state B, while the second corresponds to an emission process.

The matrix elements $\langle B|\exp(\pm i\mathbf{k}.\mathbf{r})\mathbf{p}|A\rangle$ between electron states are encountered in the semiclassical theory. Since the wavelength k^{-1} of the radiation is normally very large compared with molecular dimensions, it is usual to make the long wavelength approximation. This involves taking the origin of the electron coordinate $\mathbf{r}$ to be in the molecule and expanding the exponential as a power series in $\mathbf{k}.\mathbf{r}$. This expansion parameter is very small compared with one so that only the first term in the expansion need be retained, that is, $\exp(\pm i\mathbf{k}.\mathbf{r})$ is replaced by unity. The problem now reduces to that of evaluating $\langle B|\mathbf{p}|A\rangle$, but this may be expressed in a more convenient form by taking $|A\rangle$ and $|B\rangle$ to be eigenfunctions of the approximate electronic Hamiltonian:

$$\mathcal{H}_e = p^2/2m + U(\mathbf{r}), \tag{13.28}$$

where $U(\mathbf{r})$ is a potential function. We then have:

$$[\mathcal{H}_e, \mathbf{r}] = -(i\hbar/m)\mathbf{p}, \qquad (13.29)$$

so that

$$\begin{aligned}\langle B|\mathbf{p}|A\rangle &= (im/\hbar)\langle B|[\mathcal{H}_e, \mathbf{r}]|A\rangle \\ &= (im(E_B - E_A)/\hbar e)\langle B|e\mathbf{r}|A\rangle \\ &= im(E_B - E_A)\mathbf{P}_{BA}/\hbar e. \qquad (13.30)\end{aligned}$$

With these approximations the matrix element, Equation 13.26, becomes:

$$\langle B, n_{\mathbf{k}\lambda} - 1|(e/m)(\mathbf{A}.\mathbf{p})|A, n_{\mathbf{k}\lambda}\rangle$$

$$= i(n_{\mathbf{k}\lambda}/2\hbar\omega_{\mathbf{k}} V\epsilon_0)^{\frac{1}{2}}(E_B - E_A)\,\boldsymbol{\varepsilon}_{\mathbf{k}\lambda}.\mathbf{P}_{BA}, \qquad (13.31)$$

with a similar expression for 13.27.

For simplicity only one electron has been considered so far and, in doing this, summations and electron and nuclear subscripts have been avoided. However, the preceding development is readily extended to molecules. The perturbation $(e/m)(\mathbf{A}.\mathbf{p})$ must be summed over all electrons and nuclei, but the same results are obtained provided $|A\rangle$ and $|B\rangle$ are now eigenfunctions of the molecular Hamiltonian and $\mathbf{P}_{BA}$ becomes the matrix element of the molecular electric dipole moment between $|A\rangle$ and $|B\rangle$. Consequently these matrix elements are responsible for the so-called electric dipole transitions in molecules.

There are situations where $\mathbf{P}_{BA}$ vanishes and it is necessary to go to the second term in the expansion of $\exp(\pm i\mathbf{k}.\mathbf{r})$ to obtain non-zero matrix elements:

$$\langle B|\exp(\pm i\mathbf{k}.\mathbf{r})\mathbf{p}|A\rangle = \langle B|\mathbf{p}|A\rangle \pm i\langle B|(\mathbf{k}.\mathbf{r})\mathbf{p}|A\rangle + \ldots. \qquad (13.32)$$

The additional matrix element may be evaluated by splitting it in two:

$$\begin{aligned}\langle B|(\mathbf{k}.\mathbf{r})\mathbf{p}|A\rangle &= \tfrac{1}{2}[\langle B|(\mathbf{k}.\mathbf{r})\mathbf{p} + (\mathbf{k}.\mathbf{p})\mathbf{r}|A\rangle \\ &\quad + \langle B|(\mathbf{k}.\mathbf{r})\mathbf{p} - (\mathbf{k}.\mathbf{p})\mathbf{r}|A\rangle]. \qquad (13.33)\end{aligned}$$

By noting that for the electronic Hamiltonian (Equation 13.28):

$$[\mathcal{H}_e, r_i r_j] = -(i\hbar/m)(p_i r_j + r_i p_j) \qquad (i, j = x, y, z), \qquad (13.34)$$

the first part of Equation 13.33 may be shown to be:

$$\tfrac{1}{2}\langle B|(\mathbf{k}.\mathbf{r})\mathbf{p} + (\mathbf{k}.\mathbf{p})\mathbf{r}|A\rangle = (im(E_B - E_A)/2\hbar e^2)\langle B|\mathsf{T}|A\rangle.\mathbf{k}, \qquad (13.35)$$

where the tensor T has components $T_{ij} = e^2 r_i r_j$. On substitution into Equations 13.26 or 13.27 the scalar product of Equation 13.35 with $\boldsymbol{\varepsilon}_{\mathbf{k}\lambda}$

must be taken and, since $\boldsymbol{\varepsilon}_{\mathbf{k}\lambda}.\mathbf{k}$ vanishes (Equations 12.5), the results are unaltered if the scalar $-e^2r^2/3$ is added to each of the diagonal elements of **T**. This gives a new tensor **Q** with elements:

$$Q_{ij} = e^2(r_i r_j - \delta_{ij}r^2/3). \qquad (13.36)$$

When we generalize to a molecule, Q_{ij} is just the contribution of the electron to the molecular electric quadrupole moment. In this way matrix elements corresponding to electric quadrupole transitions are obtained.

To evaluate the second part of Equation 13.33 we again recall that we are interested in its scalar product with $\boldsymbol{\varepsilon}_{\mathbf{k}\lambda}$, so that we may use:

$$(\mathbf{k}.\mathbf{r})(\boldsymbol{\varepsilon}_{\mathbf{k}\lambda}.\mathbf{p}) - (\mathbf{k}.\mathbf{p})(\boldsymbol{\varepsilon}_{\mathbf{k}\lambda}.\mathbf{r}) = (\mathbf{k}_\wedge \boldsymbol{\varepsilon}_{\mathbf{k}\lambda}).(\mathbf{r}_\wedge \mathbf{p}); \qquad (13.37)$$

this relation also depends on the fact that $\boldsymbol{\varepsilon}_{\mathbf{k}\lambda}.\mathbf{k} = 0$. Thus:

$$\tfrac{1}{2}\boldsymbol{\varepsilon}_{\mathbf{k}\lambda}.\langle B|(\mathbf{k}.\mathbf{r})\mathbf{p} - (\mathbf{k}.\mathbf{p})\mathbf{r}|A\rangle$$
$$= (m/e)(\mathbf{k}_\wedge \boldsymbol{\varepsilon}_{\mathbf{k}\lambda}).\langle B|(e/2m)(\mathbf{r}_\wedge \mathbf{p})|A\rangle. \qquad (13.38)$$

Now $(e/2m)(\mathbf{r}_\wedge \mathbf{p})$ is the orbital magnetic moment of the electron and reference to Equation 13.5 shows that $(\mathbf{k}_\wedge \boldsymbol{\varepsilon}_{\mathbf{k}\lambda})$ is related to the magnetic field so that the matrix element (Equation 13.38) gives rise to magnetic dipole transitions.

So far only the first term in the interaction Hamiltonian (Equation 13.3) has been considered. The last term $(e\hbar/m)(\mathbf{s}.\mathbf{B})$ also leads to magnetic dipole transitions. For illustration the matrix element corresponding to absorption is quoted:

$$\langle B, n_{\mathbf{k}\lambda} - 1|(e\hbar/m)(\mathbf{s}.\mathbf{B})|A\rangle$$
$$= i(\hbar n_{\mathbf{k}\lambda}/2\omega_{\mathbf{k}} V\epsilon_0)^{\frac{1}{2}}(\mathbf{k}_\wedge \boldsymbol{\varepsilon}_{\mathbf{k}\lambda}).\langle B|(e\hbar/m)\mathbf{s}|A\rangle, \qquad (13.39)$$

where the long wavelength approximation has been used. The similarity to the other magnetic dipole matrix element (Equation 13.38) is easily seen.

The electric quadrupole and magnetic dipole contributions to the matrix elements of the interaction Hamiltonian are in general smaller than the electric dipole contribution by a factor of order kr, where k is the reciprocal of the wavelength of the radiation and r is of molecular dimensions. They are important when the electric dipole matrix element $\mathbf{P}_{BA}$ vanishes or when special techniques are employed; for example, it is the magnetic dipole contribution that is usually involved in magnetic resonance. In the rest of this chapter only the electric dipole contributions are considered, since we are interested in principles rather than details, but in practice the other contributions may be dealt with in much the same way.

There is one term in the interaction Hamiltonian that has yet to be considered, this being $(e^2/2m)A^2$. Since it is quadratic in creation and annihilation operators, it can connect states differing by zero or two photons and in general two types of photon must be considered. Now $(e^2/2m)A^2$ is a much smaller perturbation than $(e/m)(\mathbf{A}.\mathbf{p})$, so that the long wavelength approximation may again be made and only the first term in the expansions of the exponentials need be retained. The resulting matrix element connecting states with the same total number of photons is:

$$\langle B, n_{\mathbf{k}\lambda} - 1, n_{\mathbf{k}'\lambda'} + 1 | (e^2/2m)A^2 | A, n_{\mathbf{k}\lambda}, n_{\mathbf{k}'\lambda'} \rangle$$
$$= [e^2 \hbar n_{\mathbf{k}\lambda}^{\frac{1}{2}} (n_{\mathbf{k}'\lambda'} + 1)^{\frac{1}{2}} (\boldsymbol{\varepsilon}_{\mathbf{k}\lambda} . \boldsymbol{\varepsilon}_{\mathbf{k}'\lambda'}) / 2m\omega_{\mathbf{k}}^{\frac{1}{2}} \omega_{\mathbf{k}'}^{\frac{1}{2}} V\epsilon_0] \langle B|A \rangle. \qquad (13.40)$$

Since the eigenfunctions of $\mathcal{H}_e$ are orthogonal to one another, this must vanish unless A and B are the same electron state. At the end of the last section it was seen that energy is conserved in transitions so that the important matrix elements of the type 13.40 are those in which $\omega_{\mathbf{k}}$ and $\omega_{\mathbf{k}'}$ are equal. This matrix element then corresponds to Rayleigh scattering in which the incident and scattered photons have the same energy and the electron state is unaltered. The matrix elements of $(e^2/2m)A^2$ between states differing by two photons, that is those connecting $|A, n_{\mathbf{k}\lambda}, n_{\mathbf{k}'\lambda'}\rangle$ with $|B, n_{\mathbf{k}\lambda} \pm 1, n_{\mathbf{k}'\lambda'} \pm 1\rangle$, are also proportional to $\langle B|A\rangle$ and are unimportant, since energy conservation is not possible if $A = B$.

13.4 Absorption and emission

The absorption of a single photon by a molecule in state A to give the molecule in state B may be represented by the equation:

$$A + \gamma \rightarrow B, \qquad (13.41)$$

where γ is a photon. Initially just one type of photon is considered, but eventually the summation over all types must be made. Thus, the initial state i is $|A, n_{\mathbf{k}\lambda}\rangle$ and the final state f is $|B, n_{\mathbf{k}\lambda} - 1\rangle$.

The transition rate is given by Equation 13.23:

$$w_{fi} = (2\pi/\hbar)|R_{fi}|^2 \delta(E_f - E_i). \qquad (13.42)$$

Since it is a single-photon process that is under discussion, only the leading term V_{fi} in the expansion 13.24 for R_{fi} is relevant; the other terms represent many-photon processes, which are discussed later. This matrix element V_{fi} has been considered already and the major contribution to it is the electric dipole contribution given in Equation 13.31. Thus, Equation 13.42 becomes:

$$w_{fi} = (\pi/\hbar^2 V\epsilon_0)(n_{\mathbf{k}\lambda}/\omega_{\mathbf{k}})(\boldsymbol{\epsilon}_{\mathbf{k}\lambda}.\mathbf{P}_{BA})^2(E_B - E_A)^2\delta(E_B - E_A - \hbar\omega_{\mathbf{k}})$$
$$= (\pi/V\epsilon_0)n_{\mathbf{k}\lambda}\omega_{\mathbf{k}}(\boldsymbol{\epsilon}_{\mathbf{k}\lambda}.\mathbf{P}_{BA})^2\delta(E_B - E_A - \hbar\omega_{\mathbf{k}}), \qquad (13.43)$$

where $\hbar\omega_{\mathbf{k}}$, E_A and E_B are the energy of the photon and the molecular states A and B respectively.

So far just one type of photon has been considered and, in addition, Equation 13.43 is an operator equation. It is now necessary to sum over the photon polarization λ and all the possible wave vectors $\mathbf{k}$. To do this it is expedient to replace the summation over $\mathbf{k}$ by an integration over ω and the solid angle Ω. At the same time a number density $\rho(\omega)$ must be introduced and from Equation 6.75 this is just $V\omega^2/(2\pi c)^3$. Thus we make the replacement:

$$\sum_{\mathbf{k}} \rightarrow \int (V\omega^2/(2\pi c)^3)\mathrm{d}\omega\mathrm{d}\Omega. \qquad (13.44)$$

Before performing the integration it is convenient to express $n_{\mathbf{k}\lambda}$ in Equation 13.43 in terms of the radiation intensity $I(\omega)$, which is defined so that $I(\omega)\mathrm{d}\omega$ is the energy per unit volume of radiation with angular frequency between ω and $\omega + \mathrm{d}\omega$; it is assumed here that all polarizations and orientations of $\mathbf{k}$ are equally likely but the details may be modified to deal with other cases. Now each photon contributes an energy $\hbar\omega/V$ to the intensity, so that:

$$I(\omega)\mathrm{d}\omega = n_{\mathbf{k}\lambda}(\hbar\omega/V)(V\omega^2/(2\pi c)^3)8\pi\mathrm{d}\omega. \qquad (13.45)$$

Here the number density $\rho(\omega)$ has again been used and the factor 8π arises from the sum over the two possible polarizations for each $\mathbf{k}$ and integration over the orientations of $\mathbf{k}$. On rearrangement Equation 13.45 gives:

$$n_{\mathbf{k}\lambda} = I(\omega)\pi^2c^3/\hbar\omega^3. \qquad (13.46)$$

Using 13.44 and Equation 13.46 the transition rate (Equation 13.43) now becomes:

$$w_{fi} = \sum_{\lambda}\int (I(\omega)/8\epsilon_0\hbar)(\boldsymbol{\epsilon}_{\mathbf{k}\lambda}.\mathbf{P}_{BA})^2\delta(E_B - E_A - \hbar\omega)\mathrm{d}\omega\mathrm{d}\Omega. \qquad (13.47)$$

The sum over λ and the angular integration is performed first, the relevant part of Equation 13.47 being:

$$\sum_{\lambda}\int (\boldsymbol{\epsilon}_{\mathbf{k}\lambda}.\mathbf{P}_{BA})^2\mathrm{d}\Omega. \qquad (13.48)$$

The situation is illustrated in Fig. 13.1; χ_λ is the angle between $\mathbf{P}_{BA}$ and $\boldsymbol{\epsilon}_{\mathbf{k}\lambda}$ and the angular integration is over θ and ϕ. The expression 13.48

reduces to:

$$\sum_{\lambda} \int |\mathbf{P}_{BA}|^2 \cos^2 \chi_\lambda \, d\Omega = |\mathbf{P}_{BA}|^2 \int \sin^2\theta d\Omega, \tag{13.49}$$

where the trigonometric relations:

$$\cos \chi_1 = \sin \theta \cos \phi; \qquad \cos \chi_2 = \sin \theta \sin \phi \tag{13.50}$$

have been used. The angular integral is now readily evaluated and 13.48 becomes $(8\pi/3)|\mathbf{P}_{BA}|^2$, so that the transition rate is:

$$w_{fi} = \int (\pi I(\omega)|\mathbf{P}_{BA}|^2/3\,\epsilon_0\hbar)\delta(E_B - E_A - \hbar\omega)d\omega. \tag{13.51}$$

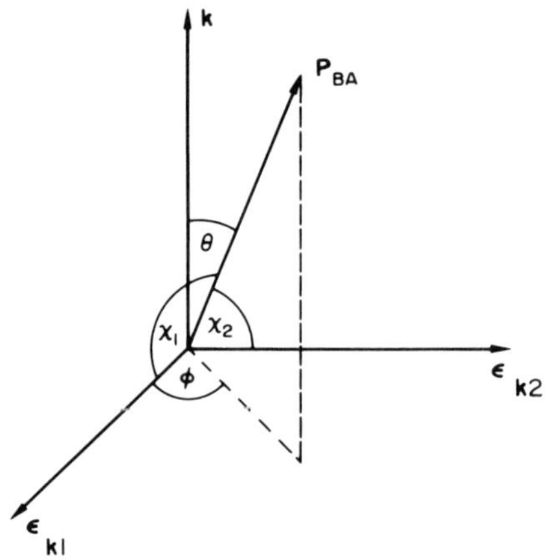

Fig. 13.1. The relative orientations of $\mathbf{P}_{BA}$ and $\boldsymbol{\epsilon}_{\mathbf{k}\lambda}$.

Finally, the integration over ω must be performed and this is readily accomplished using Equations 2.82 and 2.88:

$$\begin{aligned} w_{fi} &= \pi I(\omega)|\mathbf{P}_{BA}|^2/3\epsilon_0\hbar^2 \\ &= B_{B\leftarrow A} I(\omega), \end{aligned} \tag{13.52}$$

where $B_{B\leftarrow A}$ is the Einstein coefficient for absorption and, because energy is conserved in the transition:

$$\omega = (E_B - E_A)/\hbar. \tag{13.53}$$

The corresponding emission process, that is:

$$B \rightarrow A + \gamma, \tag{13.54}$$

may be treated in very much the same way. If just one type of photon is considered the initial and final states are $|B, n_{\mathbf{k}\lambda}\rangle$ and $|A, n_{\mathbf{k}\lambda} + 1\rangle$ respectively and from Equation 13.27 the transition rate for emission is:

$$w_{fi} = (\pi/V\epsilon_0)(n_{\mathbf{k}\lambda} + 1)\omega_{\mathbf{k}}(\boldsymbol{\epsilon}_{\mathbf{k}\lambda}.\mathbf{P}_{AB})^2\delta(E_A - E_B + \hbar\omega_{\mathbf{k}}). \quad (13.55)$$

When this is compared with the analogous equation (13.43) for absorption, it is noticed that for emission the transition rate may be split into two parts. The first is proportional to $n_{\mathbf{k}\lambda}$ and hence to the intensity of the radiation. This gives rise to stimulated emission, since it depends on radiation being present. The second part is independent of $n_{\mathbf{k}\lambda}$ and has no counterpart in absorption. It is responsible for spontaneous emission, since it does not vanish when there is no radiation present initially. These two parts are considered separately.

The stimulated emission term is similar to the absorption transition rate (Equation 13.43) and may be processed in the same way. Reference to Equation 13.52 shows that after the inclusion of all types of photons the stimulated emission rate is given by:

$$\begin{aligned} w_{fi} &= \pi I(\omega)|\mathbf{P}_{AB}|^2/3\epsilon_0\hbar^2 \\ &= B_{A\leftarrow B}I(\omega), \end{aligned} \quad (13.56)$$

where ω again satisfies the energy conservation condition (Equation 13.53). The Einstein coefficient for stimulated emission $B_{A\leftarrow B}$ is precisely the same as that for absorption $B_{B\leftarrow A}$. It is emphasized that this emission only occurs if there is radiation of the appropriate frequency already present and this is why it is said to be stimulated (or induced).

When $n_{\mathbf{k}\lambda}$ vanishes, Equation 13.55 reduces to the transition rate for spontaneous emission. This equation is for one type of photon only and must be summed over $\mathbf{k}$ and λ. The sum over $\mathbf{k}$ may be replaced by an integral using the prescription 13.44 and the sum over λ and the angular integration may be performed as for the absorption case, since all directions of emission and polarization are equally likely. This gives:

$$\begin{aligned} w_{fi} &= \int (\omega^3|\mathbf{P}_{BA}|^2/3\pi c^3\epsilon_0)\delta(E_A - E_B + \hbar\omega)\mathrm{d}\omega \\ &= \omega^3|\mathbf{P}_{BA}|^2/3\pi c^3\epsilon_0\hbar \\ &= A_{A\leftarrow B}, \end{aligned} \quad (13.57)$$

where ω is given by Equation 13.53 and $A_{A\leftarrow B}$ is the Einstein coefficient of spontaneous emission. This emission occurs whether radiation of the appropriate energy is present or not.

13.5 Comparison of the semiclassical and quantized theories

In the semiclassical theory the radiation is described classically and in particular the vector potential is a function of space coordinates. The radiation can influence a molecule but it cannot be influenced by the molecule in return. When the molecule makes a radiative transition the vector potential is not affected. This theory works whenever the radiation is sufficiently intense that the appearance or disappearance of a single photon has a negligible effect on the intensity.

The details of the use of the semiclassical theory to describe the absorption and emission of radiation is given in many texts. The absorption process may be accounted for quite readily with this theory and with some difficulty so can stimulated emission. However, the semiclassical theory is incapable of describing spontaneous emission, since classically **A** vanishes when there is no radiation and the interaction Hamiltonian disappears.

The way in which this problem is surmounted is to use Planck's radiation law, which may be derived from statistical mechanics, to relate the Einstein coefficients of stimulated and spontaneous emission. In addition, this approach shows that the Einstein coefficients of absorption and stimulated emission are equal, so that in practice only the absorption process needs to be considered in the semiclassical theory. The results are the same as those obtained in the previous section.

If the quantized theory of radiation is used the absorption and emission rates are obtained directly and there is no need to invoke Planck's radiation law. In addition, both stimulated and spontaneous emission arise in the same way. Even when there is no radiation present a molecule may interact with the radiation vacuum state and this may be attributed to the presence of the vacuum fluctuations (Section 12.5).

Not surprisingly the quantized theory may be used to derive Planck's radiation law. We consider a collection of molecules and a radiation field that can exchange energy by the emission and absorption of photons:

$$A + \gamma \leftrightharpoons B. \tag{13.58}$$

When thermal equilibrium is established the ratio of the numbers of molecules in states A and B is given by the Boltzmann law:

$$N_A/N_B = \exp(\hbar\omega/kT), \tag{13.59}$$

where $\hbar\omega = E_B - E_A$ and k is the Boltzmann constant. Now, N_A and N_B are also related by the equilibrium condition that N_A/N_B is equal to the ratio of the emission rate to the absorption rate. When it is remembered that both stimulated and spontaneous emission are possible, use of Equations 13.52, 13.56 and 13.57 shows that:

$$N_A/N_B = 1 + (\omega^3\hbar/\pi^2 I(\omega)\, c^3). \tag{13.60}$$

Combination of Equations 13.59 and 13.60 now gives:

$$I(\omega) = \omega^3\hbar/\pi^2c^3\,[\exp(\hbar\omega/kT) - 1], \tag{13.61}$$

or in terms of the frequency ν rather than the angular frequency ω:

$$I(\nu) = 8\pi h\nu^3/c^3\,[\exp(h\nu/kT) - 1]. \tag{13.62}$$

Although only one frequency of radiation has been considered this equation can immediately be extended to all frequencies if it is assumed that the radiation field is surrounded by black walls consisting of molecules capable of absorbing and emitting photons of all energies. Equation 13.62 is just Planck's radiation law, which gives the energy density of radiation in equilibrium with its surroundings at a temperature T.

Finally, it may be remarked that the quantized theory provides a more appealing physical picture for the interaction of radiation with matter, since photons are absorbed and emitted and energy is conserved. Indeed these processes may conveniently be described by diagrams. In Fig. 13.2 time increases from the bottom to the top of the page, the wavy lines represent photons and the straight lines molecular states; although there is a resemblance, these are strictly not Feynman diagrams. In the present instance these diagrams are trivial, but similar diagrams can be very useful in discussing many-photon processes, since they provide a convenient way of keeping track of all the terms in the expansion 13.24 of the reaction matrix element.

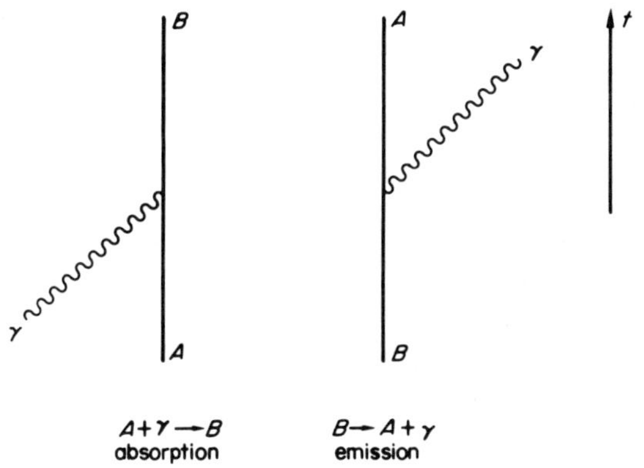

Fig. 13.2. Diagrams describing absorption and emission.

13.6 Multi-photon processes

Two-photon absorption has been observed in microwave and radio-frequency spectra and also, with the advent of intense laser sources, in the optical spectra of, for example, organic crystals. Although only the case of two-photon absorption is briefly considered here, the principles may readily be extended to emission and multi-photon processes.

In practice the two photons absorbed are usually of the same frequency, but we discuss the general case:

$$A + \gamma + \gamma' \rightarrow B. \tag{13.63}$$

Thus, the initial state is $|A, n_{\mathbf{k}\lambda}, n_{\mathbf{k}'\lambda'}\rangle$ and the final state is $|B, n_{\mathbf{k}\lambda} - 1, n_{\mathbf{k}'\lambda'} - 1\rangle$. The transition rate w_{fi} is given by Equation 13.23, so that a non-zero reaction matrix element R_{fi} is needed. The first term in the expansion 13.24 of R_{fi} is V_{fi}, but this vanishes. In the first place the perturbation $(e/m)(\mathbf{A}.\mathbf{p})$ only connects states differing by one photon. Secondly, although $(e^2/2m)A^2$ connects states differing by two photons its matrix elements are proportional to $\langle B|A\rangle$, which is zero in this case; this depends on the use of the long wavelength approximation, but higher terms in the expansion of $\exp(i\mathbf{k}.\mathbf{r})$ make negligible contributions.

For a non-zero R_{fi} it is necessary to go to the second term in its expansion and this is just:

$$\sum_n V_{fn} V_{ni}/(E_i - E_n). \tag{13.64}$$

Only the perturbation $(e/m)(\mathbf{A}.\mathbf{p})$ makes a significant contribution and then only if the state n is $|I, n_{\mathbf{k}\lambda}, n_{\mathbf{k}'\lambda'} - 1\rangle$ or $|I, n_{\mathbf{k}\lambda} - 1, n_{\mathbf{k}'\lambda'}\rangle$, where I is any state of the molecular system. Use of Equation 13.31 then gives:

$$\begin{aligned} R_{fi} = {} & -(n_{\mathbf{k}\lambda} n_{\mathbf{k}'\lambda'}/\omega_{\mathbf{k}}\omega_{\mathbf{k}'})^{\frac{1}{2}} (2\hbar V\epsilon_0)^{-1} \sum_I (E_B - E_I)(E_I - E_A) \times \\ & [(\boldsymbol{\epsilon}_{\mathbf{k}\lambda}.\mathbf{P}_{BI})(\boldsymbol{\epsilon}_{\mathbf{k}'\lambda'}.\mathbf{P}_{IA})/(E_A - E_I + \hbar\omega_{\mathbf{k}'}) \\ & + (\boldsymbol{\epsilon}_{\mathbf{k}'\lambda'}.\mathbf{P}_{BI})(\boldsymbol{\epsilon}_{\mathbf{k}\lambda}.\mathbf{P}_{IA})/(E_A - E_I + \hbar\omega_{\mathbf{k}})]. \end{aligned} \tag{13.65}$$

This may now be substituted in Equation 13.23 to give the transition rate. As for one-photon processes generalization to all types of photons must be made, but this may be accomplished in a similar manner. The details are not given here, but it may be noted that the presence of the Dirac delta function in Equation 13.23 ensures that the transition rate vanishes unless energy is conserved in the overall process:

$$E_B - E_A = \hbar\omega_{\mathbf{k}} + \hbar\omega_{\mathbf{k}'}. \tag{13.66}$$

A few words are appropriate about the molecular states I, which are called intermediate or virtual states. They are definite molecular states and as such they are eigenfunctions of the molecular Hamiltonian. However, these states cannot be observed in the two-photon process. Over the entire process energy is conserved, but it is not necessarily conserved while the molecule is in an intermediate state. The reason for this is that the lifetime of an intermediate state is so short that Heisenberg's uncertainty principle allows there to be some uncertainty in its energy. Strictly speaking, the intermediate states include continuum states and an appropriate integral over these states should be added to the sum over the discrete states in Equation 13.65; however, they are usually unimportant because the corresponding energy denominators are very large.

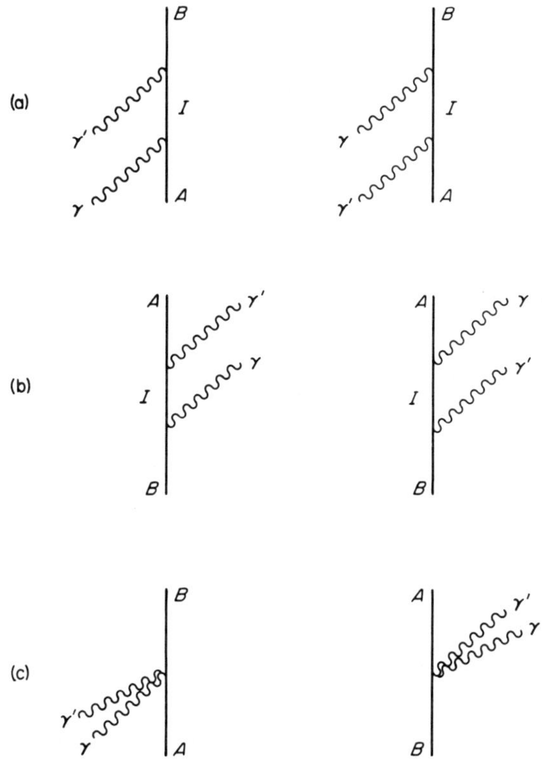

Fig. 13.3. Diagrams describing (a) two-photon absorption: $A + \gamma + \gamma' \rightarrow B$, (b) two-photon emission: $B \rightarrow A + \gamma + \gamma'$, (c) negligible two-photon absorption and emission processes.

The two-photon absorption (Equation 13.63) may be described by the diagrams given in Fig. 13.3(a). There is a diagram for each of the two perturbation terms in Equation 13.65; in one case the photon γ is absorbed first and in the other it is γ'. In Fig. 13.3(b) the corresponding diagrams for the two-photon emission process:

$$B \rightarrow A + \gamma + \gamma' \qquad (13.67)$$

are given. Finally, the negligible absorption and emission processes due to the perturbation $(e^2/m)A^2$ are described by the diagrams in Fig. 13.3(c); in this case there are no intermediate states, since the operator A^2 connects states differing by two photons and in principle can contribute to the first term V_{fi} in the expansion 13.24 of R_{fi}.

13.7 The scattering of photons by molecules

In the two-photon processes discussed in the last section both photons are either absorbed or emitted. However, it is possible for one photon to be absorbed and one to be emitted:

$$A + \gamma \rightarrow B + \gamma'; \qquad (13.68)$$

this process is referred to as scattering. If γ and γ' both have the same energy and $A = B$ we have Rayleigh scattering, while the more general phenomenon where γ and γ' are of different energy and $A \neq B$ is known as Raman scattering.

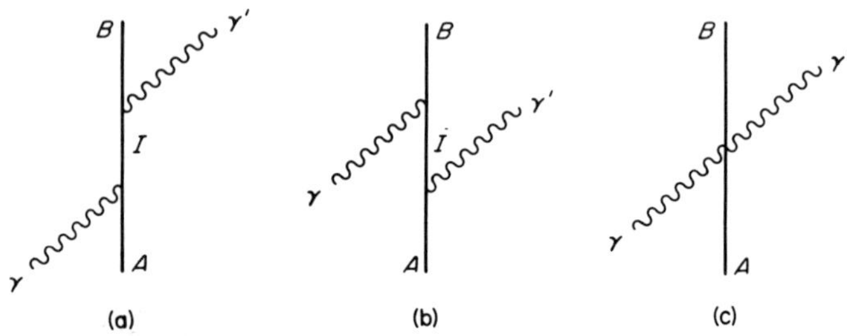

Fig. 13.4. Diagrams describing photon scattering: $A + \gamma \rightarrow B + \gamma'$.

The general process 13.68 may be described using diagrams as in Fig. 13.4. In (a) the photon γ is absorbed to give an intermediate state I which subsequently emits the photon γ'. However, there is no reason why γ' should not be emitted before γ is absorbed and this possibility is illustrated in (b). The first-order process in which absorption and emission occur simultaneously is shown in (c); this is due to the perturbation $(e^2/2m)A^2$ and is only important for Rayleigh scattering for which $A = B$.

We again consider a single bound electron, but as before the generalization to a molecule may be made without difficulty. The initial state for the scattering process may be taken as $|A, n_{\mathbf{k}\lambda}, n_{\mathbf{k}'\lambda'}\rangle$ and the final state as $|B, n_{\mathbf{k}\lambda} - 1, n_{\mathbf{k}'\lambda'} + 1\rangle$. The matrix element of the perturbation $(e^2/2m)A^2$ between these two states is given by Equation 13.40. The second-order contribution of $(e/m)(\mathbf{A.p})$ to the reaction matrix element is given by 13.64 and in this case the relevant states n are $|I, n_{\mathbf{k}\lambda} - 1, n_{\mathbf{k}'\lambda'}\rangle$ and $|I, n_{\mathbf{k}\lambda}, n_{\mathbf{k}'\lambda'} + 1\rangle$; the first occurs in diagram (a) in Fig. 13.4 and the second in diagram (b). Thus, the matrix element R_{fi} is given by:

$$\begin{aligned} R_{fi} = {} & [n_{\mathbf{k}\lambda}(n_{\mathbf{k}'\lambda'} + 1)/\omega_{\mathbf{k}}\omega_{\mathbf{k}'}]^{\frac{1}{2}}(2\hbar V\epsilon_0)^{-1} \times \\ & \{(e^2\hbar^2/m)(\boldsymbol{\varepsilon}_{\mathbf{k}\lambda}.\boldsymbol{\varepsilon}_{\mathbf{k}'\lambda'})\delta_{AB} - \sum_I (E_B - E_I)(E_I - E_A) \times \\ & [(\boldsymbol{\varepsilon}_{\mathbf{k}\lambda}.\mathbf{P}_{BI})(\boldsymbol{\varepsilon}_{\mathbf{k}'\lambda'}.\mathbf{P}_{IA})/(E_A - E_I - \hbar\omega_{\mathbf{k}'}) \\ & + (\boldsymbol{\varepsilon}_{\mathbf{k}'\lambda'}.\mathbf{P}_{BI})(\boldsymbol{\varepsilon}_{\mathbf{k}\lambda}.\mathbf{P}_{IA})/(E_A - E_I + \hbar\omega_{\mathbf{k}})]\}. \end{aligned} \tag{13.69}$$

This may now be used to calculate the transition rate, but the details of any further development depend on the experimental situation under consideration. Indeed, Equation 13.69 and similar expressions provide a starting point for a discussion of many phenomena such as optical rotation and related effects as well as Rayleigh and Raman scattering. However, except for a few more general remarks, this topic is not pursued further.

If the energy of the incident and scattered photons is much larger than the binding energy of the electrons, only the first term in the curly parentheses of Equation 13.69 is important. The scattering is largely due to the process depicted in (c) of Fig. 13.4 and is insensitive to the binding of the electron; this is the situation exploited in X-ray diffraction, for example. When the photon energy is small the resulting Rayleigh and Raman scattering is usually described in terms of molecular polarizability and this section is concluded with a demonstration that Equation 13.69 is related to a polarizability matrix element for the specific case of Rayleigh scattering, in which $A = B$ and $\omega_{\mathbf{k}} = \omega_{\mathbf{k}'}$.

The initial step is to rewrite the first term in Equation 13.69 in a similar form to those involving intermediate states. This is done by noting that, because of the commutation relations of $\mathbf{p}$ and $\mathbf{r}$:

$$\begin{aligned} i\hbar(\boldsymbol{\varepsilon}_{\mathbf{k}\lambda}.\boldsymbol{\varepsilon}_{\mathbf{k}'\lambda'}) &= (\boldsymbol{\varepsilon}_{\mathbf{k}\lambda}.\mathbf{r})(\boldsymbol{\varepsilon}_{\mathbf{k}'\lambda'}.\mathbf{p}) - (\boldsymbol{\varepsilon}_{\mathbf{k}'\lambda'}.\mathbf{p})(\boldsymbol{\varepsilon}_{\mathbf{k}\lambda}.\mathbf{r}) \\ &= \sum_I [\langle A|\boldsymbol{\varepsilon}_{\mathbf{k}\lambda}.\mathbf{r}|I\rangle\langle J|\boldsymbol{\varepsilon}_{\mathbf{k}'\lambda'}.\mathbf{p}|A\rangle \\ &\quad + \langle A|\boldsymbol{\varepsilon}_{\mathbf{k}'\lambda'}.\mathbf{p}|I\rangle\langle J|\boldsymbol{\varepsilon}_{\mathbf{k}\lambda}.\mathbf{r}|A\rangle], \end{aligned} \tag{13.70}$$

where the completeness of the intermediate states I has been used. Use of Equation 13.30 now shows that:

$$(e^2\hbar^2/m)(\boldsymbol{\varepsilon}_{\mathbf{k}\lambda}.\boldsymbol{\varepsilon}_{\mathbf{k}'\lambda'}) = \sum_I (E_I - E_A)[(\boldsymbol{\varepsilon}_{\mathbf{k}\lambda}.\mathbf{P}_{AI})(\boldsymbol{\varepsilon}_{\mathbf{k}'\lambda'}.\mathbf{P}_{IA}) + (\boldsymbol{\varepsilon}_{\mathbf{k}'\lambda'}.\mathbf{P}_{AI})(\boldsymbol{\varepsilon}_{\mathbf{k}\lambda}.\mathbf{P}_{IA})], \quad (13.71)$$

where $\mathbf{P}_{AI}$ is the matrix element of $e\mathbf{r}$ between A and I. The expression in Equation 13.69 which is enclosed in curly parentheses now becomes:

$$\hbar\omega \sum_I (E_I - E_A)[(\boldsymbol{\varepsilon}_{\mathbf{k}\lambda}.\mathbf{P}_{AI})(\boldsymbol{\varepsilon}_{\mathbf{k}'\lambda'}.\mathbf{P}_{IA})/(E_I - E_A + \hbar\omega) - (\boldsymbol{\varepsilon}_{\mathbf{k}'\lambda'}.\mathbf{P}_{AI})(\boldsymbol{\varepsilon}_{\mathbf{k}\lambda}.\mathbf{P}_{IA})/(E_I - E_A - \hbar\omega)]. \quad (13.72)$$

Since $\hbar\omega$ is assumed to be much smaller than $(E_I - E_A)$ the terms $(E_I - E_A \pm \hbar\omega)^{-1}$ may be expanded as power series in $\hbar\omega/(E_I - E_A)$ and only the linear terms need be retained. Thus, Equation 13.72 is approximated by:

$$\sum_I \hbar\omega[(\boldsymbol{\varepsilon}_{\mathbf{k}\lambda}.\mathbf{P}_{AI})(\boldsymbol{\varepsilon}_{\mathbf{k}'\lambda'}.\mathbf{P}_{IA}) - (\boldsymbol{\varepsilon}_{\mathbf{k}'\lambda'}.\mathbf{P}_{AI})(\boldsymbol{\varepsilon}_{\mathbf{k}\lambda}.\mathbf{P}_{IA})]$$
$$- \sum_I (\hbar\omega)^2[(\boldsymbol{\varepsilon}_{\mathbf{k}\lambda}.\mathbf{P}_{AI})(\boldsymbol{\varepsilon}_{\mathbf{k}'\lambda'}.\mathbf{P}_{IA}) + (\boldsymbol{\varepsilon}_{\mathbf{k}'\lambda'}.\mathbf{P}_{AI})(\boldsymbol{\varepsilon}_{\mathbf{k}\lambda}.\mathbf{P}_{IA})]/(E_I - E_A). \quad (13.73)$$

Now the first term vanishes, since it is proportional to the matrix element $\langle A|[(\boldsymbol{\varepsilon}_{\mathbf{k}\lambda}.\mathbf{r}), (\boldsymbol{\varepsilon}_{\mathbf{k}'\lambda'}.\mathbf{r})]|A\rangle$, so that the reaction matrix element R_{fi} is finally given by:

$$R_{fi} = -[n_{\mathbf{k}\lambda}(n_{\mathbf{k}'\lambda'} + 1)]^{\frac{1}{2}}(\hbar\omega/2V\epsilon_0) \times \sum_I [(\boldsymbol{\varepsilon}_{\mathbf{k}\lambda}.\mathbf{P}_{AI})(\boldsymbol{\varepsilon}_{\mathbf{k}'\lambda'}.\mathbf{P}_{IA}) + (\boldsymbol{\varepsilon}_{\mathbf{k}'\lambda'}.\mathbf{P}_{AI})(\boldsymbol{\varepsilon}_{\mathbf{k}\lambda}.\mathbf{P}_{IA})]/(E_I - E_A). \quad (13.74)$$

Although Equation 13.74 has been derived for a single bound electron R_{fi} has the same form for a molecule and can be rewritten as:

$$R_{fi} = -[n_{\mathbf{k}\lambda}(n_{\mathbf{k}'\lambda'} + 1)]^{\frac{1}{2}}(\hbar\omega/2V\epsilon_0)\boldsymbol{\varepsilon}_{\mathbf{k}\lambda}.\langle A|\boldsymbol{\alpha}|A\rangle.\boldsymbol{\varepsilon}_{\mathbf{k}'\lambda'}, \quad (13.75)$$

where $\boldsymbol{\alpha}$ is the molecular polarizability tensor. This concludes our demonstration that for low energy photons the Rayleigh scattering may be described in terms of the molecular polarizability.

13.8 Line widths and resonance fluorescence

It has already been mentioned that the time-dependent perturbation theory in Section 13.2 has deficiencies. One of these is that the theory leads to transitions that are infinitely sharp, although in practice spectral lines have a finite line width. Of course, there are a number of causes for line broadening.

For example, in high resolution gas phase studies the Doppler effect and pressure broadening are important. However, even if these contributions are eliminated there is a residual line width, known as the natural line width, which cannot be removed.

Another difficulty is that the reaction matrix element (Equation 13.69) for photon scattering involves the denominator $(E_A - E_I + \hbar\omega)$, where $\hbar\omega$ is the energy of an incident photon. If $\hbar\omega$ is equal to $(E_I - E_A)$ for an intermediate state I, then R_{fi}, and hence the transition rate w_{fi}, both become infinite. This does not, of course, occur in practice, although in the region where $\hbar\omega = E_I - E_A$ the scattering probability is observed to go through a sharp maximum. This phenomenon is called resonance fluorescence.

Both the natural line width and resonance fluorescence may be attributed to the fact that excited states have finite lifetimes and that this is not accounted for in the theory of Section 13.2. Heisenberg's uncertainty principle, which is stated in Equation 12.60, shows that the lifetime τ of a state and the uncertainty in its energy are related by:

$$\tau(\Delta E) \geqslant \hbar/2, \tag{13.76}$$

where we have used the fact that the energy operator is $i\hbar(\partial/\partial t)$; thus, ΔE is of the order of $\hbar/\tau$. For the ground state of a molecule in the absence of radiation the lifetime is infinite and the energy is well defined. However, excited states have finite lifetimes, since photon emission is always possible even if there is no radiation present. In general the average lifetime of a state is the reciprocal of the sum of the transition rates to all other states. Consequently, the energy of an excited state cannot be definite and the energy level is smeared out.

This uncertainty in the energies of excited states means that all spectral lines are broadened, since transitions always involve excited molecular states even if, as in the case of Rayleigh scattering by a molecule in its ground state, the excited states involved are intermediate states. The uncertainty in the energies of states was not accounted for in the theory of Section 13.2, since in obtaining Equation 13.10 it was assumed that the states are eigenfunctions of $\mathcal{H}_0$ and that their energies are constant and independent of time. As a result the differential equation 13.10 allows for a build up of a particular state by transitions from other states, but it does not account for the depletion of that state by transitions from it. This possibility can be allowed for phenomenologically by adding an extra term $c_k(t)\exp(-t/2\tau_k)$ to Equation 13.10 for $\dot{c}_k(t)$. Inclusion of this term then leads to finite line widths for transitions.

These are qualitative considerations, but the results arise naturally in more advanced theories. Here, we confine ourselves to an unsophisticated demonstration that finite line widths are predicted by quantitative treatments. To do this we consider a system consisting of just two states, a and b. Corresponding to Equation 13.10 there is then a pair of coupled differential equations:

$$\left.\begin{aligned} \dot{c}_a &= -(i/\hbar)V_{ab}c_b \exp\,[i(E_a - E_b)t/\hbar] \\ \dot{c}_b &= -(i/\hbar)V_{ba}c_a \exp\,[i(E_b - E_a)t/\hbar] \end{aligned}\right\}. \qquad (13.77)$$

To simplify the subsequent development it is assumed that at time $t = 0$ only state a is populated, so that

$$c_a(0) = 1; \qquad c_b(0) = 0. \qquad (13.78)$$

In Section 13.2 initial conditions of this type were deliberately avoided, since they give spurious terms in second order, but here only the first-order results are considered.

The qualitative discussion earlier suggests that state a will decay as exp $(-\,t/\tau_a)$, so that c_a will decrease as exp $(-\,t/2\tau_a)$, since the probability of the system being in state a is given by $|c_a|^2$. Consequently, the trial solution:

$$c_a(t) = \exp\,(-\,t/2\tau_a) \qquad (13.79)$$

is investigated. By substituting this in the second of Equations 13.77 a differential equation in c_b alone is obtained and the solution of this using the initial conditions (Equations 13.78) gives:

$$c_b = V_{ba}\,(\exp\,[i(E_b - E_a)t/\hbar - t/2\tau_a] \quad 1)/(E_a - E_b - i\hbar/2\tau_a). \qquad (13.80)$$

The first of Equations 13.77 may now be used with Equations 13.79 and 13.80 to determine τ_a^{-1}:

$$\tau_a^{-1} = (2i/\hbar)|\,V_{ab}|^2(1 - \exp\,[i(E_a - E_b)t/\hbar + t/2\tau_a])/(E_a - E_b - i\hbar/2\tau_a), \qquad (13.81)$$

or neglecting terms in τ_a on the right-hand side of this equation:

$$\tau_a^{-1} = (2i/\hbar)|\,V_{ab}|^2(1 - \exp\,[i(E_a - E_b)t/\hbar])/(E_a - E_b); \qquad (13.82)$$

the justification for this neglect is that $\hbar\tau_a^{-1}$ is very much smaller than $(E_a - E_b)$.

Now τ_a^{-1} is expected to be time independent and indeed this has been assumed in obtaining Equation 13.82. Consequently, the limit of Equation 13.82 is taken as $t \to \infty$ and to avoid oscillations we employ the same trick as before (Section 13.2) and insert an artificial damping factor, which is later made to vanish. Abbreviating $(E_a - E_b)/\hbar$ to x, we have:

$$\begin{aligned} \underset{t\to\infty}{Lt}\,[(1-\exp(ixt))/x] &= -i\int_0^\infty \exp(ixt)\mathrm{d}t \\ &= -\,i\underset{\eta\to 0}{Lt}\int_0^\infty \exp(ixt-\eta t)\mathrm{d}t \\ &= \underset{\eta\to 0}{Lt}\,(x+i\eta)^{-1} \\ &= \underset{\eta\to 0}{Lt}\,[x/(x^2+\eta^2)-i\eta/(x^2+\eta^2)] \\ &= 1/x - i\pi\delta(x), \end{aligned} \tag{13.83}$$

where the representation (Equation 13.21) of a Dirac delta function has been used. Thus, Equation 13.82 becomes:

$$\tau_a^{-1} = (2i/\hbar)|V_{ab}|^2/(E_a-E_b) + (2\pi/\hbar)|V_{ab}|^2\delta(E_a-E_b). \tag{13.84}$$

Reference to Equations 13.9 and 13.79 shows that the imaginary part of τ_a^{-1} leads to an energy shift. It is a term like this that is responsible for the Lamb shift (Section 11.9), the two levels being an electron plus the vacuum state and the electron plus a photon. The electron emits a photon which it subsequently absorbs and, since the emission is due to the electron interacting with the vacuum state, the vacuum fluctuations may be considered to be responsible for the Lamb shift.

The real part of τ_a^{-1} (Equation 13.84) is just the rate of transition (Equation 13.23) from state a to state b, so that τ_a is the lifetime of state a in the two-level system under consideration. This result is in agreement with the qualitative arguments given earlier.

We are now in a position to consider the line shape associated with the transition a to b. At $t = \infty$ the transition is certain to have taken place and from Equation 13.80:

$$c_b(\infty) = -V_{ba}/(E_a - E_b - i\hbar/2\tau_a). \tag{13.85}$$

The transition probability is then:

$$|c_b(\infty)|^2 = |V_{ba}|^2/[(E_a-E_b)^2 + \hbar^2/4\tau_a^2]. \tag{13.86}$$

The important thing to note is that this is not proportional to a Dirac delta function, but is an ordinary function of the energy difference $(E_a - E_b)$, so that the spectral line associated with the transition is not infinitely sharp.

If the transition from a to b involves the absorption or emission of a photon of energy $\hbar\omega$ by a molecule in state A to give the molecular state B, we can write:

$$E_a - E_b = E_A - E_B \pm \hbar\omega = \hbar(\omega_0 \pm \omega). \tag{13.87}$$

The transition probability (Equation 13.86) is then a function of the photon energy and is proportional to:

$$[(\omega - \omega_0)^2 + 1/4\tau_a^2]^{-1}. \qquad (13.88)$$

Thus, the spectral line corresponding to the transition has the Lorentz shape. It is centred on $\omega = \omega_0$ and has a width at half height of τ_a^{-1}, so that the natural line width is just equal to the transition rate from a to b.

Although only a two-state system has been considered, the extension to more complicated systems yields analogous results, the width of a line being the total transition rate to all other states. These considerations also provide an explanation for resonance fluorescence, which was mentioned earlier and which occurs in scattering processes when the incident photon energy is the same as the energy difference between the initial state of the molecule and one of the intermediate states. A more sophisticated theory is needed, since scattering involves second-order processes, but the result is that the reaction matrix element (Equation 13.69) is modified by the replacement of the denominator $(E_A - E_I + \hbar\omega_{\mathbf{k}})$ by $(E_A - E_I + \hbar\omega_{\mathbf{k}} - i\hbar/2\tau_I)$ as in the simpler example considered here. As a consequence the transition rate does not become infinite when $E_I - E_A = \hbar\omega_{\mathbf{k}}$, although it does have a maximum in agreement with experiment.

Bibliography

The interaction of radiation and matter

Grandy: pages 201-211.

Hameka: the interaction Hamiltonian is discussed in Section 6-3, absorption and emission in Chapter 7 and scattering in Chapters 11 and 12.

Sakurai: pages 36-57.

Wallace, R. (1966), 'Diagrammatic perturbation theory of multiphoton transitions', *Molec. Phys.*, **11**, 457.

Time-dependent perturbation theory

Grandy: Appendix C.

Hameka: Chapter 5.

Line widths and resonance fluorescence

Heitler, W. (1954), *The Quantum Theory of Radiation*, Clarendon Press, Oxford: pages 181-189.

Sakurai: pages 53-57 describe the phenomenological approach, which is justified in pages 65-68.

APPENDIX A

Units

This book employs the 'Système International d'Unités' (SI). (It is also known as the (rationalized) MKSA system, since four of the base units employed are the metre, kilogram, second and ampere.) A discussion of the advantages of using this system is given in Section 5.1. Here a brief summary is given, with particular reference to those units used in this text. For the details of these and other units, reference should be made to the bibliography.

A.1 SI units

There are three types of SI units: base, supplementary and derived.

(*a*) The seven **base units** are regarded as being dimensionally independent:

Physical quantity	SI unit		Dimension
	Name	Symbol	
length	metre	m	*L*
mass	kilogram	kg	*M*
time	second	s	*T*
electric current	ampere	A	*A*
thermodynamic temperature	kelvin	K	
luminous intensity	candela	cd	
amount of substance	mole	mol	

(*b*) There are two **supplementary units**:

Physical quantity	SI unit	
	Name	Symbol
plane angle	radian	rad
solid angle	steradian	sr

(*c*) **Derived units** are employed for all other physical quantities. They are said to be coherent, since they are obtained by multiplying and dividing the base units without introducing any numerical factors whatsoever. Some of these have special names and symbols:

Physical quantity	SI unit			Dimensions
	Name	Symbol	Definition	
energy	joule	J	kg m^2 s^{-2}	$M L^2 T^{-2}$
force	newton	N	kg m s^{-2}	$M L T^{-2}$
pressure	pascal	Pa	kg m^{-1} s^{-2}	$M L^{-1} T^{-2}$
power	watt	W	kg m^2 s^{-3}	$M L^2 T^{-3}$
electric charge	coulomb	C	s A	$T A$
electric potential difference	volt	V	kg m^2 s^{-3} A^{-1}	$M L^2 T^{-3} A^{-1}$
electric resistance	ohm	Ω	kg m^2 s^{-3} A^{-2}	$M L^2 T^{-3} A^{-2}$
electric capacitance	farad	F	kg^{-1} m^{-2} s^4 A^2	$M^{-1} L^{-2} T^4 A^2$
magnetic flux	weber	Wb	kg m^2 s^{-2} A^{-1}	$M L^2 T^{-2} A^{-1}$
inductance	henry	H	kg m^2 s^{-2} A^{-2}	$M L^2 T^{-2} A^{-2}$
magnetic flux density	tesla	T	kg s^{-2} A^{-1}	$M T^{-2} A^{-1}$
frequency	hertz	Hz	s^{-1}	T^{-1}

There are other derived units which have no special names or symbols. Many of these are obvious, such as that for density (kg m^{-3}), and only the pertinent electromagnetic units are given here. For completeness electric charge, electric potential and magnetic flux density are included, although their units have special names.

Physical quantity	Symbol for SI unit	Dimensions
electric charge	C	$T\ A$
electric field strength **E**	V m^{-1}	$M\ L\ T^{-3}\ A^{-1}$
electric displacement **D**	C m^{-2}	$L^{-2}\ T\ A$
permittivity ϵ ($\mathbf{D} = \epsilon\mathbf{E}$)	F m^{-1}	$M^{-1}\ L^{-3}\ T^4\ A^2$
electric potential ϕ	V	$M\ L^2\ T^{-3}\ A^{-1}$
electric charge density ρ	C m^{-3}	$L^{-3}\ T\ A$
electric dipole moment	C m	$L\ T\ A$
magnetic field strength **H**	A m^{-1}	$L^{-1}\ A$
magnetic flux density **B**	T	$M\ T^{-2}\ A^{-1}$
permeability μ ($\mathbf{B} = \mu\mathbf{H}$)	H m^{-1}	$M\ L\ T^{-2}\ A^{-2}$
vector potential **A**	T m	$M\ L\ T^{-2}\ A^{-1}$
electric current density j	A m^{-2}	$L^{-2}\ A$
magnetic dipole moment	J T^{-1}	$L^2\ A$

(*d*) **Decimal multiples** (in steps of 10^3) of SI units are indicated by prefixes:

Multiple	Prefix	Symbol	Multiple	Prefix	Symbol
10^{-3}	milli	m	10^3	kilo	k
10^{-6}	micro	μ	10^6	mega	M
10^{-9}	nano	n	10^9	giga	G
10^{-12}	pico	p	10^{12}	tera	T
10^{-15}	femto	f			
10^{-18}	atto	a			

A.2 Conversion from the mixed (Gaussian) CGS system to the SI system

(*a*) **The conversion of equations** expressed in the mixed CGS system to the corresponding SI equations only poses problems if electromagnetic quantities are involved. For the situations considered in this book the conversion may be accomplished using the following table, the quantity in the CGS column being replaced by the expression in the SI column. Before the conversion all quantities must be expressed in mixed CGS units and after the conversion all must be in SI units.

Physical quantity	Dimensions in the mixed CGS system	CGS	SI
electric charge	$M^{1/2}\ L^{3/2}\ T^{-1}$	q	$q(4\pi\epsilon_0)^{-1/2}$
electric field strength	$M^{1/2}\ L^{-1/2}\ T^{-1}$	$\mathbf{E}$	$\mathbf{E}(4\pi\epsilon_0)^{1/2}$
electric potential	$M^{1/2}\ L^{1/2}\ T^{-1}$	ϕ	$\phi(4\pi\epsilon_0)^{1/2}$
magnetic flux density	$M^{1/2}\ L^{-1/2}\ T^{-1}$	$\mathbf{B}$	$\mathbf{B}(\mu_0/4\pi)^{-1/2}$
vector potential	$M^{1/2}\ L^{1/2}\ T^{-1}$	$\mathbf{A}$	$\mathbf{A}(\mu_0/4\pi)^{-1/2}$
Bohr magneton	$M^{1/2}\ L^{5/2}\ T^{-1}$	μ_B	$\mu_B(\mu_0/4\pi)^{1/2}$

This table must not be used to provide numerical conversion factors for units, since powers of 10 must be introduced to allow for the change from centimetres and grams to metres and kilograms; this topic is considered below.
(*b*) **Conversion factors for units** are given in the following table:

Physical quantity	Name of unit	Conversion factor (in SI units)	Definition of unit
energy	erg	10^{-7}	10^{-7}J
	calorie*	4.2	4.2J
	wave number (cm^{-1})	$hc\ 10^2$	1.986×10^{-23} J
	electron volt (eV)†	1.6022×10^{-19}	1.6022×10^{-19}J
force	dyne	10^{-5}	10^{-5} N
electric charge‡	e.s.u.	$(4\pi\epsilon_0)^{1/2}\ 10^{-9/2}$	3.334×10^{-10} C
electric field strength‡	e.s.u. cm^{-1}	$c\ 10^{-6}$	2.998×10^2 Vm^{-1}
electric dipole moment	debye (10^{-18} e.s.u.cm)	$(4\pi\epsilon_0)^{1/2}\ 10^{-49/2}$	3.334×10^{-30} Cm
magnetic flux density	gauss (G)	$(\mu_0/4\pi)^{1/2}\ 10^{-1/2}$	10^{-4} T

* Only an approximate conversion factor is given, since there are various types of calories.

† The electron volt is the product of the charge on the proton and the unit V.

‡ Note that the practical CGS units for electric charge and electric potential are the coulomb and the volt, respectively, and that these are SI units.

A.3 Recommended values of fundamental constants

The uncertainties in these values (one standard-deviation) are indicated by the numbers in parentheses; for example, the uncertainty in the velocity of light in a vacuum is 0.0000010×10^8 m s^{-1} or 1.0×10^2 m s^{-1}.

Fundamental Constant	Symbol	Value
velocity of light in a vacuum	c	$2.9979250(10) \times 10^8 \, ms^{-1}$
permeability of a vacuum	μ_0	$4\pi \times 10^{-7} \, H \, m^{-1}$ (exact)
permittivity of a vacuum	$\epsilon_0 = \mu_0^{-1} c^{-2}$	$8.8541853(58) \times 10^{-12} \, F \, m^{-1}$
fine structure constant	$\alpha = \mu_0 e^2 c/2h$	$7.297351(11) \times 10^{-3}$
	α^{-1}	$137.03602(21)$
charge of a proton	e	$1.6021917(70) \times 10^{-19} \, C$
Planck constant	h	$6.626196(50) \times 10^{-34} \, J \, s$
	$\hbar = h/2\pi$	$1.0545919(80) \times 10^{-34} \, J \, s$
unified atomic mass constant	$m_u = m_a(^{12}C)/12$	$1.660531(11) \times 10^{-27} \, kg$
rest mass of electron	m_e	$9.109558(54) \times 10^{-31} \, kg$
rest mass of proton	m_p	$1.672614(11) \times 10^{-27} \, kg$
	m_p/m_e	$1836.109(11)$
Rydberg constant	$R_\infty = \mu_0^2 m_e e^4 c^3/8h^3$	$1.09737312(11) \times 10^7 \, m^{-1}$
Bohr radius	$a_0 = h^2/\pi\mu_0 c^2 m_e e^2$	$5.2917715(81) \times 10^{-11} \, m$
Bohr magneton	$\mu_B = e\hbar/2m_e$	$9.274096(65) \times 10^{-24} \, JT^{-1}$
nuclear magneton	$\mu_N = (m_e/m_p)\mu_B$	$5.050951(50) \times 10^{-27} \, JT^{-1}$
free electron g factor	g	$2.002319315(7)$
Boltzmann constant	k	$1.380622(59) \times 10^{-23} \, JK^{-1}$
Avogadro constant	L	$6.022169(40) \times 10^{23} \, mol^{-1}$
gravitational constant	G	$6.6732(31) \times 10^{-11} \, Nm^2 \, kg^{-2}$

Bibliography

The Symbols Committee of the Royal Society (1971), *Quantities, Units and Symbols,* The Royal Society, London.

McGlaschan, M. L. (1971), *Physico-Chemical Quantities and Units* (second edition), The Royal Institute of Chemistry (London), Monographs for Teachers, number 15.

Taylor, B. N., Parker, W. H., and Langenberg, D. N. (1969), 'Determination of e/h, using macroscopic quantum phase coherence in superconductors. Implications for quantum electrodynamics and the fundamental physical constants', *Rev. Mod. Phys.*, **41**, 375: with the exception of the electron g factor (see Section 9.5) the values of the fundamental constants quoted here are taken from this reference.

Coulson, C. A. (1958), *Electricity,* Oliver and Boyd, Edinburgh: Chapter 14.

APPENDIX B

Vector Relations in Three Dimensions

General relations. The scalar ϕ and the components of the vectors **A**, **B**, ... are all assumed to commute with one another; $\boldsymbol{\nabla}$ is the vector operator with components $(\partial/\partial r_i)$.

$$\mathbf{A}.(\mathbf{B}_\wedge \mathbf{C}) = \mathbf{B}.(\mathbf{C}_\wedge \mathbf{A}) = \mathbf{C}.(\mathbf{A}_\wedge \mathbf{B}),$$

$$\mathbf{A}_\wedge(\mathbf{B}_\wedge \mathbf{C}) = (\mathbf{A}.\mathbf{C})\mathbf{B} - (\mathbf{A}.\mathbf{B})\mathbf{C},$$

$$(\mathbf{A}_\wedge \mathbf{B}).(\mathbf{C}_\wedge \mathbf{D}) = (\mathbf{A}.\mathbf{C})(\mathbf{B}.\mathbf{D}) - (\mathbf{A}.\mathbf{D})(\mathbf{B}.\mathbf{C}),$$

$$\boldsymbol{\nabla}.(\phi\mathbf{A}) = (\boldsymbol{\nabla}\phi).\mathbf{A} + \phi(\boldsymbol{\nabla}.\mathbf{A}),$$

$$\boldsymbol{\nabla}_\wedge(\phi\mathbf{A}) = (\boldsymbol{\nabla}\phi)_\wedge \mathbf{A} + \phi(\boldsymbol{\nabla}_\wedge \mathbf{A}),$$

$$\boldsymbol{\nabla}(\mathbf{A}.\mathbf{B}) = (\mathbf{B}.\boldsymbol{\nabla})\mathbf{A} + (\mathbf{A}.\boldsymbol{\nabla})\mathbf{B} + \mathbf{B}_\wedge(\boldsymbol{\nabla}_\wedge \mathbf{A}) + \mathbf{A}_\wedge(\boldsymbol{\nabla}_\wedge \mathbf{B}),$$

$$\boldsymbol{\nabla}.(\mathbf{A}_\wedge \mathbf{B}) = (\boldsymbol{\nabla}_\wedge \mathbf{A}).\mathbf{B} - \mathbf{A}.(\boldsymbol{\nabla}_\wedge \mathbf{B}),$$

$$\boldsymbol{\nabla}_\wedge(\mathbf{A}_\wedge \mathbf{B}) = \mathbf{A}(\boldsymbol{\nabla}.\mathbf{B}) - (\mathbf{A}.\boldsymbol{\nabla})\mathbf{B} + (\mathbf{B}.\boldsymbol{\nabla})\mathbf{A} - (\boldsymbol{\nabla}.\mathbf{A})\mathbf{B},$$

$$\boldsymbol{\nabla}_\wedge(\boldsymbol{\nabla}\phi) = 0,$$

$$\boldsymbol{\nabla}.(\boldsymbol{\nabla}_\wedge \mathbf{A}) = 0,$$

$$\boldsymbol{\nabla}_\wedge(\boldsymbol{\nabla}_\wedge \mathbf{A}) = \boldsymbol{\nabla}(\boldsymbol{\nabla}.\mathbf{A}) - \nabla^2\mathbf{A}.$$

Special relations involving the position vector **r**; $\delta(\mathbf{r})$ is the Dirac delta function (see Section 2.13).

$$\boldsymbol{\nabla}.\mathbf{r} = 3,$$

$$\nabla_\wedge \mathbf{r} = 0,$$

$$(\mathbf{A}.\nabla)\mathbf{r} = \mathbf{A},$$

$$\nabla(1/r) = -\mathbf{r}/r^3,$$

$$\nabla^2(1/r) = -\nabla.(\mathbf{r}/r^3) = -4\pi\delta(\mathbf{r}),$$

$$\nabla_i(r_j/r^3) = (4\pi/3)\delta_{ij}\delta(\mathbf{r}) + \delta_{ij}/r^3 - 3r_ir_j/r^5.$$

The derivation of other relations is often facilitated by the use of ϵ_{ijk} (see Chapter 2). The following sum rules may be useful:

$$\epsilon_{ijk}\epsilon_{imn} = \delta_{jm}\delta_{kn} - \delta_{jn}\delta_{km},$$

$$\epsilon_{ijk}\epsilon_{ijn} = 2\delta_{kn},$$

$$\epsilon_{ijk}\epsilon_{ijk} = 6.$$

APPENDIX C

General Bibliography

The following references are cited a number of times in this book. They are referred to by the authors' names in the bibliographies at the ends of the chapters.

Bethe, H. A., and Salpeter, E. E. (1957), *Quantum Mechanics of One- and Two-Electron Atoms,* Springer-Verlag, Berlin.

Feynman, R. P., Leighton, R. B., and Sands, M. (1964), *The Feynman Lectures in Physics,* Addison-Wesley, Reading, Mass.

Foldy, L. L. (1962). 'Relativistic wave equations', *Pure Appl. Physics* (D. R. Bates, editor. Academic Press, New York), **10** (part 3), 1.

Grandy, W. T., Jr. (1970), *Introduction to Electrodynamics and Radiation,* Academic Press, New York.

Hameka, H. F. (1965), *Advanced Quantum Chemistry,* Addison-Wesley, Reading, Mass.

Landau, L. D., and Lifschitz, E. M. (1962), *The Classical Theory of Fields,* Pergamon Press, Oxford.

Rose, M. E. (1961), *Relativistic Electron Theory,* Wiley, New York.

Sakurai, J. J. (1967), *Advanced Quantum Mechanics,* Addison-Wesley, Reading, Mass.

Indexes

Author Index

(Authors cited in the bibliographies at the ends of the chapters.)

Subject Index

Little distinction is drawn between classical and quantum mechanics, since classical theory is confined to Chapters 2 to 6 inclusive (pages 13 to 103).